Biobased Composites

Processing, Characterization, Properties, and Applications

First Edition

Anish Khan
King Abdulaziz University, Jeddah, Saudi Arabia

Sanjay M. Rangappa
King Mongkut's University of Technology North Bangkok, Bangkok, Thailand

Suchart Siengchin
King Mongkut's University of Technology North Bangkok, Bangkok, Thailand

Abdullah M. Asiri
King Abdulaziz University, Jeddah, Saudi Arabia

WILEY

This edition first published 2021.

Published by Jon Wiley & Sons, Inc., Hoboken, New Jersey
Published simultaneously in Canada.

Registered Office
John Wiley & Sons, Inc., 111 River Street, Hoboken, NJ 07030, USA

Editorial Office
111 River Street, Hoboken, NJ 07030, USA

For details of our global editorial offices, customer services, and more information about Wiley products visit us at www.wiley.com.

Wiley also publishes its books in a variety of electronic formats and by print-on-demand. Some content that appears in standard print versions of this book may not be available in other formats.

For general information on our other products and services or for technical support, please contact our Customer Care Department within the United States at (800) 762-2974, outside the United States at (317) 572-3993 or fax (317) 572-4002.

Wiley also publishes its books in a variety of electronic formats. Some content that appears in print may not be available in electronic formats. For more information about Wiley products, visit our web site at www.wiley.com.

Library of Congress Cataloging-in-Publication Data

Names: Khan, Anish, author. | Rangappa, Sanjay Mavinkere, author. | Siengchin, Suchart, author. | Asiri, Abdullah M., author.
Title: Biobased composites : processing, characterization, properties, and applications / Anish Khan, King Abdulaziz University, Sanjay M Rangappa, King Mongkut's University of Technology, Suchart Siengchin, King Mongkut's University of Technology, Abdullah M. Asiri, King Abdulaziz University.
Description: First edition. | Hoboken, New Jersey : John Wiley & Sons, Inc., 2021. | Includes bibliographical references and index.
Identifiers: LCCN 2020036206 (print) | LCCN 2020036207 (ebook) | ISBN 9781119641797 (hardback) | ISBN 9781119641810 (adobe pdf) | ISBN 9781119641827 (epub)
Subjects: LCSH: Fibrous composites. | Biomass chemicals.
Classification: LCC TA418.9.C6 K478 2021 (print) | LCC TA418.9.C6 (ebook) | DDC 620.1/97–dc23
LC record available at https://lccn.loc.gov/2020036206
LC ebook record available at https://lccn.loc.gov/2020036207

Cover image: © Zukvosten / Getty Images
Cover design by Wiley

MIX
Paper from responsible sources
FSC
www.fsc.org
FSC® C013604

Contents

List of Contributors

Nikita Agrawal
Department of Pedodontics, People's College of Dental Sciences, Bhopal, Madhya Pradesh, India

Jamal Akhter Siddique
Department of Chemistry, School of Basic and Applied Sciences, Lingaya's Vidyapeeth (Formerly known as Lingaya's University), Faridabad, Haryana, India

Faris M. AL-Oqla
Department of Mechanical Engineering, Faculty of Engineering, Hashemite University, Zarqa, Jordan

Marcos Aquino
Department of Textile Engineering, Textile Engineering Laboratory, Federal University of Rio Grande do Norte, Natal, RN, Brazil

V. Arul Mozhi Selvan
Department of Mechanical Engineering, National Institute of Technology, Tiruchirappalli, India

Abdullah M. Asiri
Center of Excellence for Advanced Materials Research and Chemistry Department, Faculty of Science, King Abdulaziz University, Jeddah, Saudi Arabia

Aftab Aslam Parwaz Khan
Center of Excellence for Advanced Materials Research and Chemistry Department, Faculty of Science, King Abdulaziz University, Jeddah, Saudi Arabia

Nadir Ayrilmis
Department of Wood Mechanics and Technology, Faculty of Forestry, Istanbul University-Cerrahpasa, Istanbul, Turkey

E. Biswas
Chemistry and Biochemistry Department, Georgia Southern University, Statesboro, GA, USA

M. Chandrasekar
School of Aeronautical Sciences, Hindustan Institute of Technology and Science, Chennai, Tamil Nadu, India

Piotr Czub
Department of Chemistry and Technology of Polymers, Cracow University of Technology, Cracow, Poland

D. Divya
Research and Development Department, Pinnacle Bio-Sciences, Kanyakumari, Tamil Nadu, India

Rubens Fonseca
Department of Textile Engineering, Textile Engineering Laboratory, Federal University of Rio Grande do Norte, Natal, RN, Brazil

Brijesh Gangil
Department of Mechanical Engineering, H.N.B. Garhwal University, Srinagar, Uttarakhand, India

T. F. Garrison
Chemistry Department, King Fahd University of Petroleum and Minerals, Dhahran, Saudi Arabia

S. Hawkins
Chemistry and Biochemistry Department, Georgia Southern University, Statesboro, GA, USA

Mohit Hemath Kumar
Department of Mechanical Engineering, National Institute of Technology, Tiruchirappalli, Tamil Nadu, India

S. Indran
Department of Mechanical Engineering, Rohini College of Engineering and Technology, Kanyakumari, Tamil Nadu, India

N. B. Karthik Babu
Department of Mechanical Engineering, National Institute of Technology, Tiruchirappalli, Tamil Nadu, India

S. Karthikeyan
Department of Mechanical Engineering, Kalasalingam Academy of Research and Education, Krishnankoil, Tamil Nadu, India

Anish Khan
Center of Excellence for Advanced Materials Research and Chemistry Department, Faculty of Science,
King Abdulaziz University, Jeddah, Saudi Arabia

A. V. Kiruthika
Assistant Professor of Physics, Seethalakshmi Achi College for Women, Pallathur, Tamil Nadu, India

Dipen Kumar Rajak
Department of Mechanical Engineering, Sandip Institute of Technology and Research Centre, Nashik, Maharashtra, India

Pawan Kumar Rakesh
Department of Mechanical Engineering, NIT, Srinagar, Uttarakhand, India

J. Maniraj
Department of Mechanical Engineering, KIT-Kalaignarkarunanidhi Institute of Technology, Coimbatore, Tamil Nadu, India

Moises Melo
Department of Textile Engineering, Laboratory of Characterization of the Textile Materials, Federal University
of Rio Grande do Norte, Natal, RN, Brazil

H. Mohit
Department of Mechanical Engineering, National Institute of Technology, Tiruchirappalli, India

K. Monroe
Chemistry and Biochemistry Department, Georgia Southern University, Statesboro, GA, USA

Kátia Moreira
Department of Textile Engineering, Textile Engineering Laboratory, Federal University of Rio Grande do Norte, Natal, RN, Brazil

Durgesh D. Pagar
Department of Mechanical Engineering, K. K. Wagh Institute of Engineering Education and Research, Nashik, Maharashtra, India

Catalin I. Pruncu
Department of Mechanical Engineering, Imperial College London, London, UK
Department of Mechanical Engineering, School of Engineering, University of Birmingham, Birmingham, UK

R. L. Quirino
Chemistry and Biochemistry Department, Georgia Southern University, Statesboro, GA, USA

N. Rajesh Jesudoss Hynes
Department of Mechanical Engineering, Mepco Schlenk Engineering College, Sivakasi, Tamil Nadu, India

L. Rajesh Kumar
Department of Mechanical Engineering, KPR Institute of Engineering and Technology, Coimbatore, Tamil Nadu, India

N. Rajini
Department of Mechanical Engineering, Kalasalingam Academy of Research and Education, Krishnankoil, Tamil Nadu, India

M. Ramesh
Department of Mechanical Engineering, KIT-Kalaignarkarunanidhi Institute of Technology, Coimbatore, Tamil Nadu, India

P. Ramesh
Department of Production Engineering, National Institute of Technology, Tiruchirappalli, India

T. Ramesh
Department of Mechanical Engineering, National Institute of Technology, Tiruchirappalli, Tamil Nadu, India

Lalit Ranakoti
Department of Mechanical Engineering, NIT, Srinagar, Uttarakhand, India

M. R. Sanjay
Natural Composites Research Group Lab, Academic Enhancement Department, King Mongkut's University of Technology North Bangkok, Bangkok, Thailand

Thiago Santos
Department of Textile Engineering, Textile Engineering Laboratory, Federal University of Rio Grande do Norte, Natal, RN, Brazil

Caroliny Santos
Department of Textile Engineering, Textile Engineering Laboratory, Federal University of Rio Grande do Norte, Natal, RN, Brazil

K. Senthilkumar
Department of Mechanical Engineering, Kalasalingam Academy of Research and Education, Krishnankoil, Tamil Nadu, India
Department of Mechanical and Process Engineering, The Sirindhorn International Thai German Graduate School of Engineering (TGGS), King Mongkut's University of Technology North Bangkok, Bangkok, Thailand

T. Senthil Muthu Kumar
Department of Mechanical Engineering, Kalasalingam Academy of Research and Education, Krishnankoil, Tamil Nadu, India
Department of Mechanical and Process Engineering, The Sirindhorn International Thai German Graduate School of Engineering (TGGS), King Mongkut's University of Technology North Bangkok, Bangkok, Thailand

P. Shenbaga Velu
Department of Mechanical Engineering, P.S.R Engineering College, Sivakasi, Tamil Nadu, India

Suchart Siengchin
Department of Materials and Production Engineering, The Sirindhorn International Thai German Graduate School of Engineering (TGGS), King Mongkut's University of Technology North Bangkok, Bangkok, Thailand
Natural Composites Research Group Lab, King Mongkut's University of Technology North Bangkok, Bangkok, Thailand

Anna Sienkiewicz
Department of Chemistry and Technology of Polymers, Cracow University of Technology, Cracow, Poland

N. J. Vignesh
Department of Mechanical Engineering, Mepco Schlenk Engineering College, Sivakasi, Tamil Nadu, India

Madhu Yadav
Department of Chemistry, Bioorganic Research Laboratory, University of Allahabad, Allahabad, Uttar Pradesh, India
GGIC, Prayagraj, Uttar Pradesh, India

Preface

The objective of this book is to summarize many of the recent developments in the area of biobased composites. As the title indicates, this book emphasizes new challenges for the characterization, properties, and applications of biobased composites. This book provides an update of all the important areas of (processing, characterisation, properties and application) biobased composites in a comprehensive manner. This book covers the void for the need of one-stop reference book for the researchers. Leading researchers from industry, academy, government, and private research institutions across the globe have contributed the chapters for the book. Academics, researchers, scientists, engineers, and students in the field of biobased composites will benefit from this book, which is highly application oriented.

The editors are thankful to all the authors for their contribution. The editors also thank the editorial and publishing team for their guidance and support.

1

Introduction to Biobased Composites

Faris M. AL-Oqla

Department of Mechanical Engineering, The Hashemite University, Zarqa 13133, Jordan

1.1 Introduction

Tons of industrial wastes are dumped daily in every region of the world, making the recycling of these wastes a very critical environmental concern. For environmental-friendly industries, materials from renewable resources should be used to replace conventional materials. Hence, the sustainability will be ensured, and cheaper as well as ecological alternative products will compete with the current products from nonrenewable resources [1–3].

Biobased composite materials and green products are increasingly substituting the traditional materials and products in a wide range of applications for sustainable industries, though massive efforts are still essential to better exploit such biomaterials as well as to expand their applications [4–6]. Moreover, the necessity to advance their potentials for consistent and reliable performance is still demanding [7]. To achieve these goals, founding a well-organized and robust evaluation and selection system for the composite constituents is the most important step [8, 9]. In such techniques, several preferred fibers as well as polymer assets including mechanical, physical, economic, and environmental, have to be revealed in parallel and assessed to determine the best type of fibers for a certain application [9–11]. Further, proper capabilities and performance of new materials including the biobased ones, would enhance their industrial applications.

The weather, on the other hand, is also reaching more extremes in both hot and cold conditions and negatively affecting the available resources and the environment. Ice formation is sharply decaying over the past years due to climate changes. It is believed that the fast-increasing manner of temperature over years due to global warming leads to shorter winters, breaking the natural balance in climate resulting in the destruction of available resources. The rising of sea level due to melting of ice is also a serious problem that can lead to catastrophic disasters on the sea-neighbor lands, which would negatively affect the environment. Natural fibers are considered renewable resources and can be recycled from many industrial process wastes. These natural fibers are obtained from plant sources, such as hemp, or from animal sources [12, 13]. Natural fibers of plant sources comprise cellulose, hemicellulose, and lignin; natural fibers of animal sources consist of mainly proteins. Natural fibers can be

Biobased Composites: Processing, Characterization, Properties, and Applications, First Edition.
Anish Khan, Sanjay Mavinkere Rangappa, Suchart Siengchin and Abdullah M. Asiri.

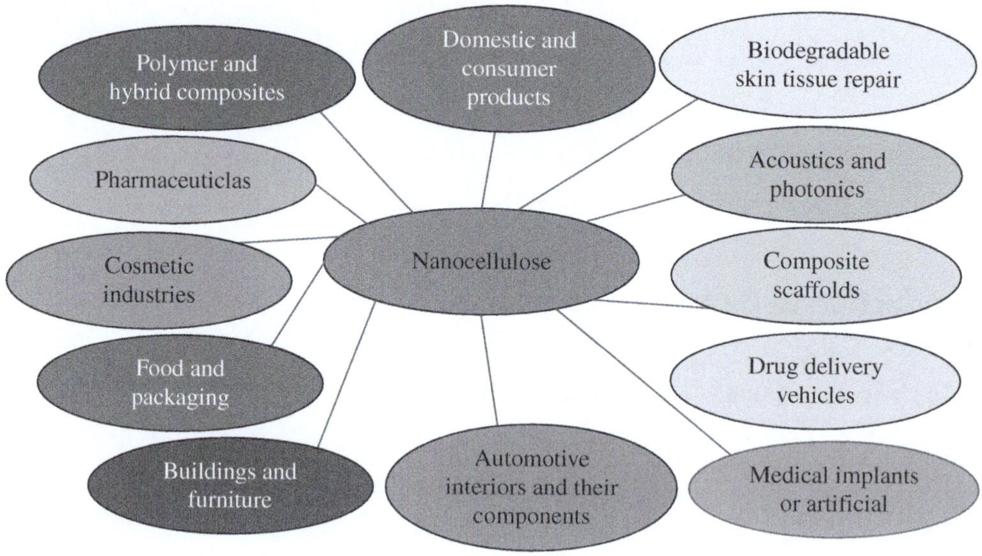

Figure 1.1 Nanocellulose applications.

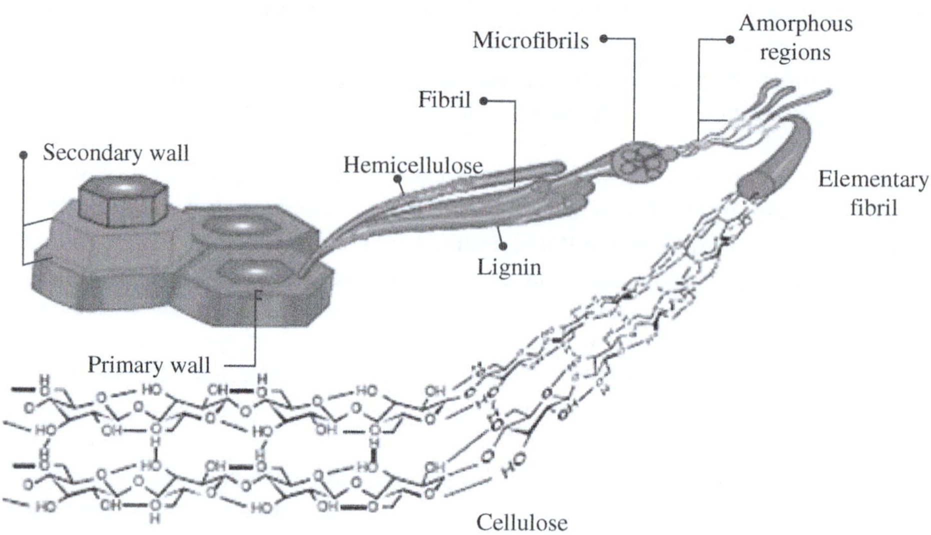

Figure 1.2 Illustration for the structure of cellulose extracted from plants.

utilized in various sizes from macro- to nanoscale fibers. For instance, the nanocellulose fibers can be utilized in a wide range of applications as indicated in Figure 1.1.

Figure 1.2 illustrates the structure of cellulose extracted from plants as a 3D illustration showing the plant cross-sectional walls, hemicellulose, lignin, fibril, microfibrils, the amorphous region, and cellulose.

In general, the natural fibers are used in polymer composites as reinforcement [14–17]. Hence, the properties of these composites will directly be influenced by the type of fibers used, their aspect ratio (length/width), their extraction processes, and their interaction with the matrix material [18–20].

1.2 Biodegradable Materials

Biodegradable materials are increasingly demanded to replace the conventional materials. Thermoplastic starch, for instance, is obtained from corn, potatoes, or other cereals. It mainly consists of amylose and amylopectin. As thermoplastic starch has highly sensitivity toward hydrolysis, and due to its low mechanical performance, it is usually used as a matrix for composites and not as a reinforcement. However, the starch phase is blended with polyesters to produce very interesting biodegradable products [21–23]. Examples of such products are commercially available in many fields such as the food packaging industry and in the manufacture of disposable items and films. Biomaterials are more and more in demand to replace the conventional materials and products in some applications. Thus, more efforts are still required to include their use in many industrial fields such as automobile, aerospace, construction, electronic, and food industries. However, before using the biodegradable materials in these industries, their reactions and performance with fire should be carefully assessed with various tools as well as with most newly established materials [24–29]. Thermoplastic starch has demonstrated good flame retardancy by including aluminum trihydroxide and coconut fibers; fire growth rate and the total heat release were significantly suppressed. This is due to the increase in carbonaceous char accompanied by the reduction of carbon content in the pyrolysis products. Also, using coconut fibers replaced a significant portion the aluminum trihydroxide causing its overall reduction in the thermoplastic starch. These results of flame retardancy opened the gate for researchers to explore alternatives from natural fibers to replace the current flame-retardant additives for the thermoplastic starch biocomposites [30]. Thermoplastic starch-blend materials are mainly produced for compostability. Hence, the organic wastes will be reduced, and the biogases will be decreased.

1.3 Polymers in Tissue Engineering

With the increasing trend toward using biomaterials in tissue engineering, great efforts are taken to develop biomaterials with desired characteristics, such as the ease of production, the compatibility with the host tissues, and the improvement of the healing rates. Polymeric-based biomaterials are increasingly replacing the metallic and ceramic-based biomaterials. As the latter type possesses high elastic modulus, the distribution of the stresses on the host bones will change completely, leading to alterations in the stress concentrations as well as relocating their locations [31]. In addition, polymeric materials are generally easier to be manufactured or shaped, and compatible with the host tissues [32]. Some classifications of polymers are illustrated in Figure 1.3. Moreover, some characteristics such as their mechanical performance and degradations can be controlled easily. Hence, they can perfectly be dedicated to a certain purpose once they are implanted in the host tissues.

For the last 50 years, many researchers have been concerned in investigating and developing potential natural and synthetic polymers for different applications in medical engineering such as biodegradable sutures, tissue scaffolds, and cardiovascular stents [33–35]. Polymers of natural resources are biocompatible and biodegradable in nature. Thus, they are readily suited for many medical applications such as tissue engineering applications and drugs production. Commonly used forms of scaffolds in tissue engineering are illustrated in Figure 1.4. Biobased composites are gaining more and more attention, and they are a concern for many researchers internationally due to their availability, recyclability,

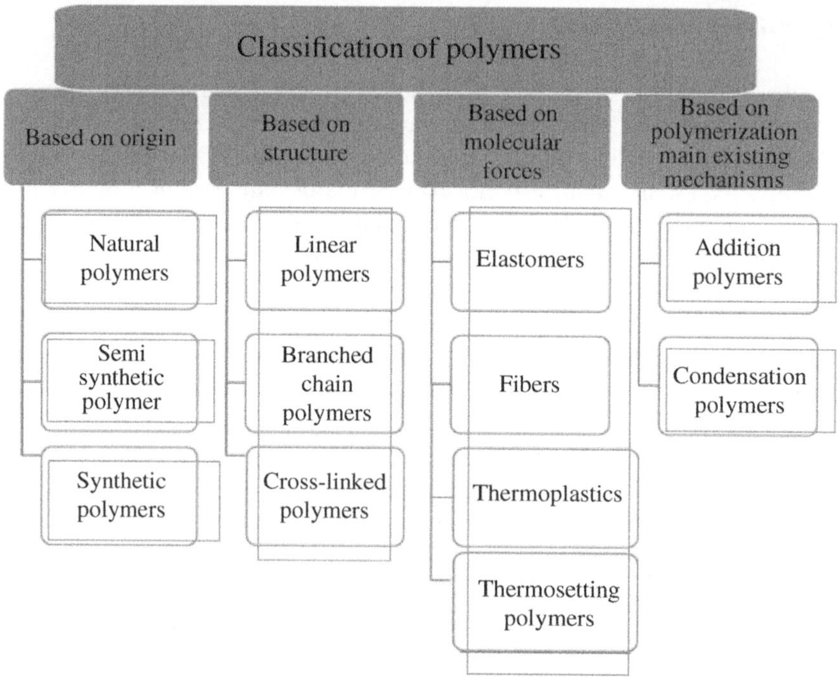

Figure 1.3 Classification of polymers.

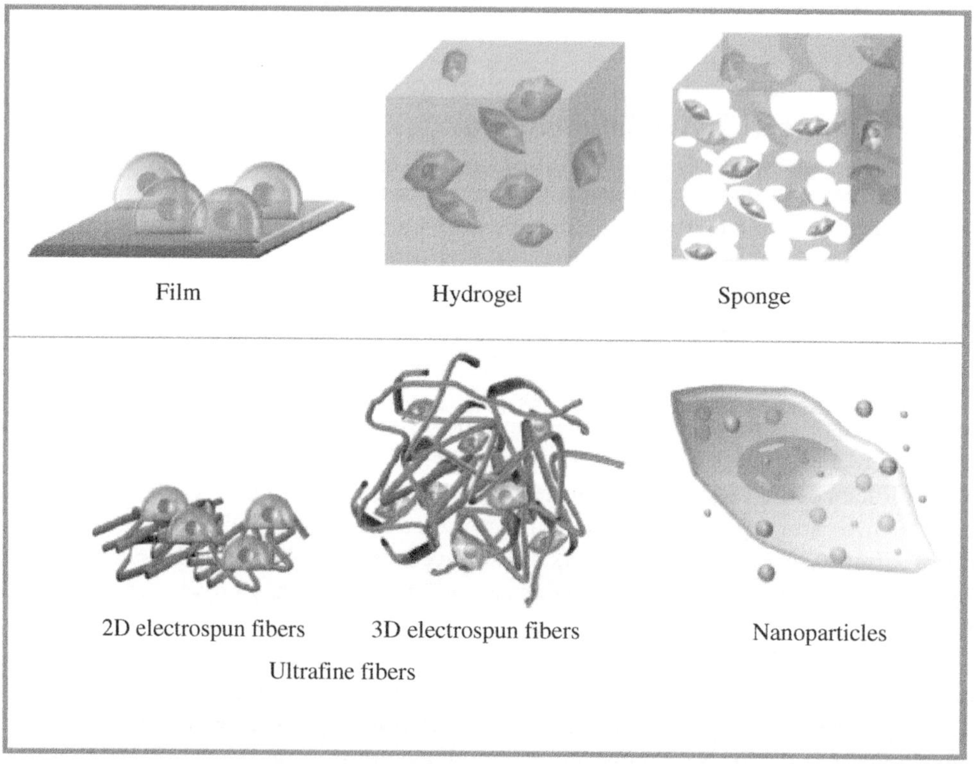

Figure 1.4 Commonly used forms of scaffolds in tissue engineering.

degradability, sustainability, low cost, light weight, and most importantly their high mechanical performance. The continuous improvements of the biocomposites will surely lead to new materials with a high potential to replace the conventional composites in the current applications or for future applications. In biocomposites, a material which is strong and similar to cement is made by the natural merging of different cells of hard plant fibers by lignin. These composites also possess high electric resistance due to the presence of cellulose fibrils embedded in lignin.

1.4 Environmental Realization

Due to the severe environmental impact of the petro-based plastics which are produced from nonrenewable resources, biobased composites of renewable resources with relatively fast biodegradability are preferred as ecofriendly alternatives. Thus, natural fibers as reinforcement of polymeric based composites can offer excellent ecofriendly alternatives and good environmental indices at a low cost. This is due to the fact that most of the natural fibers have specific reasonable mechanical properties comparable to the traditional synthetic fibers [16–19, 36–41]. Because of this, several researchers have considered the natural fibers in biomaterials as environmental alternatives of low cost materials. The number of researches that considered natural fibers as reinforced materials in polymeric based composites in three years from 2016 to 2018 is illustrated in Figure 1.5. However, the aspect of the biobased composites is challenging and huge efforts are still required to replace the petro-based composites with these biobased composites. Nevertheless, biobased composites have already been commercialized in many industrial applications such as the automotive and packaging industries.

More promising aspects and applications are under development using biobased composites, and the total weight of the biocomposite productions is expected to exceed thousands of

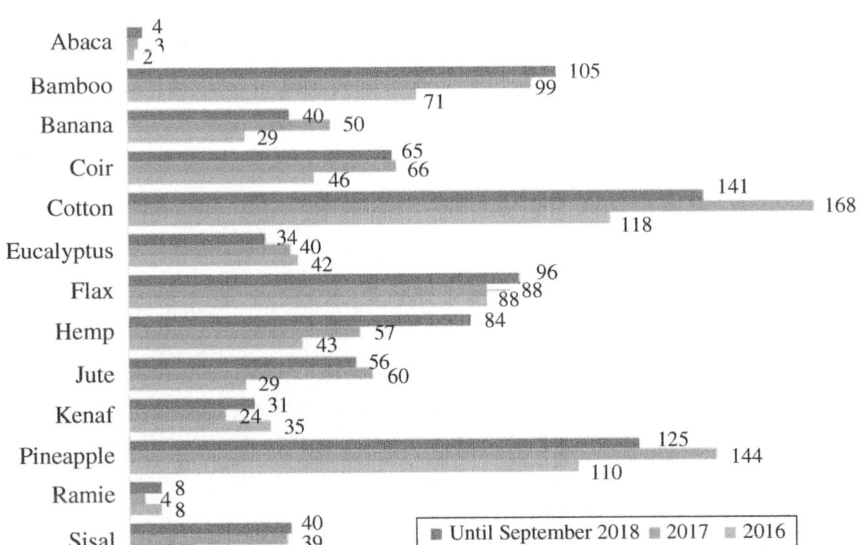

Figure 1.5 Number of publications considering natural fiber composites. *Source:* Peças et al. [42]. Licensed under CC BY 4.0.

tons annually in the few coming years [43]. It is believed that during the last three decades, the generation of solid wastes was almost doubled and about one third of these wastes were from packaging materials, especially the food packaging wastes. Hence, using biobased materials just for food packaging can yield a considerable impact on the environment by replacing the nondegradable materials with degradable and composting materials [44–46].

1.4.1 Green Biomass-based Composites

An increasing environmental awareness for the people helped in adopting a public trend toward replacing the petro-based composites with sustainable biobased ones. In general, biobased composites are a combination of fibers and polymer material (matrix), with at least one of them from a natural resource [47–49]. Hence, the combinations of natural fibers and petro-based polymers, synthetic fibers and biobased polymers, and natural fibers and biobased polymers are all called conventional biocomposites. Here, the first and second combinations are not fully ecofriendly, whereas the third combination, that is the combination of natural fibers and biobased polymers (commonly known as a green biocomposite), is more ecofriendly [6, 50–52]. A biocomposite is called biodegradable when the matrix material is biodegradable. This can be for both biobased and petro-based composites provided that they are degradable. Biopolymers are different from biodegradable polymers in their raw material. The latter can be formed from either biobased or petroleum-based composites and can be categorized as green polymeric matrices. When both the fiber and matrix are from renewable resources, they are referred to as biobased composites or biocomposites (fully biodegradable green composite). These biocomposites have less environmental impact.

1.4.2 Selection Considerations

To select the most appropriate alternative materials for various applications, several evaluation criteria have to be considered, most of which may be conflicting. Thus, the selection process and considerations may contain several aspects and may become a sophisticated issue with a manner of multicriteria decision making problem that requires proper decisions to be made via various optimizations as well as other methods like that of analytic hierarchy process, which are utilized in various engineering problems [18, 37, 53–57].

1.4.2.1 Materials Implementation Requirements

Before commercializing a new material for use in industry, it has to undergo a series of sequence steps. First, the new material is introduced. Then a possible application is identified, and the material specifications needed to meet the requirements of this application should be listed. The material specifications are mainly varieties of physical, mechanical, economical, and environmental properties such as density, appearance, strength, cost, availability, reliability, etc. Hence, the new material should meet all the required specifications through a long journey of tests and verifications. In the next step, a prototype is made to check the performance of the proposed material. In the last step, a management plan is established for studying the market needs, supplying raw materials, and ensuring the availability of raw materials in a sustainable manner. The time spent through all these preceding steps is completely unpredictable, adding great stress on the competing manufacturers. After going through all these steps, the new material is now ready for production. However, a manufacturing process must first be established.

1.4.2.2 Material Cost

Cost is very essential in the material selection process for any application, and the target cost is set in the early stages of any application development. A tradeoff between the cost and the performance is sometimes required in the decision-making stage.

1.5 Biomass Composites Characteristics and Testing

Combining fibers with polymers in biomaterials usually produces composites with totally different characteristics as well as superior desired performance over the utilized constituents. Biobased plastics are dominating the new trends in plastic industry since the petro-based plastics are nonrenewable (help in depletion of petroleum resources), nondegradable (cause shortage in landfills), and very harmful to the environment. These new trends are focusing more specifically on the renewable plants and on agro waste fiber composites. However, it is not possible to completely replace all the petro-based productions with biobased ones [25]. In such cases, the concept of combining biomaterials with petro-based products should be adopted. Natural biocomposites have become well recognized for their low cost and low density. In addition, the ease of shaping and processing due to the low abrasiveness when compared to synthetic fiber-based composites gives the biobased composites extra advantages. On the other hand, many difficulties arise in using the biobased composites in industry. One of these difficulties is the incompatibility issue between the fiber and the polymer. This is due to both the hydrophilic characteristic of the natural fibers and the hydrophobic characteristic in polymers. Reducing the incompatibility requires physical and chemical treatments for the fibers, as well as using various additives as coupling agents between the fibers and the polymers. Once the composite material is well fabricated, its characteristics are required to be tested and improved. The most critical properties are the mechanical ones, namely the tensile strength, the tensile modulus, the fatigue strength, the creep rate, and the impact strength. Agro waste natural fibers are normally suitable to reinforce polymers due to their relative high mechanical performance and their low densities.

The mechanical properties of the composite materials are the most essential characteristics even if the composites are not used in loaded applications. A certain level of strength is required for the composites to at least maintain their shapes during service. However, for some composites, it is very hard to estimate their mechanical properties as is the case with biocomposites of short natural fibers. This is due to many reasons, such as the fiber dimensions, fiber quality, fiber orientation and distribution, the fiber–matrix interface quality, as well as the matrix characteristics [41, 58]. Table 1.1 demonstrates the mechanical and physical characteristics of some natural fibers.

Improving the composite properties can sometimes be achieved by controlling some key factors, such as the fiber aspect ratio (L/D) and volume fraction of the fibers with respect to the matrix [59]. If the aspect ratio of the fiber is very small, insufficient load will transfer from the matrix to the fiber; in such cases, the fibers will work just as fillers and no considerable improvements will be achieved in the composite's mechanical performance. On the other side, high aspect ratio usually leads to poor fiber dispersion, substantially poor mechanical performance. Regarding the volume fraction, the low percentile causes discontinuities in transferring the load over the fibers; thus, the composite strength will decline. Also, the high percentile can produce the same effect due to fiber clustering.

Table 1.1 Mechanical and physical characteristics of some natural fibers.

Fiber type	Coir	Date palm	Flax	Hemp	Sisal
Density (g/cm^3)	1.15–1.46	0.9–1.2	1.4–1.5	1.4–1.5	1.33–1.5
Length (mm)	20–150	20–250	5–900	5–55	900
Diameter (μm)	10–460	100–1000	12–600	25–500	8–200
Tensile strength (MPa)	95–230	97–275	343–2000	270–900	363–700
Tensile modulus (GPa)	2.8–6	2.5–12	27.6–103	23.5–90	9–38
Specific modulus (approx.)	4	7	45	40	17
Elongation to break (%)	15–51.4	2–19	1.2–3.3	1–3.5	2–7

Source: AL-Oqla et al. [16]. © 2015, Elsevier.

Table 1.2 Informative values on the different properties of the fibers.

Fiber type	Density (g/cm^3)	Tensile strength (MPa)	Tensile modulus (GPa)	Elongation to break (%)	Cost per weight (USD/kg)
Coir	1.15–1.46 (1.31)	95–230 (162.5)	2.8–6 (4.4)	15–51.4 (33.2)	0.3
Date palm	0.9–1.2 (1.05)	97–275 (186)	2.5–12 (7.25)	2.0–19 (10.5)	0.02
Jute	1.3–1.49 (1.4)	320–800 (560)	8–78 (43)	1–1.8 (1.4)	0.3
Hemp	1.4–1.5 (1.45)	270–900 (585)	23.5–90 (56.75)	1–3.5 (2.25)	1.3
Kenaf	1.4	223–930 (576.5)	14.5–53 (33.75)	1.5–2.7 (2.1)	0.5
Oil palm	0.7–1.55 (1.13)	80–248 (164.0)	0.5–3.2 (1.85)	17–25 (21)	0.3

On the other hand, an obvious lack of research regarding the evaluation and selection processes of the natural fiber composites (NFCs) is observed. More specifically, evaluating and selecting the proper agro waste natural fibers for the NFCs is not investigated comprehensively regarding the desired features [9]. Hence, more efforts to establish sufficient comparison criteria are required in order to precisely evaluate and select the appropriate fiber type for the biobased products. The overall characteristics and capabilities of the NFCs depend on the physical, mechanical, chemical, and economic features of the composites' constituents. Therefore, in order to exploit the benefits of these materials to the full extent, comprehensive investigations of the previously mentioned features have to be completed as a primary stage in any industrial application. New techniques have been developed by AL-Oqla and Sapuan [20] for the assessment and selection of the composites. A wide range of valuable criteria has been discussed by AL-Oqla and Sapuan [20] to demonstrate that natural fibers have a primary role in natural fiber reinforced polymer composites. Another technique to evaluate various raw fibers was presented by AL-Oqla et al. [10], where six different types of natural fibers were considered in the evaluation process. These were coir, jute, hemp, kenaf, oil palm, as well as date palm. The physical, mechanical, and economic properties of these types were considered simultaneously. Informative values on the different properties of the considered fibers are listed in Table 1.2. They are obtained from literature based on experimental works, where the average values were adopted assuming that the data is uniformly distributed within the data range found in literature.

Table 1.3 Specific properties of the fibers.

Fiber type	Specific tensile strength (MPa)/(g/cm^3)	Specific tensile modulus (GPa)/(g/cm^3)	Specific elongation (%)/(g/cm^3)	Cost ratio
Coir	124.05	3.36	25.34	0.231
Date palm	177.14	6.90	10.00	0.015
Jute	400.00	30.71	1.00	0.231
Hemp	403.45	39.14	1.55	1
Kenaf	411.79	24.11	1.50	0.385
Oil palm	145.13	1.64	18.58	0.231

Table 1.4 Specific properties of the fibers with respect to the cost ratio.

Fiber type	Specific tensile strength (MPa)/(g/cm^3)/cost ratio	Specific tensile modulus (GPa)/(g/cm^3)/cost ratio	Specific elongation (%)/(g/cm^3)/cost ratio
Coir	537.53	14.55	109.82
Date palm	11 514.29	448.81	650.00
Jute	1 733.33	133.10	4.33
Hemp	403.45	39.14	1.55
Kenaf	1 070.64	62.68	3.90
Oil palm	628.91	7.09	80.53

Hence, Table 1.3 lists the specific properties for the fibers (the average values of each property divided by the average values of the density).

The obtained specific properties calculated in Table 1.3 are further calculated with respect to the cost ratio as tabulated in Table 1.4.

It is believed that a comparison of the natural fibers using the combined physical, mechanical, and economic information would result in better evaluation of the available natural fibers and resources. It can be noticed here that the specific tensile strength to cost ratio for the date palm was five times that of jute. Therefore, combined evaluations would lead to better evaluation of fibers as it can be realized that date palm fiber is better than jute once specific properties to cost ratio evaluation criterion is considered.

1.6 Life-cycle Assessment

The Life-cycle assessment (LCA), as the name implies, is an assessment method of the environmental impacts along the whole life of the product, starting from preparing the raw materials, and passing through producing, distributing, using, maintaining, and ending by disposing or recycling. Designers use this method in evaluating their products. It can widen the view of the environmental concerns through compiling an inventory of relevant energy and material inputs and environmental releases; assessing the potential effects accompanying the identified inputs and outputs; and helping in adopting more realistic decisions. In general, the LCA used an iterative process to arrive at the results. These results provide further information to the production process and guide toward the most important environmental input or output that should be highly monitored. Several case studies for the environmental impacts of the life cycle of many products were reported in the literature [60–64]. All the studies conclude that the LCA is an effective tool to assess the environmental impacts of the different products along their life, as well as it aids in selecting ecofriendly materials. Hence, using LCA in planning and management strategies helps in delivering more sustainable products and protecting the health of the people. The extra information obtained from LCA can help in improving the inventory analysis, and in adopting more informative decisions. Further environmental improvements can be achieved using LCA through evaluating the potential effects accompanying specified inputs and outputs.

There are three types of LCA; these are: conceptual, simplified, and detailed. These types can be applied in many ways where each type has its own strengths and weaknesses. The life cycle of each product passes through many stages starting from extracting and preparing the raw materials, manufacturing, distributing, storing, using, and ending by disposing or recycling the products to reduce the environmental effect.

Recycling is defined as converting the materials and the products at the end of their life to new useful products. It exploits the potentially useful materials from waste, reduce the demands for raw materials, reduce the energy consumption, and reduce the air and water pollutions. Each stage of the LCA have inputs (raw materials and energy) and outputs (gas emissions and solid wastes). Thus, the role of the LCA is to study the environmental impacts of these inputs and outputs on each stage of the product life, and trying to reduce these impacts and deriving the potential benefits from them. The LCA is composed of four phases:

- Goal and scope

The purpose of this study is to select a product and determine the objective of the study (comparison, improvement) and fix boundaries accordingly.

- Inventory analysis

Collecting as much information as possible on the inputs and outputs at each stage of the product life.

- Impact analysis

Studying the environmental impacts of the inputs and the outputs at each stage of the product life.

- Interpretation

Using the data and information collected, the product life cycle will be improved, and the environmental impact will be reduced.

The LCA is performed by passing over these phases frequently in an iterative manner.

1.7 Conclusions

Improving the different features of the biobased composites is directly reflected by the progress achieved in the evaluation and selection processes of their constituents, in addition to the preparation and treatment of fiber-polymers. Therefore, it is very important to build robust and efficient selection methods to improve the performance of the biocomposites, and to reduce the negative environmental effects. Doing this would lead to more informed decisions and save efforts and time. Application of such methods can be expanded to include more characteristics and more potential types of the composites' constituents to derive new biomaterials with optimal performance for a greener future.

References

1 AL-Oqla, F.M., Sapuan, M.S., Ishak, M.R., and Nuraini, A.A. (2015). Selecting natural fibers for bio-based materials with conflicting criteria. *American Journal of Applied Sciences* 12 (1): 64–71.

2 AL-Oqla, F.M. and Sapuan, S. (2015). Polymer selection approach for commonly and uncommonly used natural fibers under uncertainty environments. *JOM* 67 (10): 2450–2463.

3 Al-Oqla, F.M. and Sapuan, S. (2018). Natural fiber composites. In: *Kenaf Fibers and Composites*, vol. 1 (eds. S.M. Sapuan, J. Sahari, M.R. Ishak and M.L. Sanyang). CRC Press.

4 AL-Oqla, F.M. (2017). Investigating the mechanical performance deterioration of Mediterranean cellulosic cypress and pine/polyethylene composites. *Cellulose* 24 (6): 2523–2530.

5 Al-Oqla, F.M. and El-Shekeil, Y. (2019). Investigating and predicting the performance deteriorations and trends of polyurethane bio-composites for more realistic sustainable design possibilities. *Journal of Cleaner Production* 222: 865–870.

6 Alaaeddin, M., Sapuan, S., Zuhri, M. et al. (2018). Properties and common industrial applications of polyvinyl fluoride (PVF) and polyvinylidene fluoride (PVDF). In: *IOP Conference Series: Materials Science and Engineering*, vol. 409, 012021. IOP Publishing.

7 Al-Oqla, F.M., Hayajneh, M.T., and Fares, O. (2019). Investigating the mechanical thermal and polymer interfacial characteristics of Jordanian lignocellulosic fibers to demonstrate their capabilities for sustainable green materials. *Journal of Cleaner Production* 241: 118256.

8 AL-Oqla, F.M. and Omari, M.A. (2017). Sustainable biocomposites: challenges, potential and barriers for development. In: *Green Biocomposites: Manufacturing and Properties* (eds. M. Jawaid, S.M. Sapuan and O.Y. Alothman), 13–29. Cham: Springer.

9 AL-Oqla, F.M. and Salit, M.S. (2017). Material selection of natural fiber composites using the analytical hierarchy process. In: *Materials Selection for Natural Fiber Composites*, vol. 1, 169–234. Cambridge, USA: Woodhead Publishing, Elsevier.

10 AL-Oqla, F.M., Sapuan, M.S., Ishak, M.R., and Aziz, N.A. (2014). Combined multi-criteria evaluation stage technique as an agro waste evaluation indicator for polymeric composites: date palm fibers as a case study. *BioResources* 9 (3): 4608–4621. https://doi.org/10.15376/biores.9.3.4608-4621.

11 AL-Oqla, F.M., Sapuan, M.S., Ishak, M.R., and Nuraini, A.A. (2015). Decision making model for optimal reinforcement condition of natural fiber composites. *Fibers and Polymers* 16 (1): 153–163.

12 AL-Oqla, F.M. and Sapuan, S. (2018). Investigating the inherent characteristic/performance deterioration interactions of natural fibers in bio-composites for better utilization of resources. *Journal of Polymers and the Environment* 26 (3): 1290–1296.

13 Al-Oqla, F.M., Sapuan, S., Anwer, T. et al. (2015). Natural fiber reinforced conductive polymer composites as functional materials: a review. *Synthetic Metals* 206: 42–54.

14 Al-Oqla, F.M., Sapuan, S., and Fares, O. (2018). Electrical-based applications of natural fiber vinyl polymer composites. In: *Natural Fibre Reinforced Vinyl Ester and Vinyl Polymer Composites*, 349–367. Elsevier.

15 AL-Oqla, F.M., Sapuan, S., Ishak, M., and Nuraini, A. (2015). A model for evaluating and determining the most appropriate polymer matrix type for natural fiber composites. *International Journal of Polymer Analysis and Characterization* 20: 191–205. (just-accepted).

16 AL-Oqla, F.M., Sapuan, S., Ishak, M., and Nuraini, A. (2015). Predicting the potential of agro waste fibers for sustainable automotive industry using a decision making model. *Computers and Electronics in Agriculture* 113: 116–127.

17 AL-Oqla, F.M., Sapuan, S., Ishak, M., and Nuraini, A. (2016). A decision-making model for selecting the most appropriate natural fiber – polypropylene-based composites for automotive applications. *Journal of Composite Materials* 50 (4): 543–556.

18 AL-Oqla, F.M., Sapuan, S., Ishak, M., and Nuraini, A. (2014). A novel evaluation tool for enhancing the selection of natural fibers for polymeric composites based on fiber moisture content criterion. *BioResources* 10 (1): 299–312.

19 AL-Oqla, F.M., Sapuan, S., and Jawaid, M. (2016). Integrated mechanical–economic – environmental quality of performance for natural fibers for polymeric-based composite materials. *Journal of Natural Fibers* 13 (6): 651–659.

20 AL-Oqla, F.M. and Sapuan, S. (2014). Natural fiber reinforced polymer composites in industrial applications: feasibility of date palm fibers for sustainable automotive industry. *Journal of Cleaner Production* 66: 347–354. https://doi.org/10.1016/j.jclepro.2013.10.050.

21 Iannotti, G., Fair, N., Tempesta, M. et al. (2018). Studies on the environmental degradation of starch-based plastics. In: *Degradable Materials*, 425–446. CRC Press.

22 Ibrahim, H., Farag, M., Megahed, H., and Mehanny, S. (2014). Characteristics of starch-based biodegradable composites reinforced with date palm and flax fibers. *Carbohydrate Polymers* 101: 11–19.

23 Ilyas, R., Sapuan, S., Ishak, M., and Zainudin, E. (2018). Development and characterization of sugar palm nanocrystalline cellulose reinforced sugar palm starch bionanocomposites. *Carbohydrate Polymers* 202: 186–202.

24 AL-Oqla, F.M., Almagableh, A., and Omari, M.A. (2017). *Design and Fabrication of Green Biocomposites Green Biocomposites*, 45–67. Cham: Springer.

25 AL-Oqla, F.M., Alothman, O.Y., Jawaid, M. et al. (2014). Processing and properties of date palm fibers and its composites. In: *Biomass and Bioenergy*, 1–25. Cham: Springer.

26 Al-Oqla, F.M. and Omar, A.A. (2015). An expert-based model for selecting the most suitable substrate material type for antenna circuits. *International Journal of Electronics* 102 (6): 1044–1055.

27 AL-Oqla, F.M., Omar, A.A., and Fares, O. (2018). Evaluating sustainable energy harvesting systems for human implantable sensors. *International Journal of Electronics* 105 (3): 504–517.

28 Alaaeddin, M., Sapuan, S., Zuhri, M. et al. (2019). Lightweight and durable PVDF–SSPF composites for photovoltaics backsheet applications: thermal, optical and technical properties. *Materials* 12 (13): 2104.

29 Almagableh, A., Al-Oqla, F.M., and Omari, M.A. (2017). Predicting the effect of nano-structural parameters on the elastic properties of carbon nanotube-polymeric based composites. *International Journal of Performability Engineering* 13 (1): 73.

30 Toldy, A., Szolnoki, B., and Marosi, G. (2011). Flame retardancy of fibre-reinforced epoxy resin composites for aerospace applications. *Polymer Degradation and Stability* 96 (3): 371–376.

31 Li, J., Baker, B.A., Mou, X. et al. (2014). Biopolymer/calcium phosphate scaffolds for bone tissue engineering. *Advanced Healthcare Materials* 3 (4): 469–484.

32 Butcher, A.L., Offeddu, G.S., and Oyen, M.L. (2014). Nanofibrous hydrogel composites as mechanically robust tissue engineering scaffolds. *Trends in Biotechnology* 32 (11): 564–570.

33 Freeman, R., Boekhoven, J., Dickerson, M.B. et al. (2015). Biopolymers and supramolecular polymers as biomaterials for biomedical applications. *MRS Bulletin* 40 (12): 1089–1101.

34 Joung, Y.H. (2013). Development of implantable medical devices: from an engineering perspective. *International Neurourology Journal* 17 (3): 98–106. https://doi.org/10.5213/inj.2013.17.3.98.

35 Modjarrad, K. and Ebnesajjad, S. (2013). *Handbook of Polymer Applications in Medicine and Medical Devices*. Elsevier.

36 AL-Oqla, F.M. and Sapuan, S.M. (2014). Date palm fibers and natural composites. *Postgraduate Symposium on Composites Science and Technology 2014 & 4th Postgraduate Seminar on Natural Fibre Composites 2014*. Putrajaya (28 January 2014).

37 AL-Oqla, F.M. and Sapuan, S.M. (2014). Enhancement selecting proper natural fiber composites for industrial applications. *Postgraduate Symposium on Composites Science and Technology 2014 & 4th Postgraduate Seminar on Natural Fibre Composites 2014*. Putrajaya (28 January 2014).

38 Khairul, M., Faris, S., AL-Oqla, F.M., and Zainudin, E. (2019). Experimental investigation and numerical prediction for the fatigue life durability of austenitic stainless steel at room temperature. *Engineering Solid Mechanics* 7 (2): 121–130.

39 Rashid, B., Leman, Z., Jawaid, M. et al. (2017). Eco-friendly composites for brake pads from agro waste: a review. In: *Reference Module in Materials Science and Materials Engineering*. Elsevier.

40 Sadrmanesh, V., Chen, Y., Rahman, M., and Al-Oqla, F.M. (2019). Developing a decision making model to identify the most influential parameters affecting mechanical extraction of bast fibers. *Journal of Cleaner Production* 238: 117891.

41 Sapuan, S.M., Pua, F.-L., El-Shekeil, Y.A., and AL-Oqla, F.M. (2013). Mechanical properties of soil buried kenaf fibre reinforced thermoplastic polyurethane composites. *Materials & Design* 50: 467–470. https://doi.org/10.1016/j.matdes.2013.03.013.

42 Peças, P., Carvalho, H., Salman, H., and Leite, M. (2018). Natural fibre composites and their applications: a review. *Journal of Composites Science* 2 (4): 66.

43 Yousef, S., Mumladze, T., Tatariants, M. et al. (2018). Cleaner and profitable industrial technology for full recovery of metallic and non-metallic fraction of waste pharmaceutical blisters using switchable hydrophilicity solvents. *Journal of Cleaner Production* 197: 379–392.

44 Alaaeddin, M., Sapuan, S., Zuhri, M. et al. (2019). Development of photovoltaic module with fabricated and evaluated novel backsheet-based biocomposite materials. *Materials* 12 (18): 3007.

45 Majeed, K., Jawaid, M., Hassan, A. et al. (2013). Potential materials for food packaging from nanoclay/natural fibres filled hybrid composites. *Materials & Design* 46: 391–410.

46 Sanyang, M., Ilyas, R., Sapuan, S., and Jumaidin, R. (2018). Sugar palm starch-based composites for packaging applications. In: *Bionanocomposites for Packaging Applications*, 125–147. Springer.

47 AL-Oqla, F.M., Omari, M.A., and Al-Ghraibah, A. (2017). Predicting the potential of biomass-based composites for sustainable automotive industry using a decision-making model. In: *Lignocellulosic Fibre and Biomass-Based Composite Materials*, 27–43. Elsevier.

48 AL-Oqla, F.M. and Rababah, M. (2017). Challenges in design of nanocellulose and its composites for different applications. In: *Cellulose-Reinforced Nanofibre Composites*, 113–127. Elsevier.

49 Alaaeddin, M., Sapuan, S., Zuhri, M. et al. (2019). Polymer matrix materials selection for short sugar palm composites using integrated multi criteria evaluation method. *Composites Part B: Engineering* 176: 107342.

50 Abdulrahman, K.O., Abed, A.M., Bayode, A. et al. (2018). *Hierarchical Composite Materials: Materials, Manufacturing, Engineering*, vol. 8. Walter de Gruyter GmbH & Co KG.

51 Alaaeddin, M., Sapuan, S., Zuhri, M. et al. (2019). Photovoltaic applications: status and manufacturing prospects. *Renewable and Sustainable Energy Reviews* 102: 318–332.

52 Alaaeddin, M., Sapuan, S., Zuhri, M. et al. (2019). Physical and mechanical properties of polyvinylidene fluoride – short sugar palm fiber nanocomposites. *Journal of Cleaner Production* 235: 473–482.

53 Faris Mohammed Khair Faris AL-Oqla (2015). *Enhancement of Evaluation Methodologies for Natural Fiber Composites Material Selection System (Ph.D)*. UPM.

54 Al-Widyan, M.I. and Al-Oqla, F.M. (2011). Utilization of supplementary energy sources for cooling in hot arid regions via decision-making model. *International Journal of Engineering Research and Applications* 1 (4): 1610–1622.

55 Al-Widyan, M.I. and Al-Oqla, F.M. (2014). Selecting the most appropriate corrective actions for energy saving in existing buildings A/C in hot arid regions. *Building Simulation* 7 (5): 537–545. https://doi.org/10.1007/s12273-013-0170-3.

56 Dalalah, D., Al-Oqla, F., and Hayajneh, M. (2010). Application of the Analytic Hierarchy Process (AHP) in multi-criteria analysis of the selection of cranes. *Jordan Journal of Mechanical and Industrial Engineering, JJMIE* 4 (5): 567–578.

57 Dweiri, F. and Al-Oqla, F.M. (2006). Material selection using analytical hierarchy process. *International Journal of Computer Applications in Technology* 26 (4): 182–189. https://doi.org/10.1504/IJCAT.2006.010763.

58 Alves, C., Ferrão, P., Silva, A. et al. (2010). Ecodesign of automotive components making use of natural jute fiber composites. *Journal of Cleaner Production* 18 (4): 313–327. https://doi.org/10.1016/j.jclepro.2009.10.022.

59 Shah, D.U. (2016). Damage in biocomposites: stiffness evolution of aligned plant fibre composites during monotonic and cyclic fatigue loading. *Composites Part A: Applied Science and Manufacturing* 83: 160–168.

60 Arena, M., Azzone, G., and Conte, A. (2012). A streamlined LCA framework to support early decision making in vehicle development. *Journal of Cleaner Production* 41: 105–113.

61 Black, M., Whittaker, C., Hosseini, S. et al. (2011). Life cycle assessment and sustainability methodologies for assessing industrial crops, processes and end products. *Industrial Crops and Products* 34 (2): 1332–1339.

62 Luz, S.M., Caldeira-Pires, A., and Ferrao, P. (2010). Environmental benefits of substituting talc by sugarcane bagasse fibers as reinforcement in polypropylene composites: ecodesign and LCA as strategy for automotive components. *Resources, Conservation and Recycling* 54 (12): 1135–1144.

63 Milani, A., Eskicioglu, C., Robles, K. et al. (2011). Multiple criteria decision making with life cycle assessment for material selection of composites. *eXPRESS Polymer Letters* 5 (12): 1062–1074. https://doi.org/10.3144/expresspolymlett.2011.104.

64 Pegoretti, D.S., Mathieux, F., Evrard, D. et al. (2014). Use of recycled natural fibres in industrial products: a comparative LCA case study on acoustic components in the Brazilian automotive sector. *Resources, Conservation and Recycling* 84: 1–14.

2

Processing Methods for Manufacture of Biobased Composites

P. Shenbaga Velu[1], N. J. Vignesh[2], and N. Rajesh Jesudoss Hynes[2]

[1] *Department of Mechanical Engineering, P.S.R Engineering College, Sivakasi, Tamil Nadu, India*
[2] *Department of Mechanical Engineering, Mepco Schlenk Engineering College, Sivakasi, Tamil Nadu, India*

2.1 Introduction

Growing awareness among the people regarding the use of biodegradable materials instead of the great environmental killer (i.e. these plastic pollutants) has made scientists focus their research on biodegradable composites for more than a decade. These biocomposites have a wide range of applications, principally in automotive industries for parts such as door panels and inserts, rear trunk covers, side rims, tool box area, seat backs, dashboard, and parcel shelves. Nowadays, in addition to the automotive field, the aerospace industry also relies upon these biocomposites for the production of good-quality, high-strength, and thermally stable materials for making many critical parts of the aircraft. Their wide range of applications in various fields encourage the industries to focus on the various characterization techniques desirable to make these materials more commercial and also cost effective. In order to find the strength of these materials, these characterization techniques have become more prominent. Its growth in the field of biomedical engineering has fostered many practical and challenging applications in the biomedical field. The bigger challenge with these oil-based polymers is the environmental issues, which has an indirect effect on the economic growth of the country. This has led to an increased urge for the innovation of biobased materials as an alternative. Also, there is poor sustainability of the petroleum products when it is seen from an environmental standpoint [1]. The advantage of using biobased materials is that they are environmental friendly and much safer for human use/intake when compared to petroleum-based materials. Further, economic factors leading to hike in oil prices and their demand has resulted in ways of finding biobased materials of both natural and synthetic origin in order to drive away these crude oil-based polymer materials. One such example of biobased advancements in the biomedical field is the biobased topicals that enhance salicylic acid delivery [2]. Such advancements in the field of medicine have led to the exploration of gel formulations contained in natural polymers for a way of increasing the penetrating capability of the salicylic acid drug delivery to treat acne vulgaris on the skin; moreover, it is safe enough to be used as a topical on human skin. In anticancer drug delivery system, the use of calcium carbonate nanoparticles helps to enhance drug delivery [3]. In order to achieve

Biobased Composites: Processing, Characterization, Properties, and Applications, First Edition.
Anish Khan, Sanjay Mavinkere Rangappa, Suchart Siengchin and Abdullah M. Asiri.

timely delivery of the drugs in porous structures, the diffusion process through a matrix of the drug has been used in these biobased drug delivery systems [1]. In the recent years, in osteopathic medicines, polyethylene has become a replacement for platinum in joint surgery. One of the current topics for debate is that, some of the debris present in this polyethylene are causing infections in the patients.

In applications such as rotor blades of wind turbines, where controlled fiber orientation is necessary, these biobased composites provide a good and reasonable technical performance. But finding their mechanical characterization becomes a prime necessity before using these composites in real-time applications. This is because, in some cases, a highly complex loading has to be faced by the composite structure developed. Hence, a number of mechanical characterizations are carried out on the composite material before it is put to real-time application. Some of the important mechanical tests that are carried out include Tensile test, compression test, flexural test, and impact test.

Awal et al. [4] utilized polylactic acid (PLA) as matrix and cellulosic wood fibers as reinforcement for manufacturing biocomposites. They used injection molding and extrusion molding processes for the fabrication of these composite materials. In order to increase the adhesive nature in between the fiber and matrix interfaces, they used an additive named Bioadimide and this also helped to increase the process speed of the biocomposites.

Boumhaout et al. [5] developed a biocomposite for the purpose of building insulation. The method used for carrying out the manufacturing process was the hand layup technique and the authors used date palm fiber mesh with mortar as the reinforcement material. Their research findings showed that these bioreinforcements are capable of increasing the insulation properties by decreasing 70% of its thermal conductivity. Ortega et al. [6] developed a PA11 (polyamide)-based biocomposite using stone ground wood (SGW) as reinforcement. Extrusion–injection molding technique was used for the preparation of these samples. It was concluded that, when the weight percentage of SGW was about 50%, the tensile strength of the prepared biocomposites was at its maximum.

Shibata et al. [7] made a biopolymer using a biodegradable matrix (cornstarch) and kenaf/bamboo as reinforcement using press forming method. The cox model was used to predict the flexural modulus of unidirectional and random-oriented fiber composites. The values predicted were in a greater agreement with the experimental values. Arao et al. [8] used PLA as a biodegradable matrix for preparing biobased composites. They used jute fibers as a reinforcement. They suggested that, by using a well-compounded pallet in the injection molding process, the tensile strength and Young modulus of developed composites can be improved. A novel nanocomposite biodegradable material using graphene oxide as reinforcement and chitosan as the matrix has been developed by Khan et al [9]. For successfully preparing these composite films, a technique called solution casting technique was used. The developed biodegradable films are having very good tensile strength and thermal stability and hence they could be used successfully in several biomedical applications. Table 2.1 shows some of the examples of commercially available biobased thermosetting polymers.

2.2 Biobased Materials

Biobased materials are the materials which contain components of biorenewable origins. As time passes, there is no doubt that, by every chance, these novel biobased materials will replace the crude oil-based polymer materials. Some of the most common examples of these

Table 2.1 Examples of commercially available biobased thermosetting polymers.

S. no.	Manufacturer	Trade name	Raw materials	Applications
1.	DSM	Palapreg ECO P 55-01	Unsaturated polyester	SMC/BMC
2.	Bioresin	Bioresin	Castor oil	Automotive, marine
3.	Ashland	ENVIREZ 1807	Unsaturated polyester, soybean oil (18% biobased)	Tractor panels
4.	JVS Polymers Ltd	LAIT-X/POLLIT	Lactic acid based	Coatings, biomedical applications
5.	Amroy Europe Oy	EpoBiox	Natural phenols distilled from forest industry waste stream	Boats, glues, electrical cars

biorenewable resources include biofibers, biomass feedstocks, and biopolymers [10]. Considering the adverse effects of petroleum-based fuel resources, biobased materials have given a more ecofriendly and safe alternative resource, namely the Biomass feedstock. In the fabrication process of biobased composites, biofibers, such as hemp and flax, are most commonly used as reinforcements. Cellulosic plastics are one of the common examples of biopolymers [10]. One thing to be noted here is that, the biorenewable portion need not be on one side at all times during the manufacturing of biobased composites. It can be either a matrix or a filler or both.

Different types of composites require different kinds of processing methodologies. A variety of fillers and reinforcements are used for the development of these biobased composites, which make them suitable for various biomedical applications. In practical biomedical applications, both biobased and non-biobased fillers/reinforcements with either biobased or non-biobased polymers were used.

The use of biobased composites in the field of biomedicine will continue to expand as long as there is continuation of research in the field. Also, these materials will be having a direct positive impact on the society both economically and environmentally.

2.3 Processing Methods

Commercial and domestic fields have increased their usage of biobased composite materials as the demand for these biobased materials have been increasing during the past decade. A number of processing methods, developed by researchers, are now available for the successful production of these biobased components. Nevertheless, with these available processing techniques, there requires much improvement in the field in order to minimize the number of defects produced in the biobased composites. Some of the most prominent fabrication techniques used for biobased composites have been depicted in Table 2.2. The simplest and cheapest among them is the hand layup technique, which is used for producing relatively simple shapes. A method to improve the surface finish and dimensional accuracy of the products produced is the spray layup technique, which is a modified version of the hand layup technique.

Idicula et al. [11] used short randomly oriented banana and sisal fibers as reinforcing materials for developing a polyester-based hybrid biocomposite. The composites were

Table 2.2 Fabrication methods to produce biocomposites.

S. no.	Method of fabrication	Key point	Features
1.	Hand lay-up	Involves the use of an open mold in which the resin and reinforcements are placed manually and a mild pressure is applied using the roller in order to remove the voids.	Tooling cost is very low and there is no size constraint. Examples range from small tubs, medium storage tanks to larger boats, etc.
2.	Spray lay-up	A typically developed spray gun is used for spraying the mixture of chopped fiber and resin in an open mold.	This is a faster method compared to the hand lay-up technique. Also, semi-automatic in nature. Examples range from small tubs, medium storage tanks to larger boats, etc., with more accuracy.
3.	Compression molding	It is a method by which the resin and reinforcement are heated by a closed metallic mold structure in which the reinforcement and resin are initially placed in the lower half of the metallic mold and the pressure is applied by closing both the halves of the mold. The various controlling parameters include time, temperature, and pressure.	Specific features include lower postprocessing requirements and higher part design flexibility. The applications are structural components and automobile components.
4.	Injection molding	The mixture of reinforcement and resin are pushed through a screw extruder that pushes the compounds through a closed mold at the end of a nozzle. On through the way, the resin and reinforcement mixtures are heated up by a heater surrounding the extruder.	Specific features of the process include high volume production and very fast cycle time. Applications include floor pans, bumpers, automobile body panels, etc.
5.	Pultrusion method	Consists of a long continuous chain of fibers wound over a drum. It is being pulled through the path way of the resin and passed through a heated die. The shape of die controls the cross-section of the product.	Specific features include constant cross-sectional shape products. Applications are channels, long rods and extremely high fiber loading beams.
6.	Vacuum bag molding	Consists of a vacuum bag that is used to cover the mold. Used for producing laminates from prepregs or wet lay-ups. This method helps to improve the mechanical properties of laminates.	Specific features include uniform degree of consolidation and includes no void content.

developed using a compression molding machine. They found that, the composites made by using compression molding techniques are having very good impact strength when compared to those produced using resin transfer molding. Sreekumar et al. [12] developed biocomposites using a compression molding machine and investigated their mechanical properties. Polyester has been used as the matrix and sisal fiber as reinforcement. They compared the mechanical properties of these developed composites with that of the composites made using resin transfer molding. They found out that the strength of the composites made by compression molding technique is lower than that made from resin transfer molding technique. Kim et al. [13] used an injection molding machine for developing biobased composites using pineapple and cassava flour as reinforcements and PLA as matrix. Here they found out that the tensile and flexural strength has increased and it is attributed to the use of compatibilizer with flour as reinforcement. This helped to increase the interfacial adhesion between the matrix and the fillers. Feldmann et al. [14] developed biobased composites by using biorenewable polyamides as matrix and cellulosic fibers as reinforcements. The difference in this work is that, the authors have used only a single-step pultrusion process instead of the conventional two-step pultrusion process. The resultant composites from this single-step pultrusion process showed very good tensile strength when compared to the composites made from other processes. Sevkat et al. [15] developed a hybrid hollow shaft-like composite using the filament winding method. The advantage with this method is that changing the fiber loading can be achieved with much ease even when the production is running. Hence, it is concluded that hybridization could be easily achieved by using the filament winding technique. When the shapes of the composites becomes complex with higher fiber percentage, then resin transfer molding (RTM) technique could be used. Dweib et al. [16] developed biocomposites by using soybean oil as the matrix material and cellulose fiber as reinforcement. The authors, in order to manufacture composites of roof top structural applications with large dimensions, utilized a modified vacuum-assisted resin transfer molding technique. In order to eliminate the need for using a solid mold, the authors used a nylon bagging sheet for covering the entire part.

Boey et al. [17] used an autoclave process to develop thermoset composites. The study was carried out on the developed composites by considering the effect of vacuum on the voids present. Once the composites reach an initial pressurization range, there occurs a significant decrease in the void content of the composite. Some other techniques that are being used for developing these biobased composites include tube rolling, automated fiber placement (AFP), and automated tape laying (ATL).

2.4 Fabrication Techniques of Biobased Composites

Many a time, in order to improve the overall thermomechanical properties of the developed composites, either natural polymers such as cellulose are used as resins in addition to the reinforcing fibers or synthetic polymers are used for polymerization. Some of the unique advantages of these biobased materials are that they offer good biodegradability and biocompatibility. This helps them to find their application in the field of biomedical engineering and also in tissue engineering. When used in the biomedical field, inside the human body, these scaffolds that are engineered must be capable of degrading in a way which is nontoxic to the user. It should also be capable of mimicking the body's own extracellular matrix (ECM) in order to allow the proliferation of human body cells and also to avoid an immune response. The rate of gradation of these scaffolds must take place in a proper way. Hence, it becomes difficult for us to synthesize these biobased polymers alone. Nanoparticles such as hydroxyapatite (HA) could be used for nanoscale hybridization of natural and synthetic materials [18] or carbon nanotubes [19]. The following are the most common fabrication methods used for developing biobased composites.

They are:

1) Solvent Casting and Particulate Leaching (SCPL)
2) Emulsion Freeze Drying

3) Electrospinning
4) Blow Film Extrusion
5) 3D printing

2.4.1 Solvent Casting and Particulate Leaching

During Solvent Casting and Particulate Leaching (SCPL) technique, the polymer matrix is found to be dispersed with soluble particles [20]. As soon as the material is consolidated, an appropriate solvent washes away the dissolved particles, which then leads to a porous material [20]. But on the other hand, when the parts become thicker, then the solvent evaporation becomes difficult and the solvent particulate removal too becomes difficult [21]. Under these circumstances, the remaining amount of solvents or particulates available may be removed with the help of vacuum drying [22]. The orthopedic scaffolds that are used for bone replacement or repair have been fabricated by the SCPL method [23]. For avoiding the toxic nature of the solvents used with the host tissues, a highly porous biopolymer foam has been created through a method of gas foaming with the help of carbon dioxide in addition to the particulate leaching technique [24, 25]. Figure 2.1 depicts the steps involved in solvent casting and particulate leaching. For creating a homogeneous mixture, the polymer solution, prepared using an organic solvent, is mixed with a porogen particulate (i.e. NaCl). This mixture is then poured into the desired mold, after which the solvent has been made to evaporate. In order to obtain a complete porous structure, the mold has been washed by means of a particulate

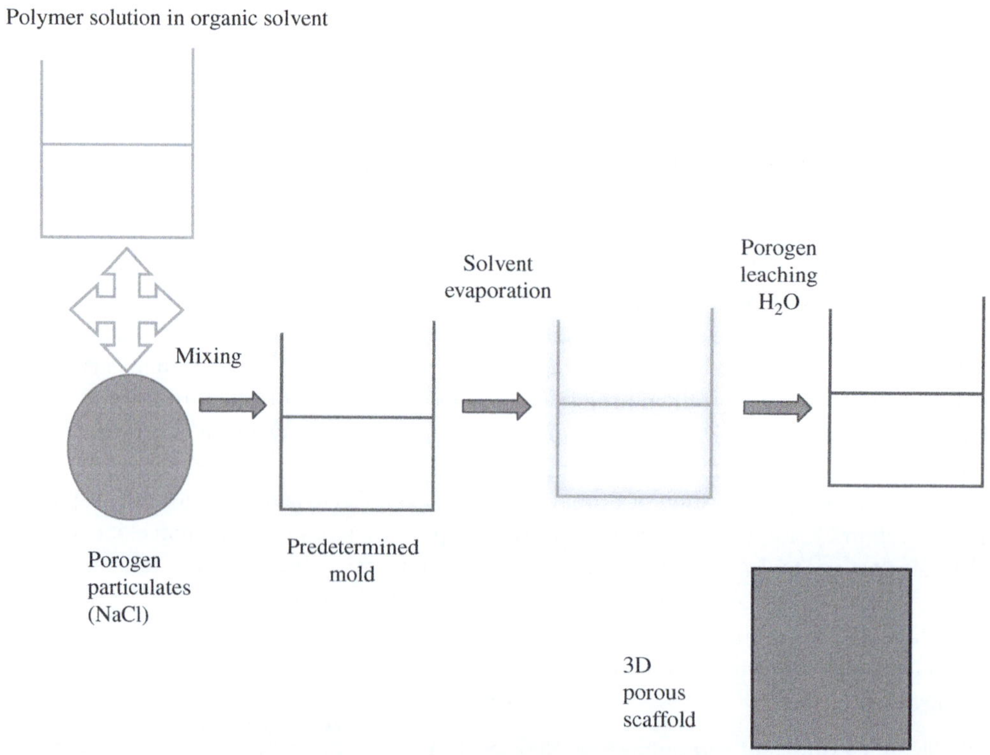

Figure 2.1 Steps involved in solvent casting and particulate leaching.

solution like water which dissolves easily for removing the residual particulates. Drying is carried out under vacuum in order to remove the residual solvent or particulate to get the newly formed scaffold.

2.4.2 Emulsion Freeze Drying

This method is particularly suitable for collagen-based scaffolds. This is because the collagen has a higher sensitivity to heat denaturation and chemicals. It has been found that aortic heart valve cells can be grown by creating these porous collagen scaffolds [26]. For entrapping the ice crystals, these collagen type 1 fibers were frozen at −30 °C. The required porous materials can be obtained by removing the crystals using a process known as lyophilization. The prominent parameters that affect the pore size include temperature, pH, and solution concentration. Emulsion freeze drying method is used for producing scaffolds having heat-sensitive bioactive molecules with pore sizes in the range 20–200 μm and with porosity greater than 90% [27, 28].

2.4.3 Electrospinning

Electrospinning is another technique which is used for the fabrication of cardiovascular scaffolds [29]. Figure 2.2 depicts the schematic representation of the electrospinning process. An electric field is applied as soon as the solution containing a conductive polymer is ejected from the needle and collected on a target. When the solvent dries out, a fibrous network exactly similar to the structure of our body's natural ECM is formed [30]. By adjusting the processing parameters (such as voltage, working distance, flow rate, and temperature) or the polymer solution conditions (such as conductivity, viscosity, and concentration), the structure as well as the diameter of the fiber can be altered [31]. Taking advantage of the synthetic polymers' mechanical strength and natural polymers' biocompatibility, serious efforts have been taken to hybridize these structures. Successful and interesting advancements have been made in the field of heart valve tissue engineering [32]. Under these circumstances, electrospinning technique has been used to develop a biohybrid scaffold, which consists of non-cross-linked decellularized bovine

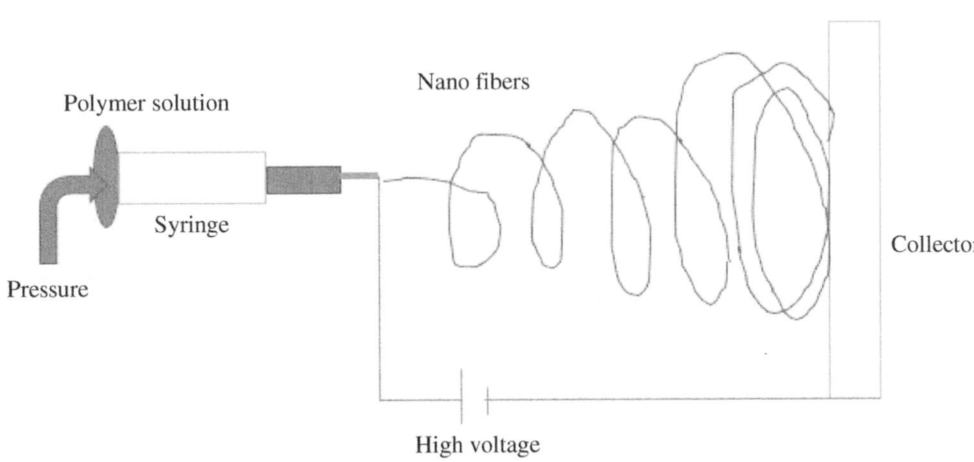

Figure 2.2 Fabrication of electrospun nanofibers under high voltage.

pericardium ECM coated with an adhesive layer of polycaprolactone (PCL)–chitosan nanofibers [33]. In order to enhance the fiber–polymer interactions, and due to the hydrophobic nature of PCL, it has been blended with a mixture of several materials (dextran, cellulose acetate, polyhydroxybutyrate, and chitosan nanofibers) [34]. Dip coating technique has been used to improve the mechanical properties of the decellularized scaffolds, without the use of cross-linkers [35]. Even though these dip-coated scaffolds resulted in having better mechanical properties than electrospinning, the structural integrity of the ECM nanofibers were disrupted with the use of organic solvents. Thus, electrospinning forms the most appropriate choice for processing.

2.4.4 Blow Film Extrusion

Blow film extrusion is used for the manufacture of polymeric films. This method involves the extrusion of molten polymeric material, which takes the shape of a die into which it is blown, while air blown into it forms a thin film. Due to their significant advantages such as biodegradability, low cost for processing, and nontoxicity, these blown films are finding their applications in the pharmaceutical industry. Chitosan [36], polysaccharides [37], cellulose [38], and their derivatives are the most commonly used biobased polymers in film industries. Increased attention has been gained for the preparation of blown films made from starch-based materials [39]. Chitosan- and starch-based thermoplastic films have been found to be suitable for applications in pharmaceutical and food processing industries. By increasing the percentage of chitosan in the film, properties such as water vapor and oxygen barrier tend to increase, while the surface hydrophilicity is reduced. Hence, by controlling the amount of chitosan in the film, the rate of degradation of the film, and consequently, its shelf life can be tuned [39].

2.4.5 3D Printing

In 3D printing technique, a preprogrammed printer head moves over the target surface and ejects a fixed quantity of molten polymer on the target in order to create the final desired shape. For attaining the final 3D structure, the movable printer head deposits the polymer material in a layer-by-layer fashion. This process is also known as Rapid Prototyping and has prominent applications in biomedical fields such as dental implants [40], orthopedic prosthetics [41, 42], 3D surgical and medical models, and also hearing aids. 3D bioprinting, a new technique, is nothing but a variation of traditional 3D printing in which 3D biofunctional structures are engineered. It involves the deposition of living cells onto a gel medium [43]. Some studies have used nanomaterials in conjunction with biohybrid scaffolds to create additional functions [44]. Some examples show that, new bone growth could be induced by adding magnesia nanoparticles to PCL–chitosan nanofibers through modulation of signal transduction and cell proliferation [44]. In another example, it has been found that this cell proliferation could also be induced by means of magnetic heating when magnetic nanoparticles are added to PCL–chitosan nanofibers [45].

In the field of tissue engineering, electrospinning and 3D printing are the most common techniques by which the scaffolds are fabricated. In addition to these techniques, common biomedical supplies are prepared by conventional techniques such as compression molding or injection molding. In recent years, in order to combine the advantages of both, a combination of advanced methods and conventional techniques is used to reduce the overall processing time and cost. Examples include particulate leaching and solvent casting [46].

2.5 Fillers and Reinforcements Used in the Preparation of Biobased Composites

Generally, there are three different methods by which biobased composites are obtained. They are as follows:

1) The reinforcement of non-biobased polymers with biobased fillers/fibers
2) The reinforcement of biobased polymers with non-biobased fillers/fibers
3) The reinforcement of biobased polymers with biobased fillers/fibers.

2.5.1 Biobased Fillers/Reinforcements with Non-biobased Polymers

Fillers/additives used to produce environmental friendly materials include waste paper sludge (WPS) and ink-eliminated sludge (IES). These materials are used in high-density polyethylene (HDPE)/wood flour composites [47]. Research shows that, the flexural properties of the HDPE/wood flour composites increased when WPS was added as an additive. This increase in flexural strength could be further increased by using a compatibilizer such as malleated anhydride grafted polyethylene (MAPE) in between the reinforcement and the matrix. Also, there occurs a decrease in the swelling and water absorption properties when this WPS and IES are added as fillers/additives. It has been found out that the overall mechanical properties of IES-added composites show better results than that of the WPS-added composites. The optimum loading condition of IES-added composites is 60 wt%. These results show a promising outcome that these composites could be used for the preparation of biobased composites.

As these HDPE/wood flour composites find its applications in the exteriors of construction industries, their amount of durability could be calculated by investigating the damage caused due to sunlight and other weathering conditions [48]. Artificial weathering experiments could be carried out on the surface of these composites and by carefully investigating the chemical changes happening on the surface, help in finding out the durability. Further, it has been proved that, when carrying out tests such as X-ray photoelectron spectroscopy (XPS) and Fourier Transform Infrared Spectroscopy (FTIR), the amount of weathering damage to the wood flour-reinforced HDPE is increased to 16 times faster than unreinforced plastic. The outcomes confirm that the degradation mechanism is triggered in the polymer chain due to the presence of carbonyl groups and leads to polymer chain scission once the surface oxidation is started. There is an overall reduction in the crystallinity of the polymer as the time increases and when the chain scission mechanism has been initiated.

2.5.2 Non-biobased Fillers/Reinforcements with Biobased Polymers

The development of biobased composites, in which non-biobased additives or fillers are reinforced with the biobased polymers has been restricted to the addition of reinforcements to PLAs [48, 49]. When PLA is reinforced with fillers such as hydroxyapatite fibers (HAF) (up to 60 wt%), there is a significant improvement in their Young's modulus under compression molding conditions at 180 °C. The maximum value of Young's modulus attained was about 10 GPa for PLA/HAF composites [49]. For heavily loaded composites, a brittle behavior was observed and a covalent fiber–matrix interaction was revealed [49]. When PLA has been reinforced with carbon nanotubes (CNTs), there occurs a significant rise in the Young's modulus and hardness of the composites [50]. Purified CNTs are able to better disperse in a PLA matrix than nonpurified CNTs, thus resulting in better thermomechanical properties.

2.5.3 Biobased Filler/Reinforcement and Biobased Polymer

Biobased composites with natural fiber reinforcements have several advantages over those with synthetic reinforcements and has become one of the most popular research topics of the decade. Further, natural polymers derived from renewable resources are ecofriendlier and less expensive compared to synthetic polymers. Hemp, keratin, pure cellulose fibers, and flax are used to produce composites with very good dampening properties and high flexibility [51, 52]. In the current research scenario, focus is on enhancing the fiber–resin interactions, neglecting fiber orientation. The poor uniformity in the fiber length and orientation in biobased composites results in huge variations in the mechanical properties of similar samples and hence represents one of the major limitations for its use as a commercial product. Researchers are carrying out both physical and chemical surface treatments for presenting them successfully in the market as a competitive alternative to synthetic fibers. Each component in the natural fiber composite has been separated and studied individually for optimizing the fiber–matrix interactions and also to completely understand their physical properties. Thermotreatment [53], calendaring [54], fiber stretching [55], and electric discharge [56] are some of the common optimization methods available. The methods indicated herein are used to improve the surface of the fibers physically on a macroscale and they do not alter or induce any chemical changes in the fibers. In some cases, they could also be used to improve the adhesion with the polymer matrix and also fiber orientation. Shorter fibers from high cellulose content fibers and plant particulates could be filtered by using Solvent extraction technique.

In some cases, in order to improve the fiber–matrix interactions, the chemical treatment of the fibers is also carried out. Alkaline treatments of fibers can be used as a pretreatment method in addition to the chemical modifications that are carried out. Alkalis like sodium hydroxide could be used for treating the fibers in which the cellulose lattice is converted into a more stable form [57]. It has been found that the mechanical properties of flax/epoxy polymers are increasing at a significant rate than the nontreated samples after alkaline treatment [58]. On the other hand, if there is no stretching of the fibers during the NaOH/water alkaline treatment, then it results in the shrinkage of the fibers [59]. Water can be substituted with ethanol for overcoming this problem [59]. Another method used for enhancing the interfacial adhesion between the reinforcements and the polymer matrix is the use of coupling agents, such as organosilanes which is a common chemical modification method [60]. Even though there occurs a decrease in the crystallinity of the reinforcements, there is still an increase in the tensile properties of the PLA/wood flour composites. Morphological analysis also proves that there is a stronger interfacial bonding that occur in the developed composites.

2.6 Conclusion

It is a well-known fact that the research activities related to the development of novel biobased polymers have been increasing drastically over the past few decades. Nowadays, the trend has changed from using simpler systems having thermoplastic materials reinforced with natural fibers to more advanced composites with carefully engineered biobased matrices or which are completely of biobased origin (i.e. both the matrix and reinforcements are biobased). Advancements in the polymer and composite processing techniques have led to significant progress in the development of novel biobased composites, which have both industrial and commercial applications. PLA and PLA-based composites are the most predominating materials used as biobased biomedical equipment, as there is only little focus or work done in other materials such as hydrogels, which are used as scaffolds in cell growth and tissue regeneration. Expectations are in such a way that the development of novel biobased polymers will continue its ladder climbing as it finds more interesting applications in the biomedical field, especially for cell scaffolding and drug delivery, as reported in this chapter.

References

1 Alghazali, K.M., Nima, Z.A., Hamzah, R.N. et al. (2015). Bone-tissue engineering: complex tunable structural and biological responses to injury, drug delivery, and cell-based therapies. *Drug Metab. Rev.* 47 (4): 431–454.

2 Langasco, R., Spada, G., Tanriverdi, S.T. et al. (2016). Bio-based topical system for enhanced salicylic acid delivery: preparation and performance of gels. *J. Pharm. Pharmacol.* 68 (8): 999–1009.

3 Render, D., Rangari, V.K., Jeelani, S. et al. (2014). Biobased calcium carbonate ($CaCO_3$) nanoparticles for drug delivery applications. *Int. J. Biomed. Nanosci. Nanotechnol.* 3 (3): 221–235.

4 Awal, A., Rana, M., and Sain, M. (2014). Thermorheological and mechanical properties of cellulose reinforced PLA bio-composites. *Mech. Mater.* 80: 87–95. https://doi.org/10.1016/j.mechmat.2014.09.009.

5 Boumhaout, M., Boukhattem, L., Hamdi, H. et al. (2017). Thermomechanical characterization of a bio-composite building material: mortar reinforced with date palm fibers mesh. *Constr. Build. Mater.* 135: 241–250.

6 Oliver-Ortega, H., Granda, L.A., Espinach, F.X. et al. (2016). Tensile properties and micromechanical analysis of stone groundwood from softwood reinforced biobased polyamide11 composites. *Compos. Sci. Technol.* 132: 123–130.

7 Shibata, S., Cao, Y., and Fukumoto, I. (2008). Flexural modulus of the unidirectional and random composites made from biodegradable resin and bamboo and kenaf fibres. *Compos. Part A* 39: 640–646.

8 Arao, Y., Fujiura, T., Itani, S., and Tanaka, T. (2015). Strength improvement in injection-molded jutefiber-reinforced polylactide green-composites. *Compos. Part B* 68: 200–206.

9 Khan, Y.H., Islam, A., Sarwar, A. et al. (2016). Novel green nano composites films fabricated by indigenously synthesized graphene oxide and chitosan. *Carbohydr. Polym.* 146: 1–466. https://doi.org/10.1016/j.carbpol.2016.03.031.

10 Mohanty, A.K., Misra, M., and Drzal, L.T. (2002). Sustainable bio-composites from renewable resources: opportunities and challenges in the green materials world. *J. Polym. Environ.* 10 (1/2): 19–26.

11 Idicula, M., Sreekumar, P.A., Joseph, K., and Thomas, S. (2009). Natural fiber hybrid composites – a comparison between compression molding and resin transfer molding. *Polym. Compos.* 30: 1417–1425. https://doi.org/10.1002/pc.20706.

12 Sreekumar, P.A., Joseph, K., Unnikrishnan, G., and Thomas, S. (2007). A comparative study on mechanical properties of sisal-leaf fibre-reinforced polyester composites prepared by resin transfer and compression moulding techniques. *Compos. Sci. Technol.* 67: 453–461.

13 Kim, K.-W., Lee, B.-H., Kim, H.-J. et al. (2012). Thermal and mechanical properties of cassava and pineapple flours-filled PLA bio-composites. *J. Therm. Anal. Calorim.* 108: 1131–1139.

14 Feldmann, M. and Bledzki, A.K. (2014). Bio-based polyamides reinforced with cellulosic fibres – processing and properties. *Compos. Sci. Technol.* 100: 113–120.

15 Sevkat, E. and Tumer, H. (2013). Residual torsional properties of composite shafts subjected to impact loadings. *Mater. Des.* 51: 956–967.

16 Dweib, M.A., Hu, B., Shenton, H.W., and Wool, R.P. (2006). Bio-based composite roof structure: manufacturing and processing issues. *Compos. Struct.* 74: 379–388.

17 Boey, F.Y.C. and Lye, S.W. (1992). Void reduction in autoclave processing of thermoset composites. Part 1: high pressure effects on void reduction. *Composites* 23: 261–265.

18 Xie, L., Yu, H., Yang, W. et al. (2016). Preparation in vitro degradability cytotoxicity and in vivo biocompatibility of porous hydroxyapatite whisker-reinforced poly(l-lactide) biocomposite scaffolds. *J. Biomater. Sci. Polym. Ed.* 27 (6): 505–528.

19 Li, X., Cui, R., Liu, W. et al. (2013). The use of nanoscaled fibers or tubes to improve biocompatibility and bioactivity of biomedical materials. *J. Nanomater.* 2013: 1–16.

20 Lu, L., Peter, S.J., Lyman, M.D. et al. (2000). In vitro degradation of porous poly (l-lactic acid) foams. *Biomaterials* 21: 1595–1605.

21 Chen, M.K. and Badylak, S.F. (2001). Small bowel tissue engineering using small intestinal submucosa as a scaffold. *J. Surg. Res.* 99 (2): 352–358.

22 Rogers, L., Said, S.S., and Mequanint, K. (2013). The effects of fabrication strategies on 3D scaffold morphology porosity and vascular smooth muscle cell response. *J. Biomater. Tissue Eng.* 3: 300–311.

23 Prasad, A., Sankar, M.R., and Katiyar, V. (2017). State of art on solvent casting particulate leaching method for orthopedic scaffolds fabrication. *Mater. Today Proc.* 4 (2): 898–907.

24 Okamoto, M. (2006). Biodegradable polymer/layered silicate nanocomposites: a review. In: *Handbook of Biodegradable Polymeric Materials and their Applications* (eds. S. Mallapragada and B. Narasimhan), 153–197. Los Angeles: American Scientific Publishers.

25 Kim, S.S., Ahn, K.M., Park, M.S. et al. (2007). A poly(lactideco-glycolide)/hydroxyapatite composite scaffold with enhanced osteoconductivity. *J. Biomed. Mater. Res. Part A* 80: 206–215.

26 Taylor, P.M., Sachlos, E., Dreger, S.A. et al. (2006). Interaction of human valve interstitial cells with collagen matrices manufactured using rapid prototyping. *Biomaterials* 27 (13): 2733–2737.

27 Aranaz, I., Gutierrez, M.C., Ferrer, M.L., and Monte, F. (2014). Preparation of chitosan nanocomposites with a macroporous structure by unidirectional freezing and subsequent freeze-drying. *Marine Drugs* 12 (11): 5619–5642.

28 Hottot, A., Vessot, S., and Andrieu, J. (2004). A direct characterization method of the ice morphology relationship between mean crystals size and primary drying times of freeze-drying processes. *Dry. Technol.* 22: 2009–2021.

29 Das, S., Hollister, S.J., Flanagan, C. et al. (2003). Freeform fabrication of Nylon-6 tissue engineering scaffolds. *Rapid Prototyp. J.* 9 (1): 43–49.

30 Xie, J., Li, X., and Xia, Y. (2008). Putting electrospun nanofibers to work for biomedical research. *Macromol. Rapid Commun.* 29: 1775–1792.

31 Xu, X. and Zhou, M. (2008). Antimicrobial gelatin nanofibers containing silver nanoparticles. *Fibers Polym.* 9: 685–690.

32 Hong, H., Dong, N., Shi, J. et al. (2009). Fabrication of a novel hybrid heart valve leaflet for tissue engineering: an in vitro study. *Artif. Organs* 33: 554–558.

33 Jahnavi, S., Kumary, T., Bhuvaneshwar, G. et al. (2015). Engineering of a polymer layered bio-hybrid heart valve scaffold. *Mater. Sci. Eng. C* 51: 263–273.

34 Malheiro, V.N., Caridade, S.G., Alves, N.M., and Mano, J.F. (2010). New poly(ε-caprolactone)/chitosan blend fibers for tissue engineering applications. *Acta Biomater.* 6 (2): 418–428.

35 Nirmal, R.S. and Nair, P.D. (2013). Significance of soluble growth factors in the chondrogenic response of human umbilical cord matrix stem cells in a porous three-dimensional scaffold. *Eur. Cells Mater.* 26: 234–251.

36 Dutta, P.K., Tripathi, S., Mehrotra, G.K., and Dutta, J. (2009). Perspectives for chitosan based antimicrobial films in food applications. *Food Chem.* 114 (4): 1173–1182.

37 Al-Hassan, A.A. and Norziah, M.H. (2012). Starch–gelatin edible films: water vaporpermeability and mechanical properties as affected by plasticizers. *Food Hydrocoll.* 26 (1): 108–117.

38 Yu, Z., Alsammarraie, F.K., Nayigiziki, F.X. et al. (2017). Effect and mechanism of cellulose nanofibrils on the active functions of biopolymer-based nanocomposite films. *Food Res. Int.* 99 (1): 166–172.

39 Dang, K.M. and Yoksan, R. (2016). Morphological characteristics and barrier properties of thermoplastic starch/chitosan blown film. *Carbohydr. Polym.* 150: 40–47.

40 Wu, J., Li, Y., and Zhang, Y. (2017). Use of intraoral scanning and 3-dimensional printing in the fabrication of a removable partial denture for a patient with limited mouth opening. *J. Am. Dent. Assoc.* 148 (5): 338–341.

41 Gross, B.C., Erkal, J.L., Lockwood, S.Y. et al. (2014). Evaluation of 3D printing and its potential impact on biotechnology and the chemical sciences. *Anal. Chem.* 86 (7): 3240–3253.

42 Zopf, D.A., Hollister, S.J., Nelson, M.E. et al. (2013). Bioresorbable airway splint created with a three-dimensional printer. *N. Engl. J. Med.* 368 (21): 2043–2045.

43 Schubert, C., Langeveld, M.C., and Donoso, L.A. (2014). Innovations in 3D printing: a 3D overview from optics to organs. *Br. J. Ophthalmol.* 98 (2): 159–161.

44 Roh, H.S., Lee, C.M., Hwang, Y.H. et al. (2017). Addition of MgO nanoparticles and plasma surface treatment of three-dimensional printed polycaprolactone/hydroxyapatite scaffolds for improving bone regeneration. *Mater. Sci. Eng. C* 74: 525–535.

45 Meng, J., Xiao, B., Zhang, Y. et al. (2013). Super-paramagnetic responsive nanofibrous scaffolds under static magnetic field enhance osteogenesis for bone repair in vivo. *Sci. Rep.* 3: 2655.

46 Okamoto, M. and John, B. (2013). Synthetic biopolymer nanocomposites for tissue engineering scaffolds. *Prog. Polym. Sci.* 38: 1487–1503.

47 Hamzeh, Y., Ashori, A., and Mirzaei, B. (2011). Effects of waste paper sludge on the physico-mechanical properties of high-density polyethylene/wood flour composites. *J. Polym. Environ.* 19 (1): 120–124.

48 Stark, N.M. and Matuana, L.M. (2004). Surface chemistry changes of weathered HDPE/wood-flour composites studied by XPS and FTIR spectroscopy. *Polym. Degrad. Stab.* 86: 1–9.

49 Kasuga, T., Ota, Y., Nogami, M., and Abe, Y. (2001). Preparation and mechanical properties of polylactic acid composites containing hydroxyapatite fibers. *Biomaterials* 22: 19–23.

50 Chiu, W., Chang, Y., Kuo, H. et al. (2008). A study of carbon nanotubes/biodegradable plastic polylactic acid composites. *J. Appl. Polym. Sci.* 108 (5): 3024–3030.

51 John, M.J. and Thomas, S. (2008). Biofibres and biocomposites. *Carbohydr. Polym.* 71: 343–364.

52 Sydenstricker, T.H., Mochnaz, S., and Amico, S.C. (2003). Pull-out and other evaluations in sisal-reinforced polyester biocomposites. *Polym. Test* 22: 375–380.

53 Ray, P.K., Chakravarty, A.C., and Bandyopadhyay, S.B. (1976). Fine structure and mechanical properties of jute differently dried after retting. *J. Appl. Polym. Sci.* 20 (7): 1765–1767.

54 Semsarzadeh, M.A. (1986). Fiber matrix interactions in jute reinforced polyester resin. *Polym. Compos.* 7: 23–25.

55 Zeronian, S.H., Kawabata, H., and Alger, K. (1990). Factors affecting the tensile properties of non-mercerized and mercerized cotton fibers. *Text. Res. J.* 60 (3): 179–183.

56 Paiva, J.M.F. and Frollini, E. (2006). Unmodified and modified surface sisal fibers as reinforcement of phenolic and lignophenolic matrices composites: thermal analyses of fibers and composites. *Macromol. Mater. Eng.* 291: 405–417.

57 Agrawal, R., Saxena, N., Sreekala, M., and Thomas, S. (2000). Effect of treatment on the thermal conductivity and thermal diffusivity of oil-palm–fiber-reinforced phenolformaldehyde composites. *J. Polym. Sci. B Polym. Phys.* 38: 916–921.

58 Vandeweyenberg, I., Chitruong, T., Vangrimde, B., and Verpoest, I. (2006). Improving the properties of UD flax fibre reinforced composites by applying an alkaline fibre treatment. *Compos. A Appl. Sci. Manuf.* 37: 1368–1376.

59 Baley, C., Busnel, F., Grohens, Y., and Sire, O. (2006). Influence of chemical treatments on surface properties and adhesion of flax fibre polyester resin. *Compos. A Appl. Sci. Manuf.* 37: 1626–1637.

60 Lee, S., Kang, I., Doh, G. et al. (2008). Thermal and mechanical properties of wood flour/talc-filled polylactic acid composites: effect of filler content and coupling treatment. *J. Thermoplast. Compos. Mater.* 21 (3): 209–223.

3

Physicochemical Analysis of Biobased Composites

N. J. Vignesh[1], *P. Shenbaga Velu*[2], *and N. Rajesh Jesudoss Hynes*[1]

[1] *Department of Mechanical Engineering, Mepco Schlenk Engineering College, Sivakasi, Tamil Nadu, India*
[2] *Department of Mechanical Engineering, P.S.R Engineering College, Sivakasi, Tamil Nadu, India*

3.1 Introduction

Nowadays researchers have started using biofibers as filler materials in the plastic composites due to their significant advantages. Some of the important properties which make them highly attractive to the manufacturers include higher specific stiffness, low cost (on a volumetric basis), and process flexibility. Plastics have been more increasingly used in this century, out of which about 80% are thermoplastics. Structural applications find more interests in reinforcing the plastic composites with biofibers.

The development of technology is directly connected with the demands of the consumers and also their expectations. These two factors continue to increase as there are issues related to environmental sustainability and material availability when the demands of the global resources become higher. There has been a remarkable transformation in the field of biofibers over the last few decades. Due to the intense research, development, and application, these biofibers have become more and more adequate. Nowadays biocomposites have become significant owing to the petroleum crisis; they have become engineering materials with a very wide range of properties. But still there is need for continuous research in this particular field especially due to competitive pressure from the global market. It is not a matter of simply mixing some natural ingredients with polymer composites. Those techniques are outdated.

Every day the concept of using biobased plastics as reinforcements is gaining much importance. There is a spectacular growth in the biobased composites from a technological point of view and their growth in the market place is awesome.

3.2 Performance of Biocomposites

In order to bring out the best possible potential of each of the natural fiber used, it is essential for us to know or have a thorough knowledge about their mechanical properties. Some of the most commonly carried out mechanical tests include flexural, tensile, impact, dynamic

Biobased Composites: Processing, Characterization, Properties, and Applications, First Edition.
Anish Khan, Sanjay Mavinkere Rangappa, Suchart Siengchin and Abdullah M. Asiri.

mechanical, and creep tests. Because of their low density, high strength, and specific stiffness, these natural fibers can be used as excellent reinforcing agents in plastics.

3.2.1 Tensile Properties

Out of the various mechanical tests, the tensile tests are the most common type of tests carried out frequently on natural fibers. The selection of a particular type of natural fiber is dependent upon the strength of the natural fiber and also its field of application. The results of the tensile tests are an outcome of the average value through the thickness of the composites or fibers, whereas flexural test results depict the properties of the specimen nearer to the top and bottom surfaces. Tensile stress values calculated from the tests are uniform throughout the specimen, whereas the flexural stress values vary in such a way that it is maximum at the extremities (i.e. top and bottom surfaces) and zero at the middle.

Chemical treatments such as silane treatment and matrix–resin preimpregnation carried out on the fibers such as HDPE/hemp fibers produced a significant increase in its tensile strength, but their tensile modulus remained unaffected [1, 2]. There occurs an increase in the longitudinal tensile strength of about 10% when the fibers are treated with silane treatments and the values increased from 71.8 to 79.3 MPa. For untreated and treated fibers, the transverse tensile strength was determined as 2.75 and 3.95 MPa, respectively. The longitudinal fiber direction is 1900% more than the transverse fiber direction and this proves that the behavior is more of fiber dominated. There is very little impact on the composites when the surface of the fiber has been modified. But for the matrix and/or interphase dominated behavior, the improvements in the fiber–matrix interactions are very significant.

A study on the effect of alkaline treatment on tensile properties of the sugar palm fiber reinforced composites was carried out [3]. The treatment was carried out by varying the concentrations of sodium hydroxide (NaOH) solutions and the soaking time periods. The results showed that, there is a decrease in the tensile strength of the treated fibers when the concentration of the alkali and also the soaking time period was increased. The results of the tensile strength improved greater for the treated fibers than that of the untreated fibers. This proves that the treatment has a positive effect on the fiber treated.

Batch mixing was used to prepare composites made of aliphatic polyesters (Bionolle) with flax fibers [4]. Investigation was carried out on the effects of processing conditions on the fiber length distribution and also the tensile properties of the fibers. As per the modified rule-of-mixtures equation, it was found out that there is a variation in the tensile properties with the variation in the fiber content. The tensile strength of the fibers tend to decrease if the load on the Bionolle/flax composites are increased. This shows that the adhesion between the fibers and the matrix is very low. Testing was also carried out by chemically modifying the surfaces of the flax fibers. A 30% increase in tensile strength was observed when flax fibers (25 vol%) were replaced by fibers having acetate groups. No significant improvement in strength was observed in composites containing fibers with valerate groups or polyethylene glycol chains grafted at the surface.

The tensile properties of the composites were investigated regarding the siloxane treatment of the jute fiber/polyester and epoxy composites [5], the temperature and loading rate effects of kenaf fiber/epoxy composites [6], the effects of differing geometry of abaca fiber/epoxy composites [7], the influence of moisture of the bamboo fiber/vinyl ester composites [8], the varying volume fraction of the oil palm/polyester composites [9], and the influence of fiber orientation and volume fraction of the alfa fiber/polyester composites [10].

3.2.2 Flexural Properties

The deformation of the composites could be predicted by a mechanical test known as flexural stiffness. The flexural stiffness of a composite structure is based on two essential properties: (i) elastic or Young's modulus (i.e. the stress per unit strain) of the material that composes it and (ii) the moment of inertia, which is a function of the cross-section of the composites.

Zampaloni et al. [11] developed kenaf fiber-reinforced PP sheets which could be easily thermoformed and used for a wide variety of applications with properties comparable to existing synthetic composites. It was found out that these composites could be best fabricated by the compression molding process. In this process, they used chopped kenaf fibers and layered sifting of a microfine PP powder. On comparison, it was found that the composites containing 40% kenaf/PP composites show higher values of flexural strengths than that of the composites having 30% kenaf fiber/PP composites. On further investigation, it was also found that, the composites with 40% kenaf/PP have flexural strength which is equal to flax/PP, larger than the hemp/PP, and almost double that of the coir/PP and sisal/PP composites. Similarly, the composites with 30% kenaf/PP showed flexural strength equal to those of the 40% hemp/PP composites, while also outperforming the coir/PP and sisal/PP composites.

Biocomposites of PP and HDPE reinforced with 20 wt% curaua fibers were developed, and the effect of screw rotation speed was calculated by measuring the flexural properties of the composites [12]. If there is increase in the speed of the screw rotation, then there will be a reduction in the yield stresses calculated during the flexural tests of the composites with PP and HDPE matrixes. The rate of decrease of the yield stress was higher for HDPE, but the flexural modulus is not affected by the screw rotation. Different types of behavior have been exhibited by the elongation during break period: the effect strengthened HDPE at screw rotations above 300 rpm, and slightly increased for PP composites at screw rotations above 350 rpm.

Researchers evaluated the effect of various malleated coupling agents (flexural properties have been found to increase around 60% of the optimum loading of coupling agent) on jute- and flax fiber-reinforced PP composites [13], and the effect of acetylation on the flexural properties of bagasse/PP composites (flexural properties decreased due to the acetylation) [14].

The flexural properties of coir fiber/polyester composites were evaluated [15]. The composites were prepared by varying the amount of coir fibers (up to 80 wt%) and also the molding pressures. Rigid composites were obtained when the coir fibers up to 50 wt% were used. Composites prepared with higher amounts of fiber content resulted in performances similar to flexible agglomerates. The comparison of the technical properties of the composites with that of the other conventional materials has been made possible due to the results obtained from the flexural strength.

Investigation was also carried out on the flexural properties of thermoset resins (such as cardanol reinforced with hemp, ramie, jute and flax fibers) [16], epoxy resin with coir fibers [17], hemp, kenaf, flax, and henequen fibers [18], polyester resin reinforced bagasse fibers [19], coir and oil palm fibers [20], and hemp and kenaf fibers [21].

Two different methods used to fabricate the biocomposites made from cellulose ester biodegradable plastic and hemp fibers are powder impregnation by compression molding and an extrusion process followed by an injection molding process [22]. The flexural strength and Young's modulus for the biocomposites of hemp (30 wt%) fiber reinforced with cellulose acetate (developed by process extrusion and then followed by injection molding process) composites were about 78 MPa and 5.6 GPa, respectively, whereas the corresponding flexural strength and Young's modulus for the biocomposites made from hemp fiber/PP-based composites were only 55 MPa and 3.7 GPa, respectively. The biocomposites fabricated by means of powder impregnated process and compression molding technique showed lower flexural strength and modulus values when compared to the same biocomposites made from extrusion and ensuing injection molding processes which showed superior values.

PCL and oil palm empty fruit bunch fiber were used to prepare biodegradable composites by using the melt blending technique [23]. The interaction between PCL and the fibers were very poor; in order to improve the interaction, polyvinyl pyrrolidone (a binder) is added. In order to improve the mechanical properties, the produced composites were irradiated with an electron. When these composites were added with 1% by weight of polyvinyl pyrrolidone and an electron beam irradiation of about 1 kGy, they are able to improve the flexural strength and modulus of the composites.

3.2.3 Impact Properties

The ability of a material to resist fracture on sudden application of a load applied at high speed is called as impact strength. Biofiber-reinforced plastic composites have properties that can compete with the properties of glass fiber-reinforced composites, especially concerning specific properties. However, it should be noted that the impact strength of the biocomposites is still a major disadvantage. The development of improved composite processing methodologies and new fiber manufacturing techniques along with the enhancement of matrix/fiber adhesion has improved the current situation.

Studies were carried out on the oil palm empty fruit bunch fiber reinforced composite. Nowadays, oil palm fiber filled-PVC composites were of greater interest as the effect of fiber loading and impact loading which are varied by the materials such as chlorinated polyethylene (CPE) and acrylic were used [24]. When the concentration of the impact modifiers was increased, there is change from ductile to brittle behavior of both the samples of the unfilled PVC with impact modifiers. There is a reduction in the value of impact strength when the fibers are added into the modified and unmodified PVC. The value of impact strength reduced to about 40% for acrylic-modified PVC and 30% for CPE-modified PVC, when there is an increase in the fiber content from 10 to 40 phr. But when the PVC composites are unmodified, then there is only a marginal reduction in impact strength.

In the automotive sector, a greater deal of value-added products are developed commercially by the biocomposites. Specific properties such as serviceability, reliability standards, durability, and high quality performance need to be improved for widening the applications of biocomposites into other commercial markets such as construction and consumer goods.

Markets that are unexplored by biocomposites could also be entered and they become very diverse, when the technology improvements in the biocomposites reinforced with natural fibers are capable of producing enhanced material and product characteristics. Automotive sectors are nowadays utilizing increased number of components produced by natural fiber reinforced biocomposites [25–29].

Burgueno et al. utilized biocomposites for developing structural components. Cellular beams and plates as load bearing structural components were manufactured from natural fibers such as jute, hemp, and flax fibers with unsaturated polyester resin [30–33]. The material and structural performance were experimentally assessed and compared with results from short-fiber composite micro-mechanics models and sandwich analyses. The cellular biocomposite materials can be used for load-bearing components by improving their structural efficiency by means of cellular material arrangements. It was then verified that these components can compete with composites made from conventional materials. This research should be taken in such a way that, conventional material composites should also develop and be used for providing a solution for improvements in future, in the field of civil engineering.

Structural applications will be able to see a greater number of biocomposite components in the near future. Further improvements will lead to various other applications. Prior to their application in a fully competitive manner with the synthetic fiber composites, a lot of problems have to be clarified and solved. Biocomposites, when developed as completely biobased and biodegradable, are more expensive. They are highly sustainable and recyclable. Their sensitivity to temperature and moisture is high. The use of an appropriate matrix will lead to make these biocomposites 100% biodegradable. However, the biodegradation is difficult to control. Even though these biocomposites give very good specific properties, still there is a higher variability in their results. Lower impact strength, nonlinear mechanical behavior, and poor long-term performance is seen in biocomposites. Advanced processing techniques could help overcome these weaknesses. Excessive processing requirements make the environmental sustainability of biocomposites questionable when compared to synthetic fiber composites. They also consume excessive energy. For developing higher performance composites, it is essential to perform careful life-cycle assessment for retaining the important advantages. Properties such as dimensional stability, moisture proof, fire resistance, and increased durability will lead to open up new and diverse markets for these biocomposites.

3.2.4 Creep

The tendency of a composite to deform slowly and permanently under the influence of mechanical stresses is known as creep. This can occur when the stress levels are high, although below the yield strength of the material due to long-term exposure of the mechanical stresses.

A flexural creep equipment is used to carry out the creep tests on biocomposites. They are carried out according to the ISO 899-2 standards. Creep properties such as creep modulus and creep strength can be measured using the creep test, as a function of temperature and time. Several natural fiber-reinforced composites ranging from jute to wood are subjected to these kinds of creep tests [34, 35]. The structural damage of a composite caused by repeated and cyclic loading is termed as fatigue damage. Fatigue damage also occurs within the yield stress limit, similar to creep. A micro crack occurs at stress concentrations that will eventually propagate and reach a size known as the critical size, leading to complete composite failure. The shape of the structure affects the fatigue properties. ISO 13003:2003 standards could be used to carry out the fatigue testing for biocomposites. Fatigue tests were carried out on the fibers such as flax, hemp, jute, and sisal and compared with other natural fiber composites [36–38]. The performance of composites can be evaluated during monotonic and cyclic loading.

3.2.5 Brittleness and Ductility

Biocomposites could be either brittle or ductile in nature when the composite breaks with or without significant deformation under stress. Usually ductile composites fail when they are subjected to tensile stress. On the other hand, brittle composites fail without giving any prior information or significant strain when subjected to stress. Before failure takes place, ductile materials absorb relatively more energy than brittle materials. Some polymers such as polystyrene are brittle and do not deform significantly before failure takes place, whereas natural fibers deform before failure takes place. The brittleness of cured resins is affected by the cross-linking available in the thermoset resins. At lower strains, these highly cross-linked resins tend to break. One of the commonly available and a commercial biocomposite is PP. This composite usually fails under tensile stress. When the glass transition temperature is attained, the properties of the polymers are changed.

Brittle and ductile fractures were used as a means to explain the sensitivity of the mechanical properties to temperature of a Kenaf fiber-reinforced polylactic acid composite [39, 40].

3.2.6 Toughness

Other important properties of biobased composites include toughness and fracture toughness. Strength and ductility of the composites are the important parameters that affect the toughness and fracture toughness of a composite. The property by which a composite absorbs energy and deforms plastically without failure is known as toughness. The property by which a composite that contains cracks resists fracture is known as fracture toughness. Izod and Charpy tests are used to determine the toughness of a composite material. Kahn test and plane strain fracture toughness (KIc) are the tests used to find out the fracture toughness of a composite material. The fracture toughness of cotton fiber-reinforced biocomposites has been calculated by researchers [40]. Silva et al. [41] carried out studies on composite-based natural fibers and vegetable oil-based polyurethane to find out their fracture toughness. Since the crack propagation could be located either in the interface or fiber or matrix, these properties are based on a numbers of factors. Hence, these fractured specimens are commonly subjected to fractographic inspections using microscopy. ASTM standards such as D5045, D6068, and D5528 could be used for carrying out the fracture toughness of a material.

3.3 Physicochemical Properties

Composite properties could also be affected by the composition of the fiber and the matrix. Lignin, cellulose, hemicellulose, pectin, moisture, and wax are the compositions of any kind of natural fibers. The composition of these components varies from one plant to other. The composition also depends on the type of species, the age or maturity of the plant, and also varies with different parts of the same plant. To find the composition and properties of these fibers becomes a tough and challenging task as these are spread in a random manner inside the cell wall of the plants. The hydroxyl groups in the cellulose fibers form inter- and intramolecular hydrogen bonds, thus making the cellulose fibers hydrophilic in nature. Cellulose of all natural fibers have an identical chemical structure. But the physicochemical and mechanical properties of the fibers vary with changes in the degree of polymerization. There is another material in plants namely lignin. It is a phenolic substance made of highly branched polymers. The plant cellulose are able to retain their structure in the plants due to the presence of lignin which acts as a binder linking the celluloses in the plant. It is also capable of acting as a resistant to microbial attacks. It can only be interrupted by either physical or chemical treatments. Despite the fact that most of the functional groups and units are determined in the lignin, its structure is still not clear. The fiber–matrix adhesion and composite processing are affected by the presence of a material namely wax in the plants which has an effect on the wettability of the matrix [42].

When sugar or lipids are fermented by bacterial reaction, polyhydroxyalkanoates are produced, which are linear polyesters. These bacteria form the basis for producing different monomers and these monomers, in turn, are combined to give polymers different characteristics. Thermoset matrices are produced when the unsaturated bonds in the triglycerides or plant oils are functionalized. Ester bonds are used to connect the three types of fatty acids that are present in the triglycerides with the glycerol. Thermoset resins that are obtained from plant oils are made based on the degree of unsaturation of the triglycerides, different fatty acids in triglycerides, and their composition in various plant oils. The amount of unsaturation in the fatty acids depends upon the iodine values and double bonds. When these values are higher, the degree of saturation is higher. When the unsaturated fatty acids are modified, then the reactive sites are introduced leading to the chemical reactions when the thermoset curing process is taking place. Fourier Transform Infrared Spectroscopy (FTIR), UV Spectroscopy, and Nuclear Magnetic Resonance Spectroscopy (NMR) were used to analyze the chemical structure of vegetable oils. Gas Liquid Chromatography (GLC) and High-Performance Liquid Chromatography (HPLC) were used to find out their compositions. Chemical analysis becomes more complex when the chemical composition contains a combination of different fatty acids with triglycerides. Individual fatty acids are usually identified by spectroscopy.

The chemical composition of the fiber and the matrix affects the composite properties and also the fiber–matrix interaction. Poor fiber–matrix interface is obtained when the polar groups in the cellulosic fibers are combined with the nonpolar matrix. In order to achieve molecular continuity along the fiber–matrix interface, fiber pretreatments are carried out on the cellulosic fibers to change the fiber surface chemistry [43].

Thermal stability of biocomposites and constituents are usually characterized over a range of temperature. Thermogravimetric analysis (TGA) is a common analytical method in which the thermal decomposition is analyzed by following the gravimetric weight loss with the function of time. The material is heated at a constant rate and the weight loss gives the information regarding physical and chemical changes. Degradation of the composite takes place commonly in several stages due to decomposition of each constituent in the composite.

Thermal degradation of natural fibers at elevated temperature is obvious and occurs in three different stages: the initial mass loss at about 110 °C is due to moisture evaporation which could also be removed before test; the second mass loss takes place between 275 and 370 °C and this is attributed to thermal decomposition of cellulose; and the third mass loss takes place above 370 °C which is attributed to decomposition and breakdown of cellulose. Swift decomposition of cellulose occurs between 300 and 370 °C. It is also known that cellulose derivatives decompose at 358 °C. Table 3.1 depicts the degradation temperatures of some natural fibers at 5, 25, 50, and 75% of total weight loss. Their kinetics is explained in detail by Yao et al. [44]. Variation in the degradation temperatures among fibers could be due to the change in the chemical composition of each fiber.

Table 3.1 Degradation temperatures of some natural fibers.

Weight loss %	*T*, 5% (°C)	*T*, 25% (°C)	*T*, 50% (°C)	*T*, 75% (°C)
Hemp	104.6	329.2	341.1	457.8
Kenaf	64.3	313.4	340.5	370.8
Henequen	61.1	301.4	337.4	391.6
Flax	88.2	322.8	338.6	441.4

3.4 Conclusion

Biocomposites reinforced with natural fibers and/or biopolymers have developed significantly over the past years due to their significant processing advantages, biodegradability, low cost, low relative density, high specific strength, and renewable nature. Interfacial adhesion between natural fibers and matrix will remain the key issue in terms of overall performance, since it dictates the final properties of the composites. Many studies are reviewed and highlighted in this chapter regarding the importance of the physicochemical properties and mechanical characterization.

Significant research is currently underway around the world to address and overcome the obstacles. This effort to develop biocomposite materials with improved performance for global applications is an ongoing process.

References

1 Herrera-Franco, P.J. and Valadez-Gonzalez, A. (2004). Mechanical properties of continuous natural fibre-reinforced polymer composites. *Composites Part A: Applied Science and Manufacturing* 35: 339–345.

2 Herrera-Franco, P.J. and Valadez-Gonzalez, A. (2005). A study of the mechanical properties of short natural-fiber reinforced composites. *Composites Part A: Applied Science and Manufacturing* 36: 597–608.

3 Bachtiar, D., Sapuan, S.M., and Hamdan, M.M. (2008). The effect of alkaline treatment on tensile properties of sugar palm fibre reinforced epoxy composites. *Materials and Design* 29: 1285–1290.

4 Baiardo, M., Zini, E., and Scandola, M. (2004). Flax fibre–polyester composites. *Composites Part A: Applied Science and Manufacturing* 35: 703–710.

5 Seki, Y. (2009). Innovative multifunctional siloxane treatment of jute fiber surface and its effect on the mechanical properties of jute/thermoset composites. *Materials Science and Engineering A* 508: 247–252.

6 Xue, Y., Du, Y., Elder, S. et al. (2009). Temperature and loading rate effects on tensile properties of kenaf bast fiber bundles and composites. *Composites Part A: Applied Science and Manufacturing* 40: 189–196.

7 Sapuan, S.M., Leenie, A., Harimi, M., and Beng, Y.K. (2006). Mechanical properties of woven banana fibre reinforced epoxy composites. *Materials and Design* 27: 689–693.

8 Chen, H., Miao, M., and Ding, X. (2009). Influence of moisture absorption on the interfacial strength of bamboo/vinylester composites. *Composites Part A: Applied Science and Manufacturing* 40: 2013–2019.

9 Yousif, B.F. (2010). Effect of oil palm fibres volume fraction on mechanical properties of polyester composites. *International Journal of Modern Physics B* 24: 4459–4470.

10 Ben Brahim, S. and Ben Cheikh, R. (2007). Influence of fibre orientation and volume fraction on the tensile properties of unidirectional alfa-polyester composite. *Composites Science and Technology* 67: 140–147.

11 Zampaloni, M., Pourboghrat, F., Yankovich, S.A. et al. (2007). Kenaf natural fiber reinforced polypropylene composites: a discussion on manufacturing problems and solutions. *Composites Part A: Applied Science and Manufacturing* 38: 1569–1580.

12 Mano, B., Araújo, J.R., Spinacé, M.A.S., and De Paoli, M.A. (2010). Polyolefin composites with curaua fibres: effect of the processing conditions on mechanical properties, morphology and fibres dimensions. *Composites Science and Technology* 70: 29–35.

13 Luz, S.M., Del Tio, J., Rocha, G.J.M. et al. (2008). Cellulose and cellulignin from sugarcane bagasse reinforced polypropylene composites: effect of acetylation on mechanical and thermal properties. *Composites Part A: Applied Science and Manufacturing* 39: 1362–1369.

14 Keener, T.J., Stuart, R.K., and Brown, T.K. (2004). Maleated coupling agents for natural fibre composites. *Composites Part A: Applied Science and Manufacturing* 35: 357–362.

15 Monteiro, S.N., Terrones, L.A.H., and D'Almeida, J.R.M. (2008). Mechanical performance of coir fiber/polyester composites. *Polymer Testing* 27: 591–595.

16 Maffezzoli, A., Calo, E., Zurlo, S. et al. (2004). Cardanol based matrix biocomposites reinforced with natural fibres. *Composites Science and Technology* 64: 839–845.

17 Harish, S., Michael, D.P., Bensely, A. et al. (2009). Mechanical property evaluation of natural fiber coir composite. *Materials Characterization* 60: 44–49.

18 Sgriccia, N., Hawley, M.C., and Misra, M. (2008). Characterization of natural fiber surfaces and natural fiber composites. *Composites Part A: Applied Science and Manufacturing* 39: 1632–1637.

19 Vilay, V., Mariatti, M., Taib, R.M., and Todo, M. (2008). Effect of fiber surface treatment and fiber loading on the properties of bagasse fiber-reinforced unsaturated polyester composites. *Composites Science and Technology* 68: 631–638.

20 Hill, C.A.S. and Abdul Khalil, H.P.S. (2000). The effect of environmental exposure upon the mechanical properties of coir or oil palm fiber reinforced composites. *Journal of Applied Polymer Science* 77: 1322–1330.

21 Medina, L., Schledjewski, R., and Schlarb, A.K. (2009). Process related mechanical properties of press molded natural fiber reinforced polymers. *Composites Science and Technology* 69: 1404–1411.

22 Mohanty, A.K., Wibowo, A., Misra, M., and Drzal, L.T. (2004). Effect of process engineering on the performance of natural fiber reinforced cellulose acetate biocomposites. *Composites Part A: Applied Science and Manufacturing* 35: 363–370.

23 Ibrahim, N.A., Ahmad, S.N.A., Yunus, W.M.Z.W., and Dahlan, K.Z. (2009). Effect of electron beam irradiation and poly (vinyl pyrrolidone) addition on mechanical properties of polycaprolactone with empty fruit bunch fibre (OPEFB) composite. *eXPRESS Polymer Letters* 3: 226–234.

24 Bakar, A.A. and Hassan, A. (2003). Impact properties of oil palm empty fruit bunch filled impact modified unplasticised poly(vinyl chloride) composites. *Jurnal Teknologi* 39: 73–82.

25 Bledzki, A.K., Faruk, O., and Jaszkiewicz, A. (2010). Cars from renewable materials. *Composites Journal* 10: 282–288.

26 Bledzki, A.K., Jaszkiewicz, A., Murr, M. et al. (2008). Natural fibre in automotive applications. *Bioplastics Magazine* 3 (2): 18–20.

27 Bledzki, A.K., Faruk, O., and Sperber, V.E. (2006). Cars from bio-fibres. *Macromolecular Materials and Engineering* 291: 449–457.

28 Suddel, B.C. and Evans, W.J. (2005). Natural fiber composites in automotive applications. In: *Natural Fibres, Biopolymers, and Biocomposites* (eds. A.K. Mohanty, M. Misra and L.T. Drzal), 231–259. Boca Raton: Taylor & Francis Group.

29 Suddel, B.C. and Evans, W.J. (2003). The increasing use and application of natural fiber composite materials within the automotive industry. In: *Seventh International Conference on Woodfiber–Plastic Composites*, 7–14. Madison, WI: Monona Terrace Community & Convention Center (19–20 May 2003).

30 Burgueno, R., Quagliata, M.J., Mohanty, A.K. et al. (2004). Load-bearing natural fiber composite cellular beams and panels. *Composites Part A: Applied Science and Manufacturing* 35: 645–656.

31 Burgueno, R., Quagliata, M.J., Mohanty, A.K. et al. (2005). Hierarchical cellular designs for load-bearing biocomposite beams and plates. *Materials Science and Engineering A* 390: 178–187.

32 Burgueno, R., Quagliata, M.J., Mohanty, A.K. et al. (2005). Hybrid biofiber-based composites for structural cellular plates. *Composites Part A: Applied Science and Manufacturing* 36: 581–593.

33 Burgueno, R., Quagliata, M.J., Mehta, G. et al. (2005). Sustainable cellular biocomposites from natural fibers and unsaturated polyester resin for housing panel applications. *Journal of Polymers and the Environment* 13: 139–149.

34 Bledzki, A.K. and Faruk, O. (2004). Creep and impact properties of wood fibre-polypropylene composites: influence of temperature and moisture content. *Composites Science and Technology* 64: 693.

35 Acha, B.A., Reboredo, M.M., and Marcovich, N.E. (2007). Creep and dynamic mechanical behavior of PP-jute composites: effect of the interfacial adhesion. *Composites Part A: Applied Science and Manufacturing* 38: 1507.

36 Shahzad, A. (2011). Impact and fatigue properties of hemp-glass fiber hybrid biocomposites. *Journal of Reinforced Plastics and Composites* 30: 1389.

37 Belaadi, A., Bezazi, A., Maache, M., and Scarpa, F. (2014). Fatigue is sisal fiber reinforced polyester composites: hysteresis and energy dissipation. *Procedia Engineering* 74: 325.

38 Shah, D.U. (2015). Damage in biocomposites: stiffness evolution of aligned plant fibre composites during monotonic and cyclic fatigue loading. *Composites Part A: Applied Science and Manufacturing* https://doi.org/10.1016/j.compositesa.2015.09.008.

39 Kaiser, M.R., Anuar, H., and Razak, S.B.A. (2013). Ductile–brittle transition temperature of polylactic acid based biocomposite. *Journal of Thermoplastic Composite Materials* 26: 216.

40 Alomayri, T., Assaedi, H., Shaikh, F.U.A., and Low, I.M. (2014). Effect of water absorption on the mechanical properties of cotton fabric-reinforced geopolymer composites. *Journal of Asian Ceramic Societies* 2: 223.

41 Silva, R.V., Spinelli, D., Bose Filho, W.W. et al. (2006). Fracture toughness of natural fibers/castor oil polyurethane composites. *Composites Science and Technology* 66: 1328.

42 Ramamoorthy, S.K., Skrifvars, M., and Persson, A. (2015). A review of natural fibers used in biocomposites: plant, animal and regenerated cellulose fibers. *Polymer Reviews* 55: 107.

43 Li, X., Tabil, L.G., and Panigrahi, S. (2007). Chemical treatments of natural fiber for use in natural fiberreinforced composites: a review. *Journal of Polymers and the Environment* 15: 25.

44 Yao, F., Wu, Q., Lei, Y. et al. (2008). Thermal decomposition kinetics of natural fibers: activation energy with dynamic thermogravimetric analysis. *Polymer Degradation and Stability* 93: 90.

4

Characterization of Biobased Composites

Anna Sienkiewicz and Piotr Czub

Department of Chemistry and Technology of Polymers, Cracow University of Technology, Cracow, Poland

4.1 Introduction

According to the regulations of US Department of Agriculture and US Department of Energy in 2020, at least 10% of all basic chemical building blocks should be obtained from renewable, plant-based sources, while by 2050 this amount should be up to 50% [1]. For the past few decades, it is noticeable that polymers have replaced many of the conventional materials used in various applications. This was possible due to the advantages of polymers over conventional materials. However, an intensive striving of comfort in human's life, as well as the ecological regulations, drive the constant need to improve polymeric materials, which are an indispensable part of our everyday reality. Composites (Figure 4.1) obtained by the introduction of various additives into the polymer matrix recently experience intensive development.

The properties of the final composite depend on the interaction of the filler and the matrix components. Composites with well-balanced mechanical properties are characterized by: (i) a homogeneous dispersion of the filler in the matrix, (ii) a well-balanced interaction between the matrix and the additive, (iii) a low porosity of the matrix, and (iv) an optimized percentage of filler [2].

4.2 The Conception of Composites

In general, the term "composites" (Latin "*compositus* = "bring together") relates to engineered materials made from two or more constituents, which significantly differ in terms of physical or chemical properties [3]. The first component (called "matrix" or "continuous phase") is described as the binding material, surrounding and holding the second material, which acts as reinforcement (also defined as "discontinuous phase") of the entire composition [4]. Typically, the discontinuous phase is harder and exhibits mechanical properties superior to those of the matrix. However, the properties of the composite material depend on: (i) individual properties of its components, (ii) each phase volume share, (iii) the arrangement of the discontinuous phase, and (iv) its geometrical features (such as size of particles,

Biobased Composites: Processing, Characterization, Properties, and Applications, First Edition.
Anish Khan, Sanjay Mavinkere Rangappa, Suchart Siengchin and Abdullah M. Asiri.

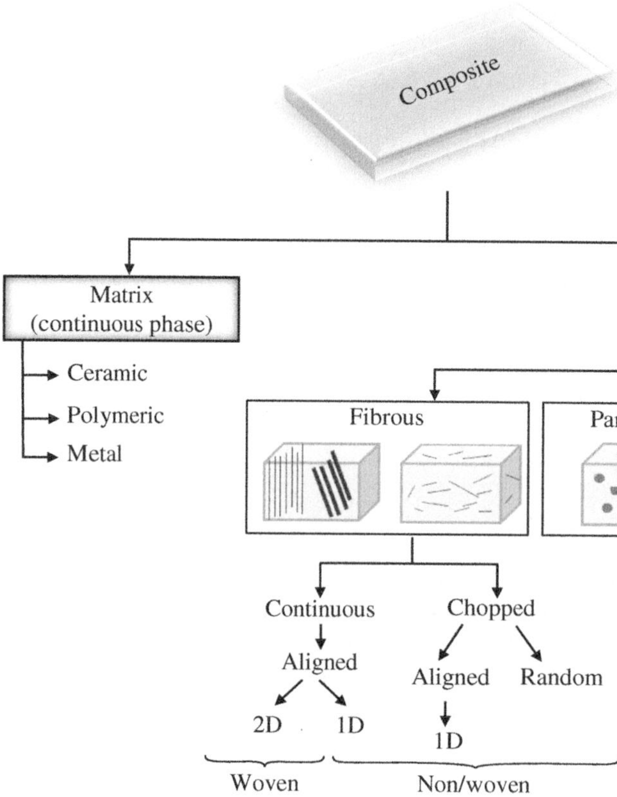

Figure 4.1 Classification of composites.

fiber length, and its orientation). The most common advantages of composite materials include lightweight, increased design flexibility, better damage tolerance (increased impact resistance, chemical and fracture toughness, as well as superior corrosion resistance), tailored strength and stiffness (anisotropic), good damping, low thermal expansion, good fatigue resistance, and increased (or decreased) electrical or thermal conductivity [3–5]. On the other hand, these composites also have some disadvantages, such as the cost of raw materials and fabrication, possible difficult reuse and disposal, and the need of the fillers' modification and use of adhesion promoters [4, 5].

4.3 Classification of Biocomposites

According to the commonly accepted definition, the terms "biocomposites," "green composites," and "ecocomposites" refer to the class of composites, which consist of at least one of the components (the matrix or the reinforcement) derived from natural resources (Figure 4.2) [6].

Natural bioreinforced biocomposites are renewable, lightweight, energy proficient, biodegradable, environmental friendly, and biocompatible. Even though biocomposites have become one of the most frequently studied materials during last few years (due to the greater ecological awareness and consumer expectation), the first application of biocomposites are dated back to ancient Egypt, 3000 years ago, when people used straw for reinforcement of mud-based walls in houses.

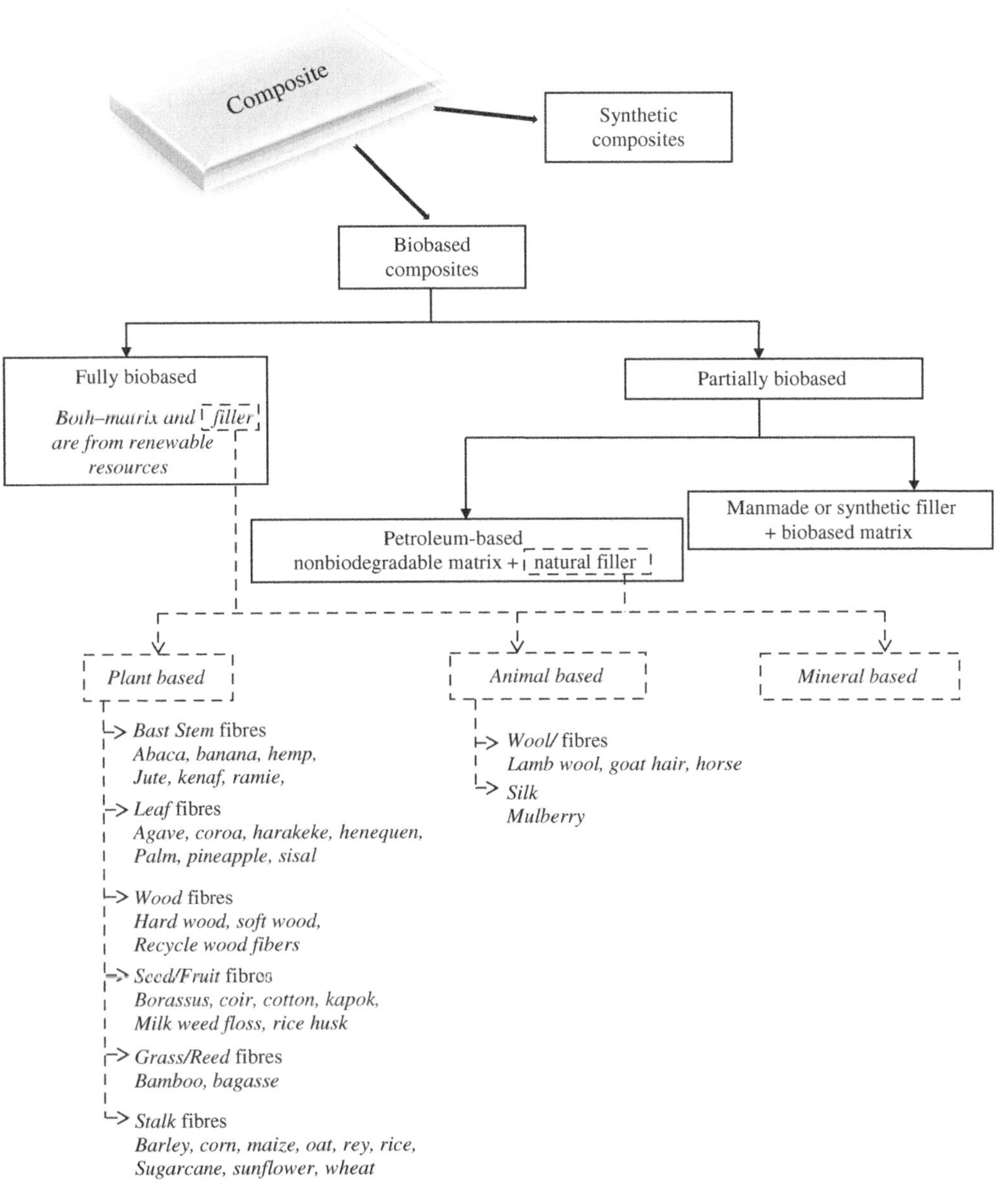

Figure 4.2 Classification of biocomposites. *Source:* Based on Netravali and Chabba [6].

4.4 Materials for the Synthesis of Biobased Composites

The overall properties of the composites depend on the properties of each component: the reinforcement, the matrix, and the interfacial shear strength (IFSS) of discontinuous–continuous phase [7]. In general, the discontinuous phase (in most cases fibers) in the composite is responsible for the overall strength of the material. On the other hand, the matrix, not only holds the reinforcement in place, but (what is more important) transfers the load from the broken fibers to their neighboring undamaged fibers through the interface. In addition, the interfacial shear strength is responsible for controlling the toughness, transferring mechanical properties, and interlaminar shear strength of the composite. It was found that the strength of the interfacial matrix–fiber adhesion results from the: (i) physicochemical interaction, (ii) mechanical interaction (interlocking), and (iii) chemical interaction or bonding [8]. Composites with good interfacial adhesion are characterized by relatively good mechanical strength (resulting from the fine structural integrity and stress transfer throughout the matrix between broken and undamaged fibers), whereas poor interfacial bonding leads to the absorption of energy through interfacial cracks (resulting in the increase of fracture toughness) [9].

4.4.1 Biopolymers as Matrix of Green Composites

As mentioned before, the final properties of the composites depend on the properties of both the matrix and the discontinuous phase. Biodegradable biopolymers, which include both thermosetting and thermoplastic polymers used as continuous phase in biocomposites, are derived from a variety of renewable (based on agricultural plants or animal products) or synthetic sources (oil by-product) [10]. Naturally derived biodegradable polymers (Figure 4.3) include polysaccharides of plant (cellulose and alginate), animal (chitin), and microbial origin (hyaluronate); proteins (e.g. casein, albumin, silk, and elastin); polyesters (e.g. poly(hydroxyalkanoate), poly(hydroxybutyrate), and poly(lactic acid)); lignin; lipids; natural rubber; some polyamides (poly(vinyl alcohol), poly(vinyl acetate)); and polycaprolactone [11]. Most of the above-mentioned polymers of natural origin degrade throughout the enzymatic reactions in suitable environments (typically, humid) [12].

Generally, major pathways to synthesize biobased polymers include: (i) application of naturally occurring polymers, which are usually modified (cellulose and starch derivatives), (ii) obtaining biobased monomers by fermentation and then polymerization (e.g. poly(lactic acid) and biobased polyamide 6), as well as (iii) biobased polymers synthesized by microorganisms (poly(hydroxyalkanoate)s) [13].

Selection of the polymer for the biocomposite matrix should be considered in terms of its mechanical properties, chemical resistance, dimensional stability, manufacturing process, and future ability to recycle or biodegrade. For example, thermosets are willingly used due to their good adhesion, relatively good thermal and chemical resistance, and excellent

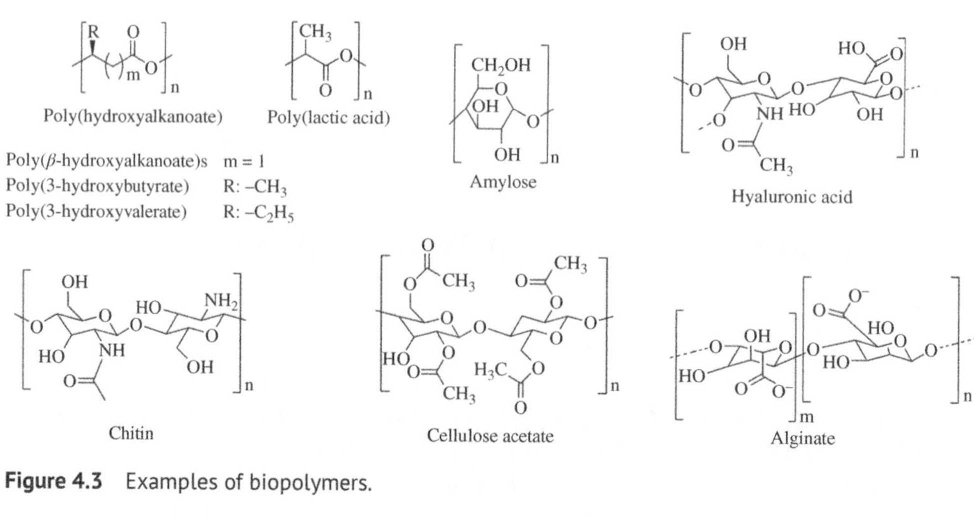

Figure 4.3 Examples of biopolymers.

mechanical properties. However, in contrast to thermoplastics, which are known for their recyclability and remoldability, thermosets are much more difficult to reuse due to temporary loosening and final breaking of the Van der Waals and hydrogen bonds in the melting process [14].

One of the frequently used biopolymer, which is already found in industrial applications is poly(lactic acid). Poly(lactic acid) is derived from renewable agricultural raw materials (such as dextrose of corn or sugar beets), fermented to form lactic acid and then polymerized by the condensation polymerization of D- or L-lactic acid or ring opening polymerization of the lactide [15]. Its relatively broad spectrum of applications is closely related to its valuable properties, such as the ability to be stress crystallized, thermally crystallized, impact modified, filled, copolymerized (employing a comonomer component of hydroxy acids or racemization of D- or L-isomer), and processed in a variety of polymer processing equipment [16]. In addition, poly(lactic acid) might be hydrolyzed to lactic acid, and eventually to water and carbon monoxide [17].

An interesting plant-derived polymer is starch, which occurs naturally as a mixture of linear amylose (poly-α-1,4-D-glucopyranoside) and branched amylopectin (poly-α-1,4-D-glucopyranoside and α-1,6-D-glucopyranoside). It is one of the least expensive biodegradable materials available in the world market today, and can be derived from corn, potatoes, wheat, and rice. Although starch might be converted to thermoplastic materials (starch plastics), through the restructuration in the presence of plasticizers (water and/or poly-alcohols), biodegradable products based on starch still show many disadvantages, mainly related to its high hydrophilic properties.

4.4.2 Fibers as Natural Reinforcement

On the market one can observe a constant pursuit of maintaining a competitive advantage, as well as exploration for new materials and improved processes to create increasingly better products and thus maintaining competitive advantage and servicing the margin [18]. For a relatively long time, man-made fillers (such as glass, aramid, and carbon fibers) were the predominant reinforcing fibers for polymeric materials, but since they feature environmental problems both during the production and disposal processes, researchers turned toward ecofriendly and nature derived fillers, which are cheaper, lighter, and advanced in terms of environmental protection when compared to synthetic fibers in polymer composites. Nowadays, most modifications of traditional polymers are made in line with the tendency to introduce and apply technologies, keeping in mind not only the better functional properties of synthesized material, but also the need to protect the environment. The reason for the growing interest in natural fibers is, among others, the method of their production, which has a less harmful effect on the environment, when compared with the process of obtaining synthetic fibers. In addition, the content of natural fillers in composites is usually higher when compared to materials with glass fibers, thus resulting in the need to use less polymer matrix in the natural fiber materials [19]. Modern chemistry and industry are focused on the applications of different natural and renewable raw materials in various branches of the synthesis and modification of already well-known industrial materials. The lower weight of composites with natural fiber results in lower fuel consumption and thus reduces carbon dioxide emissions when the component is used properly. At the end of the life cycle of natural fibers, they can be burned leading to the energy recovery and allows for financial benefits from reduced carbon dioxide emissions [20]. This is especially important in an aspect of the

European Community regulations, which states that members are obliged to reuse or recover about 95% of materials from all vehicle wrecks [21]. In addition, it has been estimated that the energy demand for natural fiber production is only 30–40% of the energy required to produce mats of fiberglass, along with 40–60% less energy consumption during the entire life cycle. Finally, an important economic aspect: the price of synthetic glass fibers is about 1–2 USD/kg, while the prices of lignocellulosic fibers, depending on the species, are lower by about one order of magnitude.

Natural fillers, as shown on Figure 4.2, are categorized into three groups based on their origin: plant, animal, or mineral based. Generally, animal fibers consist of proteins and they include hair (from sheep, goats, rabbits, alpaca, horses, etc.) and silk. On the other hand, plant fibers primarily consist of cellulose in a lignin matrix with slight amounts of additional low-molecular-weight organic compounds (extractives) and inorganic matter [3]. Plant fillers usually contain cellulose, hemicellulose, pectin, and lignin; the content, mutual ratio, and individual properties of components contribute to the overall properties of the fibers. It is worth nothing that the proportion and the composition of plant-based fillers are closely related to the plant species, growth conditions, and method of fiber extraction. Moreover, properties of the fillers depend on the fiber cell geometry of each type of cellulose and its degree of polymerization (Figure 4.4) [18]. For example, hemicellulose is responsible for the biodegradation, moisture absorption, and thermal degradation of the fiber, whereas lignin is responsible for the UV-light-degradation [14].

The main component of natural fibers, cellulose (Figure 4.5), is usually crystalline in about 60–90% and contains monoclinic and triclinic elemental cells within its structure [22]. Cellulose forms a microcrystalline structure with regions of high packing density called crystalline regions and regions of low packing density called amorphous regions [23]. Each repeating cellulose unit contains three hydroxyl groups, resulting in its hydrophilic properties. Cellulose is relatively resistant to oxidizing agents and strong alkali. Under the influence of acids, it easily hydrolyzes to water-soluble sugars. Hydroxyl groups, along with their ability to form a hydrogen bond, have a major influence on the degree of crystal packing and the physical properties of cellulose [22]. Hemicellulose is a branched polymer consisting of various five- and six-carbon sugars including glucose, mannose, galactose, and xylose, which

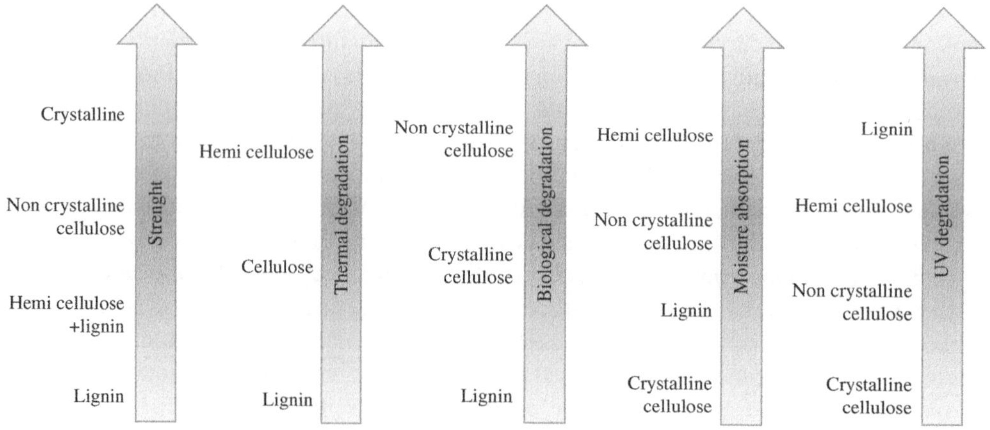

Figure 4.4 The influence of plant's components on the properties of plant fibers. *Source:* Based on Ho et al. [18].

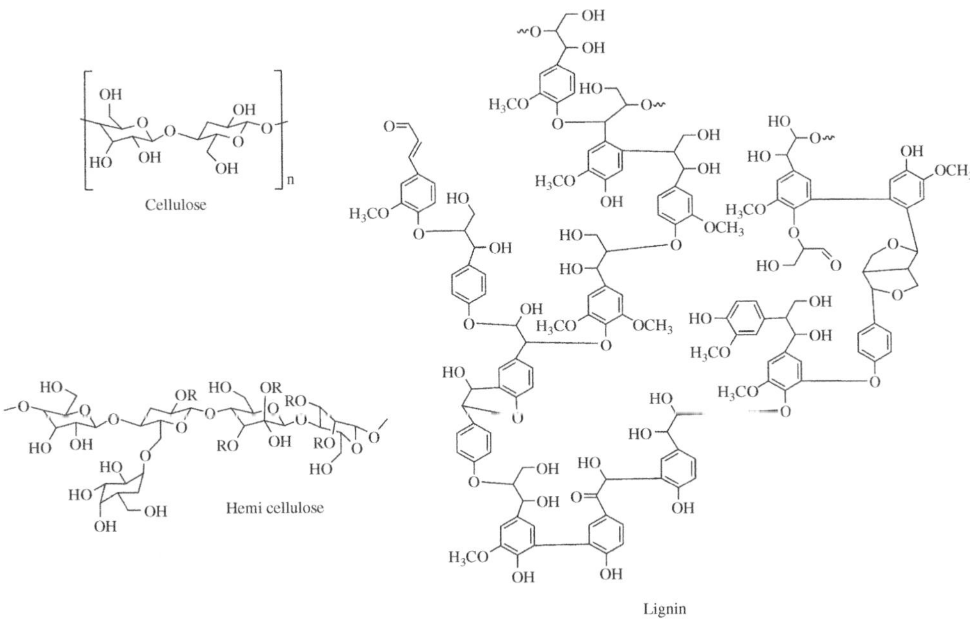

Figure 4.5 Chemical structure of cellulose, hemicellulose, and lignin.

form an assist matrix for cellulose microfibrils. In addition, due to its open structure containing many hydroxyl and acetyl groups, hemicellulose is hygroscopic in nature. It is also partly soluble in water, completely soluble in alkali, and easily hydrolyzed in acids [22, 23]. Lignin is amorphous polymer and its polymer network is formed from an irregular arrangement of various connected hydroxyl- and methoxy-substituted phenylpropanes. Lignin consists of both aliphatic and aromatic segments and acts as a barrier against the chemical and physical damage of the natural fiber. It also stiffens the cell wall and fills the spaces between cellulose fibers joining them together and forming bundles, which are then attached to the trunk with pectin, responsible for the flexibility of the plant. It is worth nothing that, compared to cellulose, both lignin and pectin are characterized by worse mechanical strength, and they must be removed from the fiber in order of achieve an effective reinforcement for the biocomposite [24]. Unfortunately, lignin is insoluble in most solvents nonhydrolyzed by acids, but dissolves in hot alkalis and easily condensates with phenol.

Another important component of natural fibers is extractives (3–10 wt% of natural filler), which include fats, waxes, resins, proteins, terpenes, and simple sugars, that provide an energy reserve, protection against microbes, and decide about properties such as color, smell, and durability of the fiber.

Natural fiber reinforcement may also be classified in accordance with the length, dimension, and orientation (Figure 4.1). They form continuous or discontinuous fiber or particle. In addition, fiber-reinforced phase arrangement may be classified as woven (continuous interlacing of perpendicular yarns of several interlocked fibers, in a regular pattern) or nonwoven. It is worth highlighting that the increase of fiber twist angle results in a decrease of fiber–matrix bond strength, lower permeability, and finally poor mechanical properties [25]. Unidirectional composites, due to the one-dimensional architecture of fiber, are generally weaker in the diagonal directions [26]. The mechanical and physical properties of

Table 4.1 Properties of natural fibers.

Type	Fiber	Cellulose (%)	Hemicellulose (%)	Lignin (%)	Density (g/cm^3)	Tensile strength (MPa)	Young's modulus (GPa)	Elongation at break (%)
Grass	Bamboo	26–43	30	21–31	0.6–1.1	270–862	17–89	1.3–8.0
Leaf	Banana	63–64	10–24	5	1.35	529–914	27–32	5–6
Bast (stem)	Hemp	60–77	15	3–10	1.47	580–1110	30–60	1.6–4.5
Fruit	Coir	30	15–25	40–45	1.2	175	6	15–25
Seed	Cotton	82.7–92	5.7–6	0	1.21	287–597	6–10	2–10

composites are closely related to the (i) orientation, (ii) strength, and (iii) physical properties of fibers, as well as (iv) interfacial adhesion property of fibers [27, 28].

Natural organic fillers include rice husk ash, nutshells, oil palm empty fruit bunch fibers, and corn plant-extracted fibers [11]. It is worth nothing that the chemical composition of natural fibers varies depending on the type of fiber (Table 4.1).

Among a large number of applied biofillers, wood flour and fibers are the mostly used. Wood flour can be easily and cheaply obtained from sawmill wastes and it is usually used after proper sieving. Wood fibers are produced by thermomechanical processes on wood waste. Both wood flour and fibers are quite interesting due to their low cost, dimensional stability, and increase in elastic modulus, without the decrease in the value of tensile strength of composites. On the other hand, their main drawbacks are the poor adhesion between the filler particles and the polymer matrix, low impact strength of composites, thermal decomposition of fibers at temperatures over 200 °C [29], the tendency to form aggregates during the preparation of composites, and poor resistance to moisture.

Novel and ecological composites are obtained using green fillers, which are naturally available from agriculture, industrial waste, or industrial recycled materials. Here, it is important to highlight that the application of natural-organic fillers, mostly derived from plants, brings several advantages, such as: (i) reduction in the use of nonbiodegradable polymers and nonrenewable resources, (ii) relatively low price of composites, (iii) less abrasive to processing instruments, when compared to inorganic-mineral fillers, and (iv) increased safety for the production employees (in case of inhalation) [30].

4.5 Challenges of the Introduction of Natural Fiber

As mentioned above, the properties of biocomposites with natural fibers depend on length, orientation, diameter, and content of filler [18]. The main drawbacks of the application of biofillers are the poor adhesion between the filler particles and the polymer matrix. Cellulose, hemicellulose, and lignin (Figure 4.5) – the chemical components of natural fibers – contain hydroxyl groups that are capable of forming a hydrogen bond with moisture. Hydroxyl groups within the structure of natural fibers make the fibers hydrophilic. Weaker interfacial bonding in composites results in worse mechanical and physical properties [28]. Hygroscopicity may induce difficulties both during the process of the biocomposite synthesis, as well as in the final properties of the obtained product. Absorbed moisture affects hydrogen bonds between the components of the cell wall leading to a change in its mechanical strength [22]. In addition, bonding between matrix and the filler in the presence of moisture is much smaller because the water molecules bind with hydroxyl groups from the surface of the fiber, thereby reducing their availability for binding to the polymer matrix. In turn, when the water evaporates voids are created, which reduce the mechanical strength of the composite. Therefore, appropriate initial drying of the natural fiber is necessary, leading to improvement in the bonding of the filler with the matrix [24]. The adhesion between reinforcement and polymeric matrix is sustained by the interfacial forces. These forces can be broadly divided into primary forces and secondary forces. Primary forces, also known as short-range forces, arise from chemical bonding. Chemical bonding comprises interlinking between molecules of the substrate by covalent, ionic, or metallic bonds. The secondary or intermolecular forces, such as Van der Waals forces, are caused from the physical interactions between two substrates.

In order to improve the interfacial interaction between filler particles (hydrophilic) and polymer macromolecules (generally hydrophobic), as well as their dispersion in the matrix, various modification techniques are performed (Figure 4.6). These techniques include modification of: (i) the surface of fibers, (ii) the polymer, or (iii) both of them at the same time [31]. Modifications of the surface of natural fibers might be done via chemical, physical, or biological treatment. Chemical treatments result in the permanent modification of the nature of fibers by grafting polymer onto the fibers or by using coupling agents.

Physical treatments change the structural and surface properties of the fiber by introducing surface crosslinking, modifying the surface energy, and/or generating reactive free radicals and groups, and thereby influence the mechanical bonding to the matrix. In general, although physical methods are an effective, environmental friendly way to modify natural fibers, they only affect the outer cell walls and do not change the chemical composition of the fibers. So often they are used as an initial stage for further chemical treatments.

One of the most effective physical modification processes toward the separation of lignocellulosic fibers is steam explosion [32]. Subjecting the lignocellulosic materials to saturated steams leads to the separation of their main components: cellulose, lignin, and hemicellulose. On the one hand, the method of steam explosion requires low energy consumption and has a low impact on the natural environment, whereas, on the other hand, steam explosion leads to obtain relatively short fibers, which are not as effective reinforcement as long fibers.

Thermomechanical processes are an example of another physical way of separating lignocellulosic materials into individual filaments [33]. Heating the fibers to a temperature close to the lignins' glass transition temperature results in the splitting of the filler bundles into individual fibers. During the process, noncellulosic chemical components of the fiber, which are characterized by lower (than lignin) glass transition temperatures, are released from the fiber bundles or depolymerized. Single fibers are characterized by greater strength and stiffness than the fiber bundles. Postmodification fibers are also characterized by a higher crystallinity index and better mechanical parameters. In addition, compared to poststeam explosion filaments, during the heat treatment, lignin and/or extractives migrate to the fiber surface, resulting in the increased hydrophobicity of the surface [33, 34].

One of the frequently applied methods of physical modification of various lignocellulosic fibers is plasma treatment. Modified fibers are characterized by a rougher surface, resulting in a better adhesion to the polymer matrix [35].

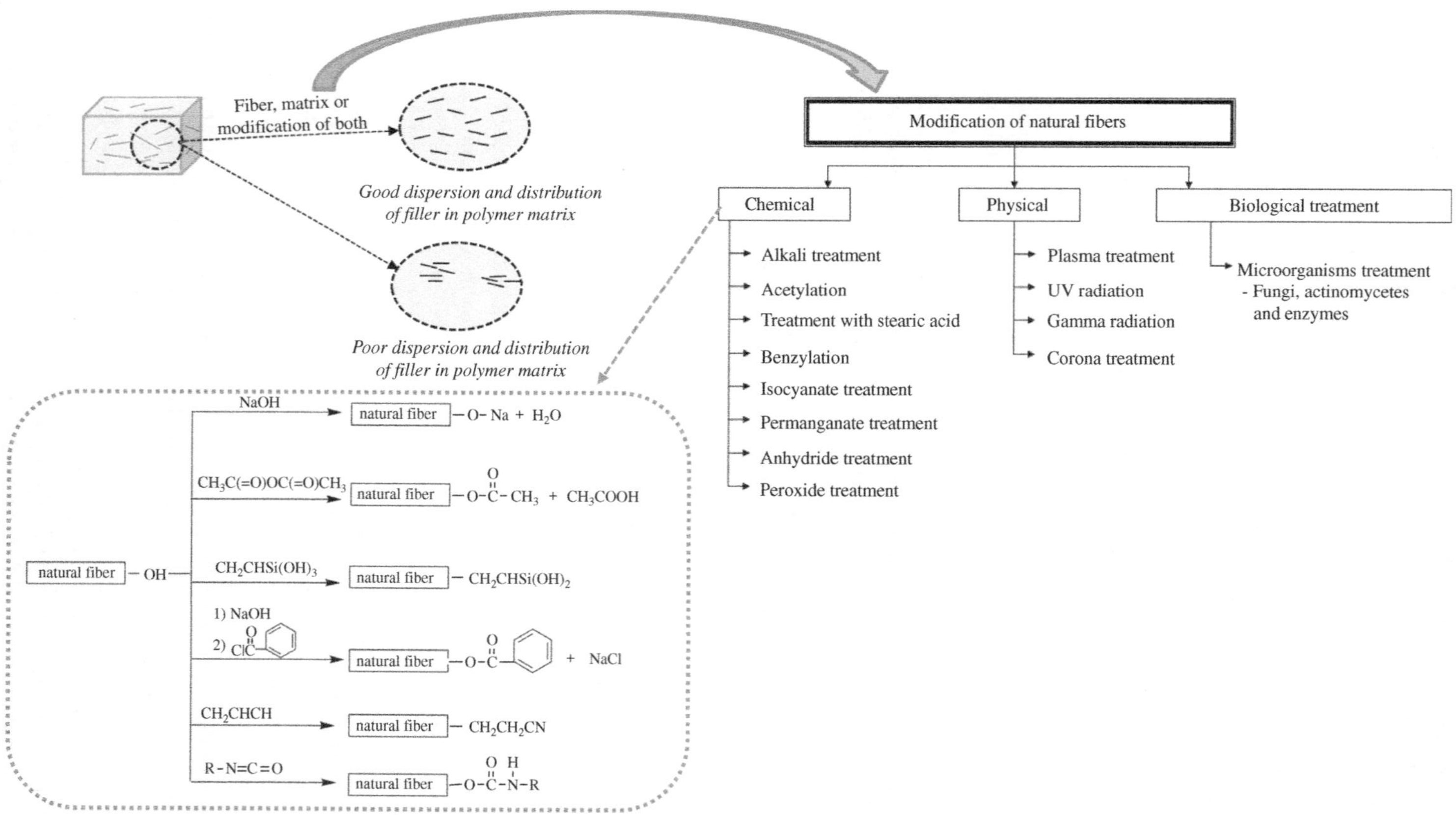

Figure 4.6 Methods of natural fiber modifications.

Biological treatments employ microorganisms, mainly fungi, actinomycetes, and enzymes and allow for the removal of lignin and hemicellulose from bundles of natural fibers with low energy expenditure. Fungi or bacteria produce enzymes that can destroy or remove pectinase glue from the fiber bundle and, consequently, release cellulose from fibers. Biological modification has an influence on the structure, chemical composition, final quality, and properties of fibers, often leading to the improvement of both moisture resistance and their mechanical properties. Despite the positive results of this method, its price, and ecological aspects, when compared to chemical treatments, the small scale of the process, the difficulties associated with its control, and the price of enzymes are the main drawbacks [36].

Methods of chemical modification of natural fillers include alkalization [37], acetylation [38], modification with benzoyl chloride [22], potassium permanganate, isocyanates, peroxides [39], or fatty acids [40]. Among the chemical methods of modification of natural fibers, alkalization is one of the most commonly used. During the process, a certain amount of lignin, waxes, and oils, covering the outer layer of cell walls is removed, cellulose depolymerized, and crystallites of small sizes are exposed [41]. Modification is based on the treatment of fibers with sodium hydroxide (Figure 4.6), which ionizes hydroxyl groups to alcoholates, thereby eliminating free hydroxyl groups that tend to bind water molecules [24]. As a result, hydrogen bonds occurring in the structure of the cell wall break, at the same time increasing roughness of the fibers' surface. The application of alkali during the modification process leads to an increase in the content of amorphous cellulose at the expense of crystalline cellulose. In addition, it is worth nothing that the concentration of alkali solutions strongly affect the final properties of the fibers. It was found that the strength of the plant fiber drastically deteriorates after exceeding a certain optimal alkali concentration – at higher concentrations of alkali, excessive delignification occurs, leading to obtain fibers with worse mechanical properties [24, 42].

Acetylation, another method of modification of natural fillers, is a well-known method of esterification leading to the plasticization of cellulose fibers as a result of replacing hydroxyl groups in the cell wall (mainly from amorphous lignin and hemicellulose) with acetyl groups [43]. The hydroxyl groups of crystalline cellulose, due to their close packing resulted from strong hydrogen bonding, are not available for the chemical reagents used in that method. In most cases, this modification is carried out using acetic anhydride in the presence of sulfuric (VI) acid [44], pyridine [45], or potassium acetate [46] as catalyst or without one. It was found that acetylation of natural fibers is effective in reducing moisture absorption and increasing the dimensional stability of biocomposites filled with acetylated fibers. The reduction of moisture sorption for acetylated jute fibers, when compared to unmodified filler, increases by about 15% [47]. In addition, studies conducted show that composites reinforced with acetylated natural fibers are characterized by greater biological resistance [22].

Another group of chemical modification of natural fillers involves the introduction of a third component into the system, which remains in the resulting polymer composite, acting as a compatibilizer between hydrophilic discontinuous phase and hydrophobic matrix. Coupling agents can improve the bonding between components of the composition by reducing the boundary layers characterized by low mechanical strength, forming highly cross-linked regions, improving wettability, creating covalent bonds of the natural filler with the matrix or by changing the acidity of its surface [48, 49]. Aminosilanes, especially 3-aminopropyltriethoxysilane, are the most often mentioned and described group of coupling agents applied for thermoplastic or thermosetting polymers [47]. Vinyl and acrylic silanes form covalent bonds with polymeric matrices in the presence of peroxide initiators, while methacrylosilanes exhibit relatively high reactivity with unsaturated polyester matrices [50]. As a result of hydrolysis of silane coupling agents, reactive silane groups and the corresponding alcohols are formed. The number of alkoxy groups of the coupling agent determines the amount of water needed for complete hydrolysis and affects the adhesion between silanes and filler. It was found that bifunctional and trifunctional alkoxysilanes exhibit better adhesion than monofunctional derivatives, because during the hydrolysis relatively more possible places to form bonds are created. After obtaining the silanol groups, the process of their condensation occurs, leading to the formation of a polymer siloxane network. It is worth nothing that reactions of alkoxy hydrolysis and silanol condensation take place in a competitive manner depending on the process conditions (used solvent, temperature, pH and silane concentration) [51].

The main disadvantage of the necessity of the biofillers' modification before the main synthesis of composites is definitely the generation of additional cost, which is undesirable in the industrial applications. That is why, a more preferable technique is the introduction of an adhesion promoter between polymer matrix and biofillers. This additional component, due to containing both a hydrophilic and a hydrophobic part, is able to create bonds with the polar groups of the filler and interact with the macromolecular chains of the polymer matrix.

References

1 Mohanty, A.K., Misra, M., Drzal, L.T. et al. (2005). Natural fibers, biopolymers, and biocomposites: an introduction. In: *Natural Fibers, Biopolymers, and Biocomposites*, 1–3. Boca Raton: Taylor & Francis.

2 Lau, K.T., Hung, P.Y., Zhu, M.H., and Hui, D. (2018). Properties of natural fibre composites for structural engineering applications. *Composites Part B: Engineering* 136: 222–233.

3 Thakur, V.K. (2013). *Overview of Composites in Green Composites from Natural Resources*, 2–8. Boca Raton: CRC Press.

4 Strong, A.B. (2008). Introduction to composites. In: *Fundamentals of Composites Manufacturing: Materials, Methods and Applications*, 1–2. Michigan: Society of Manufacturing Engineers.

5 Peters, S.T. (2013). Introduction, composite basics and road map. In: *Handbook of Composites*, 2nde, 3–30. Tonbridge: Springer Science & Business Media.

6 Netravali, A.N. and Chabba, S. (2003). Composites get greener. *Materials Today* 4 (6): 22–29.

7 Sawpan, M.A., Pickering, K.L., and Fernyhough, A. (2011). Effect of fibre treatments on interfacial shear strength of hemp fibre reinforced polylactide and unsaturated polyester composites. *Composites Part A: Applied Science and Manufacturing* 42 (9): 1189–1196.

8 Nardin, M. and Ward, I.M. (1987). Influence of surface treatment on adhesion of polyethylene fibres. *Materials Science and Technology* 3 (10): 814–826.

9 Waldman, D.A., Zou, Y.L., and Netravali, A.N. (1995). Ethylene/ammonia plasma polymer deposition for controlled adhesion of graphite fibers to PEEK. *Journal of Adhesion Science and Technology* 9 (11): 1475–1503.

10 Dicker, M.P., Duckworth, P.F., Baker, A.B. et al. (2014). Green composites: A review of material attributes and complementary applications. *Composites Part A: Applied Science and Manufacturing* 56: 280–289.

11 La Mantia, F.P. and Morreale, M. (2011). Green composites: A brief review. *Composites Part A: Applied Science and Manufacturing* 42 (6): 579–588.

12 Arora, S., Lal, S., Kumar, S. et al. (2011). Comparative degradation kinetic studies of three biopolymers: chitin, chitosan and cellulose. *Archives of Applied Science Research* 3 (3): 188–201.

13 Mitra, B.C. (2014). Environment friendly composite materials: biocomposites and green composites. *Defence Science Journal* 64 (3): 244–261.

14 Peças, P., Carvalho, H., Salman, H., and Leite, M. (2018). Natural fibre composites and their applications: a review. *Journal of Composites Science* 2 (4): 66–86.

15 Cheng, Y., Deng, S., Chen, P., and Ruan, R. (2009). Polylactic acid (PLA) synthesis and modifications: a review. *Frontiers of Chemistry in China* 4 (3): 259–264.

16 Henton, D.E., Gruber, P., Lunt, J., and Randall, J. (2005). Polylactic acid technology. In: *Natural Fibers, Biopolymers, and Biocomposites*, 527–578. Boca Raton: Taylor & Francis.

17 Anderson, J.M. and Shive, M.S. (2012). Biodegradation and biocompatibility of PLA and PLGA microspheres. *Advanced Drug Delivery Reviews* 64: 72–82.

18 Ho, M.P., Wang, H., Lee, J.H. et al. (2012). Critical factors on manufacturing processes of natural fibre composites. *Composites Part B: Engineering* 43 (8): 3549–3562.

19 Joshi, S.V., Drzal, L.T., Mohanty, A.K., and Arora, S. (2004). Are natural fiber composites environmentally superior to glass fiber reinforced composites? *Composites Part A: Applied Science and Manufacturing* 35 (3): 371–376.

20 Jacob, M., Joseph, S., Pothan, L.A., and Thomas, S. (2005). A study of advances in characterization of interfaces and fiber surfaces in lignocellulosic fiber-reinforced composites. *Composite Interfaces* 12 (1–2): 95–124.

21 Ashori, A. (2008). Wood–plastic composites as promising green-composites for automotive industries! *Bioresource Technology* 99 (11): 4661–4667.

22 Li, X., Tabil, L.G., and Panigrahi, S. (2007). Chemical treatments of natural fiber for use in natural fiber-reinforced composites: a review. *Journal of Polymers and the Environment* 15 (1): 25–33.

23 John, M.J. and Anandjiwala, R.D. (2008). Recent developments in chemical modification and characterization of natural fiber-reinforced composites. *Polymer Composites* 29 (2): 187–207.

24 Dittenber, D.B. and Gangarao, H.V.S. (2012). Critical review of recent publications on use of natural composites in infrastructure. *Composites Part A: Applied Science and Manufacturing* 43 (8): 1419–1429.

25 Shah, D.U., Schubel, P.J., and Clifford, M.J. (2013). Modelling the effect of yarn twist on the tensile strength of unidirectional plant fibre yarn composites. *Journal of Composite Materials* 47: 425–436.

26 AL-Oqla, F.M. and Salit, M.S. (2017). Material selection of natural fiber composites. In: *Materials Selection for Natural Fiber Composites*, 107–168. Amsterdam: Elsevier.

27 Shalwan, A. and Yousif, B.F. (2013). In state of art: mechanical and tribological behaviour of polymeric composites based on natural fibres. *Materials & Design* 48: 14–24.

28 Mohammed, L., Ansari, M.N., Pua, G. et al. (2015). A review on natural fiber reinforced polymer composite and its applications. *International Journal of Polymer Science* 2015: 1–15.

29 Carroll, D.R., Stone, R.B., Siringano, A.M. et al. (2001). Structural properties of recycled plastic/sawdust lumber decking planks. *Resources, Conservation and Recycling* 31: 241–251.

30 Canche-Escamilla, G., Rodriguez-Laviada, J., Cauich-Cupul, J.I. et al. (2002). Flexural, impact and compressive properties of a rigid-thermoplastic matrix/cellulose fiber reinforced composites. *Composites Part A: Applied Science and Manufacturing* 33 (4): 539–549.

31 Zhou, Y., Fan, M., and Chen, L. (2016). Interface and bonding mechanisms of plant fibre composites: an overview. *Composites Part B: Engineering* 101: 31–45.

32 Mukhopadhyay, S. and Fangueiro, R. (2009). Physical modification of natural fibers and thermoplastic films for composites – a review. *Journal of Thermoplastic Composite Materials* 22 (2): 135–162.

33 Felix, J.M. and Gatenholm, P. (1991). The nature of adhesion in composites of modified cellulose fibers and polypropylene. *Journal of Applied Polymer Science* 42: 609–620.

34 Prasad, B.M. and Sain, M.M. (2003). Mechanical properties of thermally treated hemp fibers in inert atmosphere for potential composite reinforcement. *Materials Research Innovations* 7 (4): 231–238.

35 Cruz, J. and Fangueiro, R. (2016). Surface modification of natural fibers: a review. *Procedia Engineering* 155: 285–288.

36 Lu, N., Oza, S., and Tajabadi, M.G. (2015). Surface modification of natural fibers for reinforcement in polymeric composites. In: *Surface Modification of Biopolymers*. Hoboken: Wiley.

37 Mwaikambo, L.Y. and Ansell, M.P. (2002). Chemical modification of hemp, sisal, jute, and kapok fibers by alkalization. *Journal of Applied Polymer Science* 84 (12): 2222–2234.

38 Lisperguer, J., Droguett, C., Ruf, B., and Nunez, M. (2007). Physical modification of natural fibers and thermoplastic films for composites – a review. *Journal of the Chilean Chemical Society* 52 (1): 1073–1075.

39 Hariharashayee, D., Hanif, M.A.A., Aravind, S. et al. (2018). Investigation of mechanical properties of $KMnO_4$ treated Sansevieria cylindrical fiber reinforced polymer composite. *International Journal for Advance Research and Development* 3 (4): 294–298.

40 Kumar, S. (2007). Chemical modification of wood. *Wood and Fiber Science* 26 (2): 270–280.

41 Rowell, R.M. (2012). Chemical modification of wood. In: *Handbook of Wood Chemistry and Wood Composites*, 537–599. Boca Raton: Taylor & Francis.

42 Väisänen, T., Haapala, A., Lappalainen, R., and Tomppo, L. (2016). Utilization of agricultural and forest industry waste and residues in natural fiber–polymer composites: a review. *Waste Management* 54: 62–73.

43 Rowell, R.M. (2004). Acetylation of natural fibers to improve performance. *Molecular Crystals and Liquid Crystals* 418 (1): 153–164.

44 Johnson, A.M., Kim, H., Ralph, J., and Mansfield, S.D. (2017). Natural acetylation impacts carbohydrate recovery during deconstruction of *Populus trichocarpa* wood. *Biotechnology for Biofuels* 10 (48): 1–12.

45 Papadopoulos, A.N. (2006). Pyridine-catalyst acetylation of pine wood: influence of mature sapwood vs juvenile wood. *Holz als Roh-und Werkstoff* 64 (2): 134–136.

46 Obataya, E. and Minato, K. (2009). Potassium acetate-catalyzed acetylation of wood at low temperatures II: vapor phase acetylation at room temperature. *Journal of Wood Science* 55 (1): 23–26.

47 Kalia, S., Kaith, B.S., and Kaur, I. (2009). Pretreatments of natural fibers and their application as reinforcing material in polymer composites – a review. *Polymer Engineering & Science* 49 (7): 1253–1272.

48 Kushwaha, P.K. and Kumar, R. (2014). Studies on performance of acrylonitrile-pretreated bamboo-reinforced thermosetting resin composites. *Journal of Reinforced Plastics and Composites* 29 (5): 718–724.

49 Chan, C.M., Vandi, L.J., Pratt, S. et al. (2018). Mechanical properties of poly(3-hydroxybutyrate-co-3-hydroxyvalerate)/wood flour composites: Effect of interface modifiers. *Journal of Applied Polymer Science* 135 (43): 46828. (pp. 1–10).

50 Xie, Y., Hill, C.A.S., Xiao, Z. et al. (2010). Silane coupling agents used for natural fiber/polymer composites: a review. *Composites Part A: Applied Science and Manufacturing* 41 (7): 806–819.

51 Sadhan, A.P. and Jana, C. (2002). Natural fiber composites of high-temperature thermoplastic polymers: effects of coupling agents. *Journal of Applied Polymer Science* 86 (9): 2168–2173.

5

Mechanical, Thermal, Tribological, and Dielectric Properties of Biobased Composites

T. Senthil Muthu Kumar[1,2], K. Senthilkumar[1,2], M. Chandrasekar[3], S. Karthikeyan[1], Nadir Ayrilmis[4], N. Rajini[1], and Suchart Siengchin[2]

[1] Department of Mechanical Engineering, Kalasalingam Academy of Research and Education, Krishnankoil, Tamil Nadu, India
[2] Department of Mechanical and Process Engineering, The Sirindhorn International Thai German Graduate School of Engineering (TGGS), King Mongkut's University of Technology North Bangkok, Bangkok, Thailand
[3] School of Aeronautical Sciences, Hindustan Institute of Technology and Science, Chennai, Tamil Nadu, India
[4] Department of Wood Mechanics and Technology, Istanbul University-Cerrahpasa, Istanbul, Turkey

5.1 Introduction

Composites with synthetic fiber reinforcements such as carbon, glass, and aramid have been employed in high performance applications such as aircraft, automobile, and aerospace sectors. However, serious threats to the environment in the form of pollution, waste aggregation due to the limited recycling abilities of synthetic reinforcements, and depletion of petroleum-based resources push the need for the development of biobased environment friendly materials. Over the last few decades, research has been focused on replacing the synthetic fibers with natural fibers obtained from the renewable resources such as plants and trees [1]. Natural fibers have been emerging as a potential material for reinforcement in the polymer composites due to their easy availability, renewable nature, lower density, good thermal insulation and mechanical properties, biodegradability, reduced tool wear, low price, etc. [2–4]. Hence, many industrial sectors such as automotive, construction, biomedical, packaging, military, and aerospace are considering the development of biocomposites or biobased composites [5].

5.2 Characterization of Biobased Composites

Suitability of the biobased composites for a particular application is validated with the help of various characterization techniques. Among the various characterization techniques, mechanical, thermal, tribological, and water absorption properties of the biobased composites were of particular interest to this study. Figure 5.1 shows the characterization techniques that can be used to assess the performance of the biobased composites within the scope of this chapter.

Mechanical properties of the biobased composites can be split into two categories: static properties (such as tensile, flexural, compression, and shear) and dynamic properties (such as fatigue and ballistic impact). Thermal behavior of the composites under elevated

Biobased Composites: Processing, Characterization, Properties, and Applications, First Edition.
Anish Khan, Sanjay Mavinkere Rangappa, Suchart Siengchin and Abdullah M. Asiri.

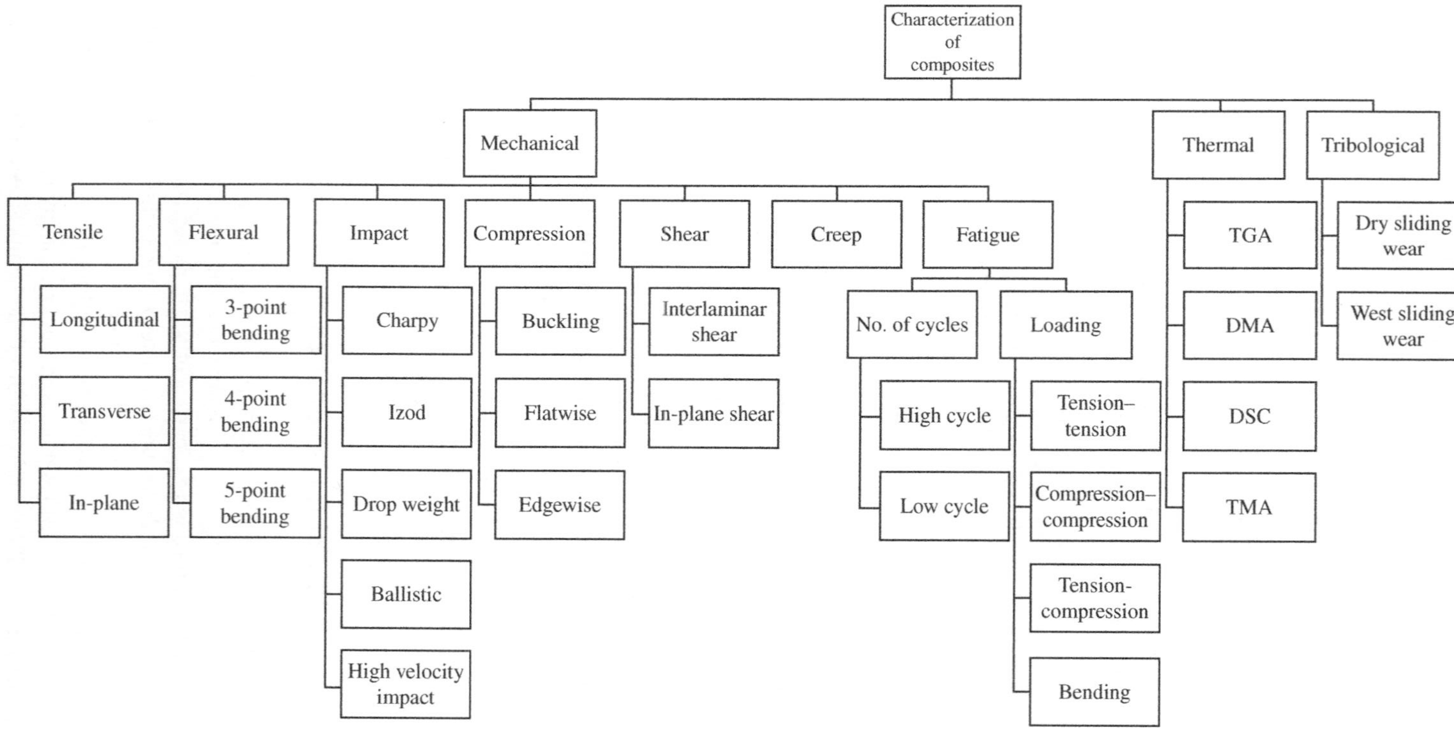

Figure 5.1 Characterization techniques commonly used to assess the properties of biocomposites.

Table 5.1 Thermal characterization techniques.

Technique	Frequently measured parameters	Remarks
Thermo-Gravimetric Analysis (TGA)	Glass transition temperature Mass loss vs. Temperature Final char residue	Higher the char residue, better the thermal stability. Detect changes in glass transition temperature. Degradation temperature of various constituents of the composite in the temperature range studied. Stages of decomposition of fiber and matrix in the temperature range studied.
Dynamic Mechanical Analysis (DMA)	Glass transition temperature Storage modulus Loss modulus Phase angle – Tan delta	DMA highlights the viscoelastic properties of the biocomposite under elevated temperatures. Elastic and viscous properties are represented by storage modulus and loss modulus, respectively.
Differential Scanning Calorimetry (DSC)	Glass transition temperature Crystallization temperature Melting temperature	Thermal transition before curing, during curing, and post curing of the polymers.
Thermo-Mechanical Analysis (TMA)	Dimensional change vs. Temperature Coefficient of Thermal Expansion (CTE)	Evaluates the dimensional changes of the biocomposite over a range of temperature.

temperature is characterized by techniques such as Thermogravimetric Analysis (TGA), Dynamic Mechanical Analysis (DMA), Differential Scanning Calorimetry (DSC), and Thermomechanical Analysis (TMA) [6]. All these techniques are aimed at determining the thermal stability of a biocomposite under elevated temperature as shown in Table 5.1. Pin-on-disc test is the commonly used technique to determine the tribological properties or wear resistance of a biocomposite. The setup consists of a rotating disc with an abrasive paper on the top. As the composite specimen comes into contact with the abrasive paper, it is subjected to sliding wear. The wear performance of a biocomposite can be evaluated under dry and wet conditions at different velocities, sliding distance, and normal loads. Parameters such as mass loss due to wear, specific wear rate, and coefficient of friction provide information on the performance of the material under wear.

5.3 Factors Influencing Various Properties of the Biobased Composites

Major factors influencing the material properties of the biobased composites is given in Figure 5.2.

5.3.1 Constituents of Biobased Composites

Fiber, matrix, and fillers are the essential constituents of a composite material. Additives such as color pigments, antioxidants, moisture absorbers, and UV stabilizers are also added

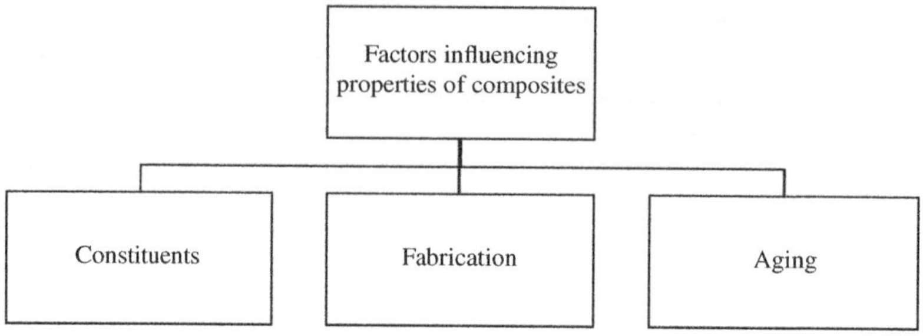

Figure 5.2 Major factors influencing the properties of biocomposites.

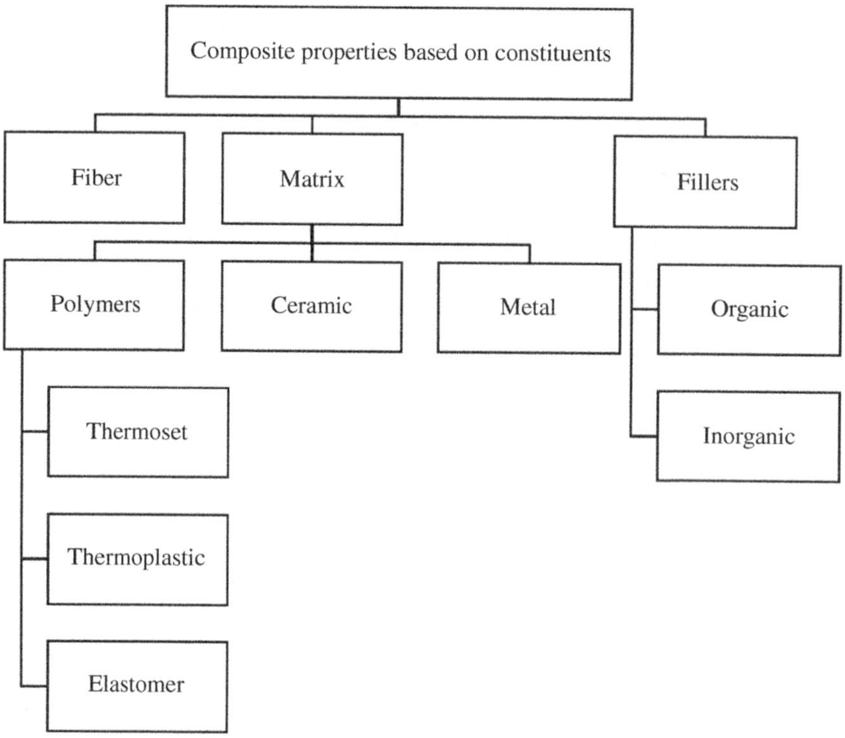

Figure 5.3 Properties of biobased composite based on the constituents.

to the composition. Composites are heterogeneous material and can retain their physical identities after their fabrication. Variants of composites can be obtained with the help of different constituents shown in Figure 5.3. Fibers are the primary load carriers in the composite and their properties are highly dependent on factors such as fiber type, fiber architecture, fiber weight or fiber loading, stacking sequence, use of two or more fibers in the hybrid configuration, and surface modification (Figure 5.4).

5.3.2 Fabrication Techniques of Biobased Composites

Fabrication techniques significantly influence the strength, physical properties, and dimensional stability of the composite. Composites with synthetic fiber reinforcements used in the

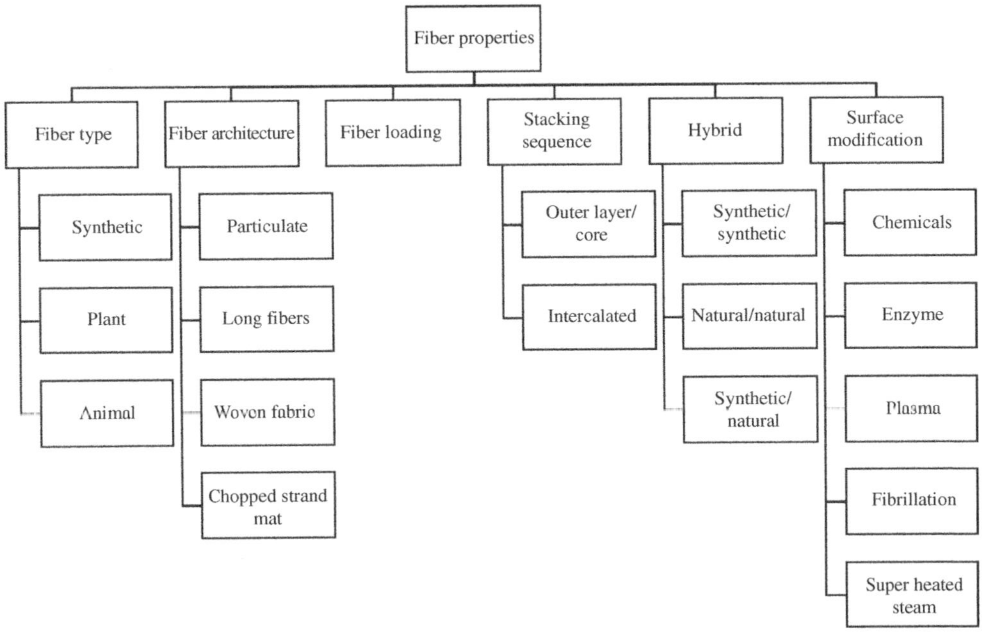

Figure 5.4 Properties of the composites influenced by the fiber parameters.

aircraft structures and aerospace applications have been cured using autoclave till date. The curing process consists of stacking the required number of prepregs (fibers impregnated with epoxy resin in a partially cured state and stored in sub-zero temperature) in a mold setup followed by the application of vacuum through vacuum bagging and curing under high temperature.

In case of biobased composites, Out of Autoclave (OOA) techniques such as compression molding, resin transfer molding, pultrusion, etc. are preferred over the autoclave as shown in Figure 5.5. This is because of the following reasons:

- Curing process using autoclave is quite expensive and requires skilled labor.
- Unavailability of the natural fiber-based prepregs. To date, there are few companies like Lineo, Procotex Corporation SA, and Lingrove Ekoa which have developed flax-based prepreg for commercial use. According to our knowledge, prepreg with other natural fibers is not available in the market. Developing prepreg with natural fibers can be challenging due to the variation in material properties such as diameter over the fiber length and inner constituents based on geographical location. Hence, development of natural fiber-based prepreg offers a lot of room for research.
- Curing requirement for thermoplastics: Thermoplastic requires melting of the resin into the mold followed by solidification under pressure to form the composite.

Additive manufacturing, also known as 3D printing (shown in Figure 5.6), developed in the recent years has proven to be a highly reliable technique for fabrication of composites. 3D printing is classified as an additive manufacturing (AM) process, where the material is added in layers, allowing users to create a real product directly from a 3D computer model. It involves the deposition of the material layer by layer in a sequence based on the computer input of a particular design. This feature allows complex shapes to be built unlike traditional

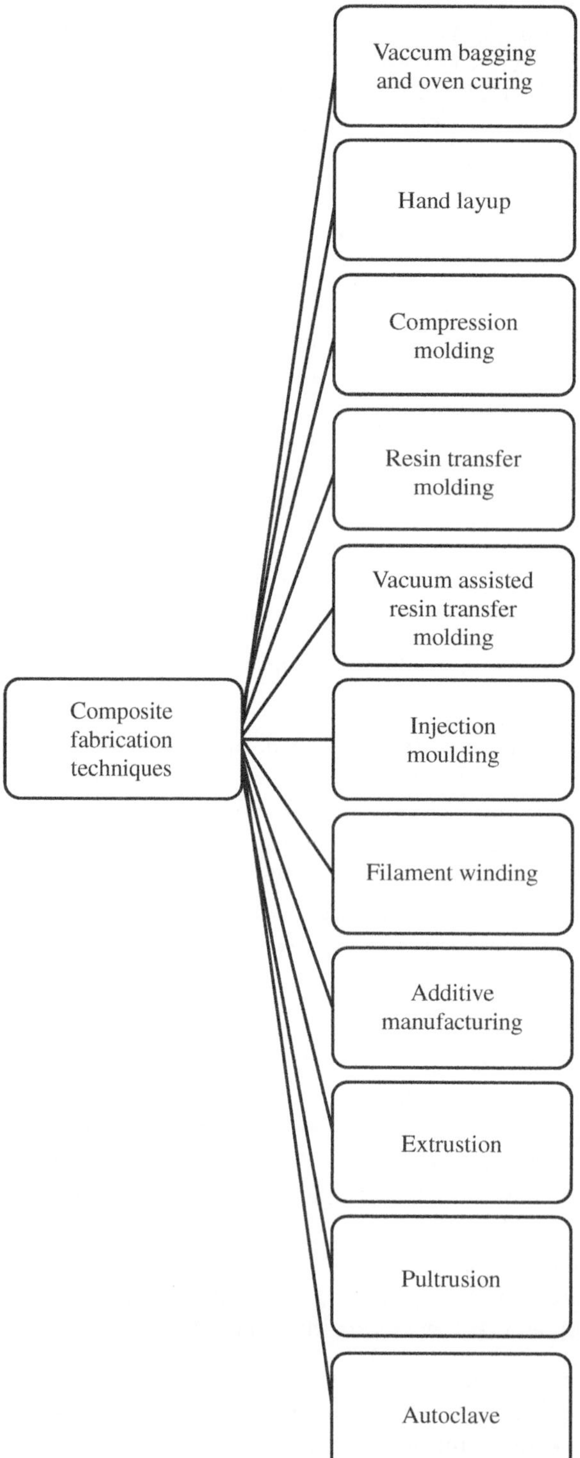

Figure 5.5 Composite fabrication techniques.

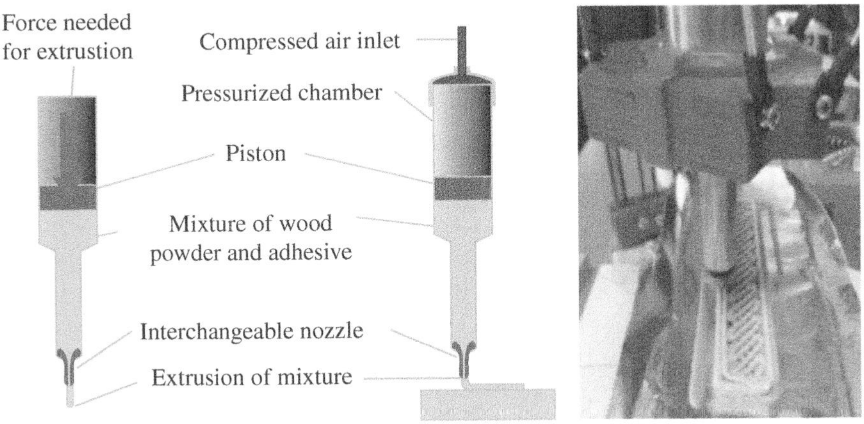

Figure 5.6 Additive manufacturing technology using a 3D printer.

curing methods which induce residual stress during the machining and postprocessing. This technique is compatible with thermoset and thermoplastic matrices. The 3D printer enables the consumers the possibilities to fabricate the complex shape products without special tools or molds. The unique one piece products or low-volume customized products can be efficiently produced using the raw material, which leads to low or zero waste.

5.3.3 Aging and Their Impact on the Composite Properties

Natural fibers extracted from the plants and trees are hydrophilic in nature while the thermoset or thermoplastic matrix used in the structural applications are hydrophobic in nature. Thus, in case of a biobased composite, the matrix is naturally resistant to moisture in the environment. However, moisture could diffuse into the composites through microvoids or defects. Moisture absorbed by the natural fiber causes thickness swelling and induces residual stress within the composite. Another environmental factor which adds to material degradation is matrix plasticization due to the sub-zero or elevated temperature. The prolonged exposure to such conditions causes degradation and leads to structural weakening. Hence, the biocomposite could fail at comparatively lower loads than the original load. The biocomposites are designed to withstand structural weakening. Therefore, the assessment of degradation effects in the biocomposite under such environmental conditions is inevitable. Performance of biocomposites under environmental conditions is studied with the help of natural weathering and simulated conditions in the laboratory through accelerated weathering and water immersion tests as shown in Figure 5.7.

5.4 Mechanical Properties of Biobased Composites

Nowadays, diverse alternative methods are used to attain enhanced properties of the materials which include the blending of polymers, using additives or reinforcements to make composites with enhanced mechanical properties. Many researchers have worked on the development of new materials which will be more effective in terms of performance and environmental aspects. Researchers proved that the reinforcements greatly influenced the

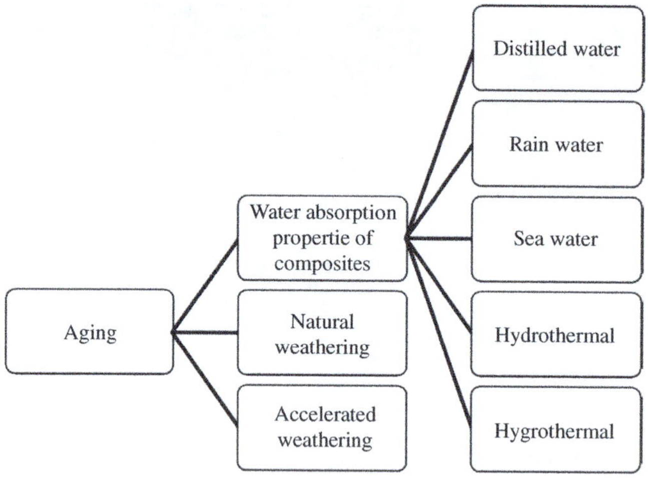

Figure 5.7 Aging test for biocomposites.

improvement in mechanical properties of the ensuing composites [7, 8]. In particular, the natural fiber reinforced composites possess better properties since the presence of cellulosic components makes the composites stiffer, thus enhancing its mechanical properties. It is well known that the mechanical properties of natural fiber reinforced composites are governed by the interfacial bonding between the matrix and the filler material. In this scenario, many natural fibers like banana, sisal, hemp, jute, coconut, sugar palm, flax, and bamboo [9, 10] have been utilized as reinforcements in composite materials for applications in consumer goods, low-cost housing, and other civil structures. A number of research works have been carried out with different natural fibers to study the reinforcing effects on the performance.

In a research, environmental friendly composites were derived using a blend of recycled agricultural plastic (RAP) and HDPE compounded with waste cellulose fibers. The inclusion of cellulose fibers greatly influenced the enhanced mechanical properties, especially tensile strength and modulus. This was due to the presence of polar acetate groups of EVA, which enhanced the physiochemical compatibility of the matrix material and the reinforcement [11].

In another study, kenaf fiber was reinforced with PLLA to form biobased composites [12]. It was found that the tensile modulus of the biobased composites was superior to that of the pure PLLA. In another study, effects of hybridizing glass fiber and biofiber on the performance of polyester composites were determined [13]. The report revealed that addition of relatively small amounts of glass fiber enhanced the mechanical properties of the hybrid composites.

Fabricating methods and processing parameters can also affect the mechanical properties of composites. In a study, the tensile strength reduced with increasing injection pressure in the jute fiber reinforced composites made by resin transfer molding process. When the injection pressure was increased from 4 to 8 bar, the tensile strength of the composites reduced. This could be attributed to the damage of the jute fiber due to increased pressure [14].

In recent times, more attention is given to the infusion of agricultural wastes as fillers in the polymer matrix. These agro-wastes are not only used as biomass but are also seen as cheaper lignocellulosic fillers for polymer-based composites. Agricultural wastes such as egg shells, tamarind nuts, brewed coffee beans and tea leaves, coconut shells, banana peels, rice husk, etc. have been used as particle fillers in polymer composites [5, 8, 15–19]. Addition of

Table 5.2 Mechanical properties of biocomposites with agro-based filler reinforcement.

Polymer matrix	Lignocellulosic filler	Tensile strength (MPa)	Tensile modulus (MPa)	Elongation at break (%)
Cellulose	Chemically treated spent coffee bean powder	149 (103)*	2.18 (2.95)	4.9 (1.1)
Polypropylene Carbonate	Tamarind nut powder	21.5 (12.2)	—	345 (11)
Polypropylene Carbonate	Banana peel powder	22.8 (10.5)	—	268 (12.5)
Poly Lactic Acid	Egg shell powder	22.5 (16.8)	1964 (1152)	286 (402)
Poly(3-hydroxybutyrate-co-3-hydroxyvalerate)	Spent coffee bean powder	9.01 (29.14)	—	5.53 (2.32)
Cellulose	Spent tea leaf powder	119.5 (97)	—	3 (1.8)

* Numericals in brackets indicate the values of pure matrix.

these fillers into polymer matrices improved the mechanical properties of the composites. For instance, chemically treated spent coffee bean powder reinforced with cellulose matrix increased the tensile properties of the biocomposites from 103 to 149 MPa [5]. Similarly, the infusion of egg shell powder in polypropylene carbonate enhanced the tensile strength from 12.8 to 23.2 MPa and tensile modulus from 823 to 1378 MPa [16]. Comparison of tensile properties of the composites with agro-based fillers is presented in Table 5.2.

As already discussed, it is very important to have better matrix and fiber or filler bonding in order to obtain better mechanical properties. Due to the hydrophilic nature of the natural fibers, the interaction between the fibers and the hydrophobic polymer matrix is very weak. Hence, chemical modification of the fillers are done to rectify this problem. For instance, untreated spent coffee bean powder reinforced with cellulose matrix reduced the tensile strength of the composites. This could be due to the incompatibility of the lignocellulosic filler and polymer matrix. When the surface of the filler is modified by different techniques such as physical modification (electric discharge [corona, cold plasma]), chemical modification (coupling agents, alkaline treatment, acetylation), thermal-treatment modification, and other chemical treatments (i.e. stearic acid in ethyl alcohol solution and permanganate treatment), the surface becomes more compatible with the matrix. This improves the impact, bending, and tensile properties of the composites [5].

5.5 Thermal Properties of Biobased Composites

Notably, new regulations formed by the governments and environmental issues play a major concern for developing a biocomposite. Besides, due to industrialization, the effects of greenhouse gas emissions are evident; hence, these are driving toward the usage of renewable and biodegradable materials in many applications [9]. Such materials can be called biocomposites or green composites, in which the matrix and/or reinforcement are biodegradable and ecofriendly [20].

Many researchers are fascinated by biocomposites, as these can be produced from agricultural resources such as wastes and other products [21]. Besides, the biocomposites are assisted in improving the manufacturing speed and recycling with improved environmental

Table 5.3 Popular thermal analysis techniques with their property measured.

Name of the technique	Acronym	Property measured
Dynamic mechanical analysis	DMA	Viscoelastic properties
Thermogravimetric analysis	TGA	Mass
Thermomechanical analysis, Thermodilatometry	TMA, TD	Length or volume
Differential scanning calorimetry	DSC	ΔT, differential power input
Dielectric analysis	DEA	Dielectric properties
Micro/nano thermal analysis	μ/n – TA	Penetration, ΔT
Differential thermal analysis	DTA	ΔT

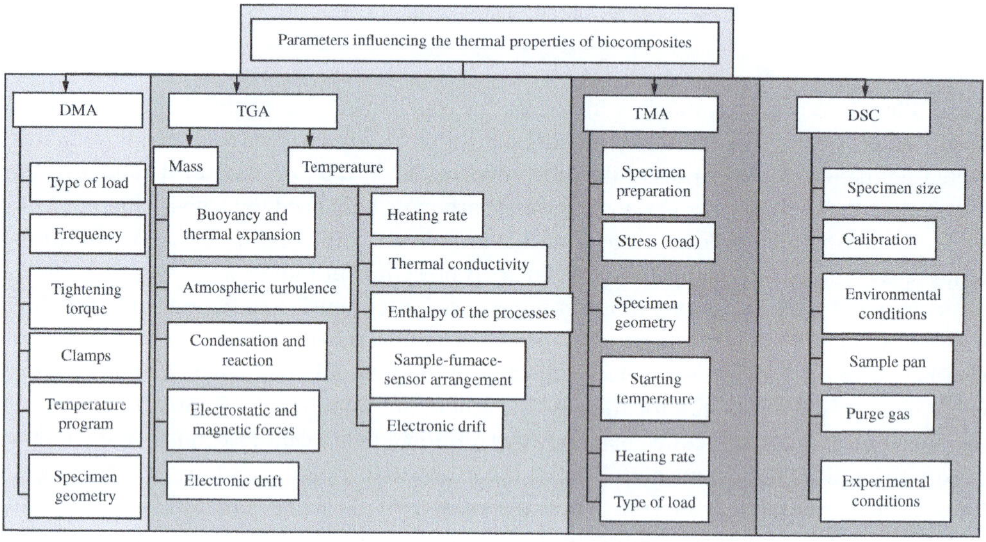

Figure 5.8 Factors influencing the thermal properties of biocomposites.

compatibility. Hence the use of biocomposites is introduced in many industries such as aerospace, construction, and building industries, especially in automotive industries [22]. Before the composites are used in any real structure, they undergo some characterization techniques such as mechanical, physical, thermal, chemical, etc. concerning their applications.

Thermal analysis is a technique used to measure the material's response, whether the material is heated or cooled (sometimes, the materials held isothermally). The main goal of this characterization is to establish a connection between the temperature and specific properties of the materials. Some of the popular thermal analysis techniques are (i) DMA, (ii) TGA, (iii) TMA, (iv) DSC, (v) dielectric analysis (DEA), and (vi) micro/nano thermal analysis (μ/n – TA) [6]. Table 5.3 shows different types of techniques, along with their property measured. The influencing factors that affect the thermal properties of composites are depicted in Figure 5.8.

In this chapter, a particular emphasis is on the thermal properties of thermosets, thermoplastics, and biopolymer-based composites. This review highlights the basics and the literature on TGA and DMA of (i) single fiber reinforced composites, (ii) hybrid fiber reinforced composites, (iii) chemically modified fiber reinforced composites (for both single and hybrid fibers), (iv) nanocomposites, and (v) filler-filled composites.

5.5.1 Thermogravimetric Analysis of Biobased Composites

In TGA, the mass of the polymer is measured as a function of time or temperature while the sample is subjected to a controlled temperature program in a controlled atmosphere. During the measurement of any polymeric materials, the temperature is fixed from ambient to 1000 °C or sometimes more concerning their intended applications. A purge gas (e.g. nitrogen, helium, or organ) can be used as an atmosphere that could act as an inert gas.

In addition to the mass loss, the polymers are sometimes subjected to mass gain in an oxidizing atmosphere at slow heating rates whereby the mass gain is observed before the degradation of the polymers.

Mass loss can be considered as volatile components such as (i) residual solvents and (ii) absorbed moisture. These are generally evaporated between atmospheric temperature and 300 °C. The reaction products (e.g. water and formaldehyde) could form due to the curing of phenolic and amino resins, whereby the temperatures are generally between 100 and 250 °C. Volatile degradation products are formed by chain scission (200–800 °C). All of the above said mass loss processes could be used to provide information such as (i) composition, (ii) extent of cure, and (iii) thermal stability by interpreting the mass loss processes [23].

Jumaidin et al. [24] have recently fabricated cogon grass fiber/thermoplastic cassava starch (TPCS) biocomposites by varying the fiber loadings (1, 3, and 5 wt%). The TPCS was found to be a promising replacement for many petroleum-based polymers. However, the TPCS exhibited lower mechanical properties. Hence, the cogon grass fibers were incorporated with TPCS using a compression molding technique. The results showed that the addition of cogon grass fiber in the TPCS improved the onset degradation temperatures (292.14 °C/5 wt%) of starch. It was ascribed by the existence of higher lignocellulose elements in the fiber. Besides, the addition of cogon grass fiber in the TPCS matrix was found to be responsible for obtaining a higher degradation rate. In another study, the onset degradation temperature of single fiber reinforced composites was improved by producing hybrid composites, in which agar/cotton/thermoplastic waxy rice starch hybrid composites exhibited higher onset degradation temperature (168 °C) as described in Ref. [25]. Moreover, the hybrid composites showed higher thermal stability than the pure fiber reinforced composites whereas the fiber ratio was maintained at 2 : 8 (Agar : Cotton). It was ascribed to the higher thermal stability of agar and cotton fibers. Likewise, Curvelo et al. [26], Ma et al. [27], and Prachayawarakorn et al. [28] improved the thermal stability of composites using higher thermal stability of fibers such as Eucalyptus fiber, winceyette fiber, and cotton fiber, respectively.

In the continuing research on thermal stability, Muthu Kumar et al. [1] used a waste filler (e.g. spent coffee bean powder) as reinforcement in poly(3-hydroxybutyrate-*co*-3-hydroxyvalerate) (PHBV) matrix. They reported that the fillers exhibited higher thermal stability than the PHBV matrix owing to the existence of polyphenols in coffee bean powder. Besides, increasing the filler content from 5 to 25% in PHBV significantly improved the thermal stability of these composites.

It is well known that reducing the particle size plays a significant role in enhancing the material properties, especially toughness and stiffness. Reduction of particle size to a nanoscale may be called nanofillers, and these fillers added with polymers are so-called as nanocomposites. The nanobased composites are expected to replace (i) the traditional composites fabricated by the melt processing method, (ii) polymer blends, and (iii) polymers [29]. Nanocalcium carbonate/pineapple leaf fiber reinforced epoxy matrix composites exhibited higher thermal stability than the unfilled matrix composites [30]. It was ascribed by the improved interaction between the nanoparticle and the epoxy matrix molecules. A similar kind of improvement had also been shown by Yang et al. [31].

5.5.2 Dynamic Mechanical Analysis of Biobased Composites

DMA is one of the most popular techniques for characterizing the viscoelastic property of the materials since it can be readily adapted to study both the polymeric liquids and solids. Thermal engineers regularly use DMA techniques for liquids as rheology measurements. In DMA, "strain" is the controlled input to the instrument, while the output "stress" is measured or observed [6]. The basic principle of DMA operation is given in Figure 5.9, and the essential findings of the DMA are given in Table 5.4.

Sreenivasan et al. [32] analyzed the effect of fiber lengths of Sansevieria cylindrica (SC) reinforced polyester matrix composites. The DMA results showed that the storage modulus was decreased by increasing the fiber lengths owing to the curling effect of fibers. The curling effect did not allow to transfer the load from fiber to matrix. Besides, the authors reported that the damping characteristics of SC reinforced composites depended on many factors, such as (i) fiber breakage, (ii) fiber/matrix adhesion, (iii) interphase zone, (iv) frictional resistance, and (v) matrix breakage.

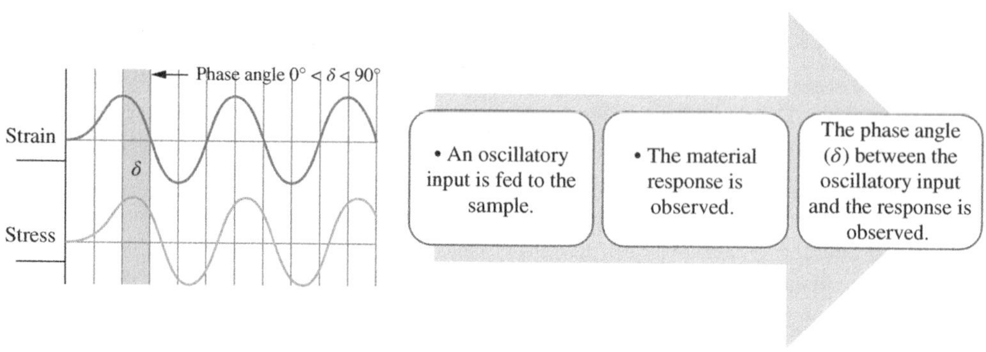

Figure 5.9 Basic principle of DMA operation.

Table 5.4 Essential findings of the DMA test.

Findings	Response
Storage modulus (E')	It measures the stiffness or elastic character of a material
Loss modulus (E'')	It measures the viscous character of a material
Tan delta (tan δ)	• It measures the damping characteristics of the material and classifies the elastic or viscous property of the composite. • It is defined by taking the ratio of $\frac{\text{Loss modulus}}{\text{Storage modulus}} = \frac{\sin(\delta)}{\cos(\delta)} = \tan\delta$

Senthilkumar et al. [6] reported that the viscoelastic properties of composites could be improved by fiber surface modification instead of incorporating fiber reinforcements in composites. Recently, Wang et al. [33] carried out the DMA on abutilon natural straw/polylactic acid (PLA) composites, whereby the straw content was varied from 1 to 5%. Before fabrication, the abutilon natural straws were crushed to a powder form (diameter: 45 μm), and these powders were surface modified using NaOH. The DMA data showed that the storage modulus and tan delta were significantly improved by increasing the straw content. The enhanced properties were attributed to the better adhesion between the treated abutilon natural straw and the PLA.

The influence of the layering sequence of sisal and hemp fiber reinforced green epoxy composites on viscoelastic properties was analyzed [10]. The results revealed that changing the layering sequence by introducing fibers in hybrids was not effective in the thermal properties. In another work, Syafiqa et al. [34] investigated the viscoelastic properties of surface-modified (benzoyl treated) sugar palm and glass fiber reinforced epoxy composites. Treated sugar palm/glass fiber composites exhibited better storage and loss modulus compared to the untreated fiber composites. Furthermore, it was observed that the properties were further improved by increasing the content of glass fibers.

Essabir et al. [35] studied the DMA property of oil palm fiber reinforced high-density polyethylene (HDPE) composites as a function of clay (bayada) content. The authors reported that the complex modulus was improved by increasing the bayada content until 12.5% whereas the ratio of optimum fiber and bayada was found to be 12.5 : 12.5. However, the damping factor showed fewer changes compared to the complex modulus.

The main aim of this chapter was to demonstrate the basics of TGA and DMA techniques. The TGA technique is a prevalent method in the area of polymers, in particular, to understand the thermal stability of any polymeric system composites. The DMA can be a beneficial technique to understand the dynamic mechanical properties such as storage modulus and damping of viscoelastic of any polymers and polymer matrix composites.

5.6 Tribological Properties of Biobased Composites

The natural fibers for biocomposites have attracted many researchers for the past three decades due to its ease of availability and processing. Most of the biocomposites show better results in terms of mechanical, thermal and tribological properties. Since these materials become a good replacement for most of the engineering applications, the research needs to be expanded in terms of finding new materials and validating the existing results in the real-time environment [36]. The need of environmental friendly materials also increased due to numerous pollution control norms and the industrial need of cost effectiveness in all aspects compelled us to find alternatives in terms of low cost material [37]. At the same time, the strength and the efficacy of the material should not be compromised. By considering all the above aspects, a lot of biobased materials are explored by various researchers (Table 5.5). In this chapter, some of the major biocomposites under various tribological studies were discussed in detail.

Based on the current industrial need, a lot of researchers have started exploring new materials for tribological applications. Nath et al. [38] made a hybrid composite using HDPE and hydroxyapatite (HAp) along with alumina as a replacement to HAp. The composite was hot pressed and tested using the ball-on-flat type testing machine. The fretting wear study

Table 5.5 Different kinds of composites made up of biobased materials and their manufacturing methods.

Reinforcement	Matrix	Manufacturing method	Testing machine	References
Hydroxyapatite (HAp)	High-density polyethylene	Hot pressed	Ball-on-flat type fretting wear tester	[38]
Nickel-free stainless steel	Hydroxyapatite (prepared by heat treating bone ash)	Cold pressing and sintering	Abrasive wear test on pin-on-disc	[39]
Seashell nanopowder	Poly(Methyl methacrylate) – PMMA	Poured into the molding	Sliding against the steel disc counterpart	[40]
Jute, hemp, and flax	Epoxy	Hand-layup technique	Pin-on-Disc	[41]
Banana/hemp hybrid	Vinyl ester resin	Compression molding	Abrasive wear test on pin-on-disc	[42]
Corn stalk fiber	Water slag and tire powder	Hot pressed	Special friction and wear tester	[43]
Ramie fiber	Phenolic resin	High temperature molding and oven curing	Chase machine	[44]
Wood flour Glass fiber	Epoxy	Hand-layup technique	Pin-on-Disc	[45]
Bauhinia-vahlii-weight/sisal	Epoxy rice husk fillers	Hand-layup technique	Pin-on-Disc	[46]

revealed the better tribological property where the COF value was about 0.09. The composition of 20% of HDPE, HAp, and Alumina offered better wear resistance even at high contact pressure. Younesi et al. [39] invented a novel biocomposite to study the tribological property for ortho applications. The material was made up of nickel-free stainless steel and hydroxyapatite which was cold pressed and sintered to make a required specimen. The wear study revealed that the wear rate decreased after a long sliding distance and the volume loss increases at high load, but these had no severe effect on the wear property with these kind of biocomposites than the ordinary material. The tribological property of the seashell-made biocomposites using poly(methyl methacrylate) was studied. The composite was molded into the required shape after melting it to the required temperature. It was made to slide against the steel disc to study the tribological property of the composite. After the addition of seashell powder, the wear properties improved with less CoF value. It was also observed that the 12 and 8% mixed composite exhibited better mechanical and tribological properties [40].

Chaudhary et al. [41] studied the wear properties of the jute, hemp, and flax fiber. The composite was fabricated using hand layup method and tested using the pin-on-disc apparatus (Figure 5.10). The result unveiled that the jute/epoxy combination gave the highest wear rate and CoF. Among all the hybrid combinations fabricated, jute/hemp hybrid delivered the better performance. Suresh et al. [42] investigated hybrid composites made up of the banana and hemp fiber. The matrix was chosen as a vinyl ester and the reinforcement was chosen from biowaste material. These materials were fabricated using the compression molding machine and abrasive property was conducted using the pin-on-disc apparatus. The result exhibited that the 20% mix of banana and hemp fiber with the matrix showed better results.

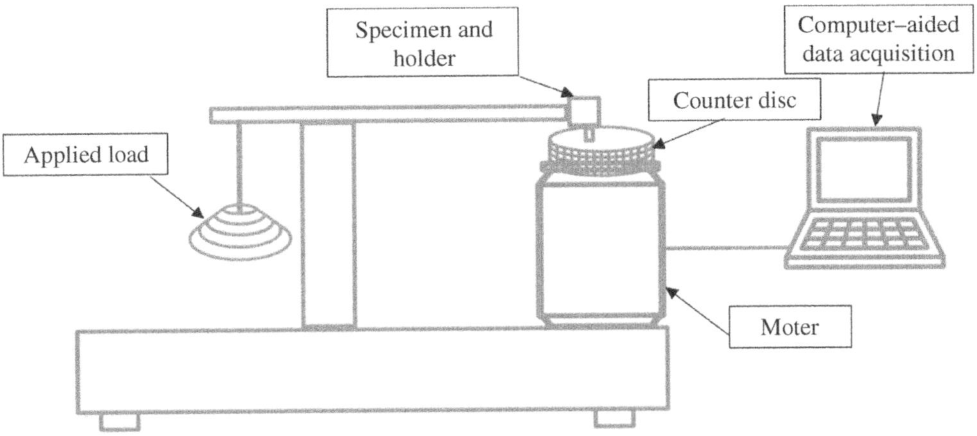

Figure 5.10 Pin-on-disc apparatus with computer-aided data acquisition.

Liu et al. [43] fabricated a composite using silane-treated corn stalk fiber and tire powder in addition with water slag. The specimen was hot pressed and tested using a special wear testing machine. The best wear-resistant values were obtained in the specimens having the 5 wt% and nine fiber contents. Ramie fiber-made composites for automobile brake pad application were fabricated and tested [44]. This composite was made with phenolic resin and fabricated at high temperature molding process. After curing, the samples were made as per the standards and tested using chase machine. The results revealed that 5 wt% of fiber reinforced composites performed well in the tribological studies and the performance decreased with increasing fiber content in the composite. The tribological properties of the hybrid composites were tested to understand the suitability of the composite for various industrial applications. Ranakoti et al. [45] used wood flour to make a composite combined with glass fiber reinforcement and epoxy polymer matrix. This composite was also fabricated using hand layup method and tested using the pin-on-disc apparatus. Along with mechanical property, the tribological property was also studied, which revealed that the combination of 10 wt% of wood and 0 wt% of glass exhibited a higher wear rate. Kumar et al. [46] investigated the natural fiber reinforced composite fabricated using Bauhinia-vahlii-weight and sisal fiber. The biowaste material, rise husk, was used as a filler material for this composite. The tribological study of these composites exhibited better results at the speed of 5.5 m/s and 20 N load with 5 wt% of the filler content. Since the study of these kinds of fibers has to be extended further to explore the possibility of the replacement of the conventional materials to recyclable materials and it has to be end up in the biodegradable materials in the future.

5.7 Dielectric Properties of Biobased Composites

The recent advancements in electronic industry makes all devices in smaller size and the circuits incorporated inside those devices need advanced materials to store more power and work efficiently even in those miniature size. The development of the low constant dielectric material plays a major role in the electronic integrated circuits and ultrascale integration. The need for low coefficient of thermal expansion, low dielectric constant, high thermal conductivity at elevated temperatures, and low cost plays a major role in the development of new dielectric materials [47]. All the above things have become possible by the advancements in polymer materials and recently with the advancement of biopolymers. The enormous availability of biopolymer materials and the ease of processing attracts researchers to work toward this area. It is not only the availability of the materials, but also the strength of the natural polymer materials that have reached level of synthetic polymer materials. The recyclable factor has also encouraged the industry to use these kinds of materials due to stringent emission norms all over the world [48]. Change of composition of the polymer materials as well as the addition of organic fillers in those materials make them suitable for dielectric application. But the hydrophilic nature of the fiber and the hydrophobic nature of the matrix add some difficulties in the processing of the materials. It can be rectified by various chemical treatments. In this chapter, the study of dielectric materials using different types of biopolymer composites and the suitable filler materials and its recent advancements were explored. The measurement of the dielectric constant and the volume resistivity can be done through the formula:

$$\rho = \frac{RA}{t} \tag{5.1}$$

$$\varepsilon' = \frac{Ct}{\varepsilon_0 A} \tag{5.2}$$

where, ρ = volume resistivity; R = resistance; A = cross-sectional area; t = thickness; ε' = dielectric constant; C = capacitance, and ε_0 = permittivity of air.

The search for alternative materials for good dielectric property was started by many researchers a decade before. Jacob et al. [49] used natural rubber as a matrix with sisal and oil palm fiber as a reinforcement to fabricate a biocomposite and studied the dielectric property. The constant values of the composite was much higher than those of the pure rubber due to the presence of lignocellulose fibers. It also increased with the addition of more fiber to the composite. But in the case of chemical modification, the results seem to be opposite and the dissipation factor of the biocomposite increased with the fiber loading and it implies the retention of the electrical charge for a long period of time. A composite fabricated using banana fiber and polyester using hand layup method, the fiber was treated with silane and NaOH. The effect of chemical treatment was studied and found out that the treated composite showed lower results of dielectric constant than the untreated one. It was also observed that the increase of the fiber content also increased the dielectric constant value [50].

Joseph et al. [51] did a comparative study of the electrical property of banana fiber reinforced composite and glass fiber reinforced composite along with various chemical treatments. The study was conducted at various frequencies and the results showed that the ε value decreased with frequency and also with banana fiber loading. The hydrophobicity of the fiber decreased after the chemical treatment thereby reduced the dielectric constant value. Blendzki et al. [52] used flax fiber in the polypropylene to understand the dielectric property of the composite under aging condition. The results revealed an interesting correlation with the aging factor and the dielectric property and this property played a significant role in the change of the dielectric property.

Cavallaro et al. [53] prepared a film-based biopolymer to study the thermal and dielectric properties of the composites. The material was made up of hydroxypropyl-cellulose along with low methoxyl pectin in addition with the halloysite nanotubes. The effect of nanotube addition was observed using the dielectric spectroscopy. Some of the recent research started with using natural materials like chicken feather, human hair, and some agricultural waste. Zhan et al. [54] used chicken feather as a reinforcement and epoxy matrix to form a composite; they studied the suitability of these composites for the printed circuit boards (PCBs). They also compared the composites with epoxy composites with E-glass fiber composite. The resistance of the chicken feather fiber was higher than the E-glass fiber composite by two to four times. The hybrid composite of the feather and glass fiber exhibited a dielectric constant of 4.1, which was lower than that of the commonly used material in the PCBs.

George et al. [55] studied the jute yarn and polypropylene composite to understand the dielectric property. The composites were fabricated using the compression mold and the fibers were modified using maleic anhydride modified polypropylene, toluene diisocyanate, and stearic acid. The dielectric constant value was observed for the treated fiber and it was found that there was a reduction of the constant value when compared with the untreated one. In case of the study on the temperature, after a particular temperature, the constant reduced and the moisture content of the composite increased the constant value. Singha et al. [56] studied the Grewia optiva fibers having modified surface and determined their mechanical, dynamical, and dielectric properties. The benzoyl chloride-, silane-, and alkali-treated fiber were used in the composite fabrication. The polyester resin was used to make the composite using the hand layup method. The dielectric property of the silane-treated material increased after the surface modification and benzoylated fibers show low dielectric strength.

El-Sayed et al. [57] used a chitosan-based nanocomposite biopolymer to study the dielectric properties of the composite. The composite film was studied at elevated temperatures and frequency. 5 wt% magnesia-filled nanocomposites showed a better result in terms of dielectric property. Agrebi et al. [58] analyzed the impact of the silane-treated pineapple leaf and kenaf fiber reinforced composite. This hybrid fiber was treated initially and fabricated using the hand layup method using phenolic resin as the matrix. Then the dielectric property showed that the chemical treatment reduced the dielectric constant value. Since the dielectric property has gained an increasing interest in the electronic sector, its exploration needs to be widened further with other biomaterials.

5.8 Conclusions

Keeping in mind the recent advancements in the field of composite materials, it is very clear that the challenges with the synthetic-based composites are increasingly being approached with biobased materials. Biobased composites are gaining more attention in recent years due to their ecofriendly nature and comparable performance with synthetic materials. However, there are some doubts about the performance of these composites and the higher costs of biodegradable polymers still prevent their wide range of commercial application. Exploration of natural fibers from waste materials can propel the growth potential for the biobased composites as light weight sustainable materials. Hybridization of different biobased fibers or fillers has been identified as a promising technique to produce economically viable composites with enhanced functional properties such as the mechanical properties, thermal properties, tribological properties, and dielectric properties. Further, the identification of cheaper biologically derived polymers can also serve this purpose. This will also have the tremendous potential to replace the synthetic polymer-based composites and fillers. Hence, the identification of the polymer and the fiber for the particular application becomes vital. Engineering, scientific, and commercial societies will have confidence on these materials only when all the major technical considerations are addressed before going for the real-time application.

References

1 Muthu Kumar, S.T., Yorseng, K., Siengchin, S. et al. (2019). Mechanical and thermal properties of spent coffee bean filler/poly(3-hydroxybutyrate-co-3-hydroxyvalerate) biocomposites: effect of recycling. *Process Saf. Environ. Prot.* 124: 187–195.

2 Atiqah, A., Chandrasekar, M., Senthil Muthu Kumar, T. et al. (2019). *Characterization and Interface of Natural and Synthetic Hybrid Composites*. Elsevier Ltd.

3 Chandrasekar, M., Ishak, M.R., Sapuan, S.M. et al. (2017). A review on the characterisation of natural fibres and their composites after alkali treatment and water absorption. *Plast. Rubber Compos.* 46 (3): 119–136.

4 Shahroze, R.M., Ishak, M.R., Salit, M.S. et al. (2018). Effect of organo-modified nanoclay on the mechanical properties of sugar palm fiber-reinforced polyester composites. *BioResources* 13 (4): 7430–7444.

5 Thiagamani, S.M.K., Nagarajan, R., Jawaid, M. et al. (2017). Utilization of chemically treated municipal solid waste (spent coffee bean powder) as reinforcement in cellulose matrix for packaging applications. *Waste Manag.* 69: 445–454.

6 Senthilkumar, K., Kumar, T.S.M., Chandrasekar, M. et al. (2019). Recent advances in thermal properties of hybrid cellulosic fiber reinforced polymer composites. *Int. J. Biol. Macromol.* 141: 1–13.

7 Senthil Muthu Kumar, T., Rajini, N., Tian, H. et al. (2017). Development and analysis of biodegradable poly(propylene carbonate)/tamarind nut powder composite films. *Int. J. Polym. Anal. Charact.* 22 (5): 415–423.

8 Senthil Muthu Kumar, T., Rajini, N., Jawaid, M. et al. (2018). Preparation and properties of cellulose/tamarind nut powder green composites: (green composite using agricultural waste reinforcement). *J. Nat. Fibers.* 15 (1): 11–20.

9 Thiagamani, S.M.K., Krishnasamy, S., Muthukumar, C. et al. (2019). Investigation into mechanical, absorption and swelling behaviour of hemp/sisal fibre reinforced bioepoxy hybrid composites: effects of stacking sequences. *Int. J. Biol. Macromol.* 140: 637–646.

10 Krishnasamy, S., Thiagamani, S.M.K., Muthukumar, C. et al. (2019). Effects of stacking sequences on static, dynamic mechanical and thermal properties of completely biodegradable green epoxy hybrid composites. *Mater. Res. Express* 6 (10): 105351.

11 Urreaga, J.M., González-Sánchez, C., Martínez-Aguirre, A. et al. (2015). Sustainable eco-composites obtained from agricultural and urban waste plastic blends and residual cellulose fibers. *J. Clean. Prod.* 108: 377–384.

12 Pan, P., Zhu, B., Kai, W. et al. (2007). Crystallization behavior and mechanical properties of bio-based green composites based on poly(l-lactide) and kenaf fiber. *J. Appl. Polym. Sci.* 105 (3): 1511–1520.

13 Mishra, S., Mohanty, A.K., Drzal, L.T. et al. (2003). Studies on mechanical performance of biofibre/glass reinforced polyester hybrid composites. *Compos. Sci. Technol.* 63 (10): 1377–1385.

14 Ashworth, S., Rongong, J., Wilson, P., and Meredith, J. (2016). Mechanical and damping properties of resin transfer moulded jute–carbon hybrid composites. *Compos. Part B Eng.* 105: 60–66.

15 Li, R., Cai, J., Zhang, L. et al. (2014). Properties of cellulose/turmeric powder green composite films. *Cellul. Chem. Technol. Cellul. Chem. Technol.* 48 (4): 321–324.

16 Feng, Y., Ashok, B., Madhukar, K. et al. (2014). Preparation and characterization of polypropylene carbonate bio-filler (eggshell powder) composite films. *Int. J. Polym. Anal. Charact.* 19 (7): 637–647.

17 Xia, G., Reddy, K.O., Maheswari, C.U. et al. (2015). Preparation and properties of biodegradable spent tea leaf powder/poly(propylene carbonate) composite films. *Int. J. Polym. Anal. Charact.* 20 (4): 377–387.

18 Senthil Muthu Kumar, T., Senthilkumar, K., Chandrasekar, M. et al. (2019). Characterization, thermal and dynamic mechanical properties of poly(propylene carbonate) lignocellulosic *Cocos nucifera* shell particulate biocomposites. *Mater. Res. Express* 6 (9): 096426.

19 Senthil Muthu Kumar, T., Rajini, N., Siengchin, S. et al. (2019). Influence of *Musa acuminate* bio-filler on the thermal, mechanical and visco-elastic behavior of poly(propylene) carbonate biocomposites. *Int. J. Polym. Anal. Charact.* 24 (5): 439–446.

20 Thiagamani, S.M.K., Krishnasamy, S., and Siengchin, S. (2019). Challenges of biodegradable polymers: an environmental perspective. *Appl. Sci. Eng. Prog.* 12 (3): 149.

21 Senthilkumar, K., Rajini, N., Saba, N. et al. (2019). Effect of alkali treatment on mechanical and morphological properties of pineapple leaf fibre/polyester composites. *J. Polym. Environ.* 27 (6): 1191–1201.

22 Chandrasekar, M., Siengchin, S., Senthilkumar, K. et al. (2018). Mechanical properties evaluation of sisal fibre reinforced polymer composites: a review. *Constr. Build. Mater.* 174: 713–729.

23 Arthur, J. and Redfern, J. (1992). Some observations on high resolution thermogravimetry: a warning. *J. Therm. Anal. Calorim.* 38 (7): 1645–1651.

24 Jumaidin, R., Khiruddin, M.A.A., Asyul Sutan Saidi, Z. et al. (2019). Effect of cogon grass fibre on the thermal, mechanical and biodegradation properties of thermoplastic cassava starch biocomposite. *Int. J. Biol. Macromol.* 146: 746–755.

25 Prachayawarakorn, J., Limsiriwong, N., Kongjindamunee, R., and Surakit, S. (2012). Effect of agar and cotton fiber on properties of thermoplastic waxy rice starch composites. *J. Polym. Environ.* 20 (1): 88–95.

26 Curvelo, A.A.S., De Carvalho, A.J.F., and Agnelli, J.A.M. (2001). Thermoplastic starch–cellulosic fibers composites: preliminary results. *Carbohydr. Polym.* 45 (2): 183–188.

27 Ma, X., Yu, J., and Kennedy, J.F. (2005). Studies on the properties of natural fibers-reinforced thermoplastic starch composites. *Carbohydr. Polym.* 62 (1): 19–24.

28 Prachayawarakorn, J., Ruttanabus, P., and Boonsom, P. (2011). Effect of cotton fiber contents and lengths on properties of thermoplastic starch composites prepared from rice and waxy rice starches. *J. Polym. Environ.* 19 (1): 274–282.

29 Karger-Kocsis, J. and Zhang, Z. (2005). Structure–property relationships in nanoparticle/semicrystalline thermoplastic composites. *Mech. Prop. Polym. Nanostruct. Morphol.* 71: 547–596.

30 Mahadevaswamy, H.S. and Suresha, B. (2019). Role of nano-$CaCO_3$ on mechanical and thermal characteristics of pineapple fibre reinforced epoxy composites. *Mater. Today Proc.* 22: 572–579.

31 Yang, G., Heo, Y.-J., and Park, S.-J. (2019). Effect of morphology of calcium carbonate on toughness behavior and thermal stability of epoxy-based composites. *Processes* 7 (4): 178.

32 Sreenivasan, V.S., Rajini, N., Alavudeen, A., and Arumugaprabu, V. (2015). Dynamic mechanical and thermo-gravimetric analysis of *Sansevieria cylindrica*/polyester composite: effect of fiber length, fiber loading and chemical treatment. *Compos. Part B Eng.* 69: 76–86.

33 Wang, H., Memon, H., Hassan, E.A.M. et al. (2019). Rheological and dynamic mechanical properties of abutilon natural straw and polylactic acid biocomposites. *Int. J. Polym. Sci.* 2019: 1–8. Article ID 8732520.

34 Safri, S.N.A., Sultan, M.T.H., Jawaid, M., and Abdul Majid, M.S. (2019). Analysis of dynamic mechanical, low-velocity impact and compression after impact behaviour of benzoyl treated sugar palm/glass/epoxy composites. *Compos. Struct.* 226: 111308.

35 Essabir, H., Boujmal, R., Bensalah, M.O. et al. (2016). Mechanical and thermal properties of hybrid composites: oil-palm fiber/clay reinforced high density polyethylene. *Mech. Mater.* 98: 36–43.

36 Nirmal, U., Hashim, J., and Ahmad, M.M.H.M. (2015). A review on tribological performance of natural fibre polymeric composites. *Tribol. Int.* 83: 77–104.

37 Azwa, Z.N., Yousif, B.F., Manalo, a.C., and Karunasena, W. (2013). A review on the degradability of polymeric composites based on natural fibres. *Mater. Des.* 47: 424–442.

38 Nath, S., Bodhak, S., and Basu, B. (2007). Tribological investigation of novel HDPE-HAp-Al_2O_3 hybrid biocomposites against steel under dry and simulated body fluid condition. *J. Biomed. Mater. Res. Part A* 83A (1): 191–208.

39 Younesi, M., Bahrololoom, M.E., and Fooladfar, H. (2010). Development of wear resistant NFSS-HA novel biocomposites and study of their tribological properties for orthopaedic applications. *J. Mech. Behav. Biomed. Mater.* 3 (2): 178–188.

40 Karthick, R., Sirisha, P., and Sankar, M.R. (2014). Mechanical and tribological properties of PMMA-sea shell based biocomposite for dental application. *Procedia Mater. Sci.* 6: 1989–2000.

41 Chaudhary, V., Bajpai, P.K., and Maheshwari, S. (2018). An investigation on wear and dynamic mechanical behavior of jute/hemp/flax reinforced composites and its hybrids for tribological applications. *Fibers Polym.* 19 (2): 403–415.

42 Suresh, S., Sudhakara, D., and Vinod, B. (2019). Investigation of bio-waste natural fiber – reinforced polymer hybrid composite: effect on mechanical and tribological characteristics of biodegradable composites. *Mech. Soft Mater.* 1: 12.

43 Liu, Y., Xie, J., Wu, N. et al. (2019). Influence of silane treatment on the mechanical, tribological and morphological properties of corn stalk fiber reinforced polymer composites. *Tribol. Int.* 131: 398–405.

44 Singh, T., Kumar, N., Ashok Raj, J. et al. (2019). Natural fiber reinforced non-asbestos brake friction composites: influence of ramie fiber on physico-mechanical and tribological properties. *Mater. Res. Express* 6 (11): 0–10.

45 Ranakoti, L., Gupta, M.K., and Rakesh, P.K. (2019). Analysis of mechanical and tribological behavior of wood flour filled glass fiber reinforced epoxy composite. *Mater. Res. Express* 6 (8): 085327.

46 Kumar, S., Mer, K.K.S., Gangil, B., and Patel, V.K. (2019). Synergy of rice-husk filler on physico-mechanical and tribological properties of hybrid Bauhinia-vahlii/sisal fiber reinforced epoxy composites. *J. Mater. Res. Technol.* 8 (2): 2070–2082.

47 Carr, S.H., Davies, D.K., and Fischer, P. (1982). *Electrical of Polymers Properties*, 379. Academic Press.

48 Zhao, X.Y. and Liu, H.J. (2010). Review of polymer materials with low dielectric constant. *Polym. Int.* 59 (5): 597–606.

49 Jacob, M., Varughese, K.T., and Thomas, S. (2006). Dielectric characteristics of sisal-oil palm hybrid biofibre reinforced natural rubber biocomposites. *J. Mater. Sci.* 41 (17): 5538–5547.

50 Pothan, L.A., George, C.N., Jacob, M., and Thomas, S. (2007). Effect of chemical modification on the mechanical and electrical properties of banana fiber polyester composites. *J. Compos. Mater.* 41 (19): 2371–2386.

51 Joseph, S. and Thomas, S. (2008). Electrical properties of banana fiber-reinforced phenol formaldehyde composites. *J. Appl. Polym. Sci.* 109 (1): 256–263.

52 Bledzki, A.K., Lucka, M., Al-Mamun, A., and Michalski, J. (2009). Biological and electrical resistance of acetylated. *BioResources* 4: 111–125.

53 Cavallaro, G., Donato, D.I., Lazzara, G., and Milioto, S. (2011). Films of halloysite nanotubes sandwiched between two layers of biopolymer: from the morphology to the dielectric, thermal, transparency, and wettability properties. *J. Phys. Chem. C* 115 (42): 20491–20498.

54 Zhan, M., Wool, R.P., and Xiao, J.Q. (2011). Electrical properties of chicken feather fiber reinforced epoxy composites. *Compos. Part A Appl. Sci. Manuf.* 42 (3): 229–233.

55 George, G., Joseph, K., Nagarajan, E.R. et al. (2013). Dielectric behaviour of PP/jute yarn commingled composites: effect of fibre content, chemical treatments, temperature and moisture. *Compos. Part A Appl. Sci. Manuf.* 47 (1): 12–21.

56 Singha, A.S., Rana, A.K., and Jarial, R.K. (2013). Mechanical, dielectric and thermal properties of *Grewia optiva* fibers reinforced unsaturated polyester matrix based composites. *Mater. Des.* 51: 924–934.

57 El-Sayed, S. and El Sayed, A.M. (2019). Synthesis, characterisation, dielectric, and optical properties of the chitosan/poly(ethylene glycol)/magnesia biopolymer nanocomposites. *Mater. Technol.* 34 (10): 602–614.

58 Agrebi, F., Hammami, H., Asim, M. et al. (2019). Impact of silane treatment on the dielectric properties of pineapple leaf/kenaf fiber reinforced phenolic composites. *J. Compos. Mater.* 54 (7): 937–946.

6

Flame Retardancy of Biobased Composites

N. B. Karthik Babu, T. Ramesh, and Mohit Hemath Kumar

Department of Mechanical Engineering, National Institute of Technology, Tiruchirappalli, Tamil Nadu, India

6.1 Introduction

The fire hazard of biocomposites is often defined by their flame-retardant properties, which tell us their rate of reaction against fire. The behavior of a material in a flammable environment (from the start of ignition to flashover) is studied by a fire reaction which also describes the smoke toxicity of flammable polymers. The four factors shown in Figure 6.1 are the main fire reaction properties which influence the fire growth. Among these properties, heat release rate (HRR) is the single most significant fire reaction property which can be measured using cone calorimetry since it is the best indicator of the fire hazard of a combustible polymer or any material. The HRR is a defined as a quantitative measure of the amount of heat energy released (in kW) by a polymer or composite per unit area (in m^2) when exposed to a fire radiating at a constant heat flux (or temperature) and measured in terms of kW/m^2. Composites contain many elements and the value of HRR of a composite material is not constant; it varies with time when the material is subjected to combustion as the composite material is ignited, progressively decomposed, and burnt fully. Therefore, the HRR is characterized by two parameters, namely *average HRR* and *peak HRR*. Average HRR is the averaged heat release over a particular period of time. Peak HRR is the maximum heat released by a material during the fire, and it often happens over a very short time period. Peak HRR is considered a very important property as it indicates the *maximum temperature* and *flame spread rate*. Polymer composite materials that possess low peak and average HRR values are the best suitable flame-resistant applications to reduce the evolution and spread of fire.

The *time-to-ignition* defines how quickly the flaming combustion occurs initially when the material is exposed to a heat source at a given heat flux and in an oxygen-controlled atmosphere. The flammability resistance of a composite material can be assessed by its ignition time. A material or composite that takes a long time to ignite (possess large ignition time) is highly suitable for flame-resistant applications.

The speed at which the flame travels on the surface of the burning material is known as *flame spread rate*. Vertical flammability tester, horizontal flammability tester, and inclined flammability tester are the three different experimental setups widely used to study the flame spread rate and its pattern. *Limiting Oxygen Index (LOI)* expresses the minimum

Biobased Composites: Processing, Characterization, Properties, and Applications, First Edition.
Anish Khan, Sanjay Mavinkere Rangappa, Suchart Siengchin and Abdullah M. Asiri.

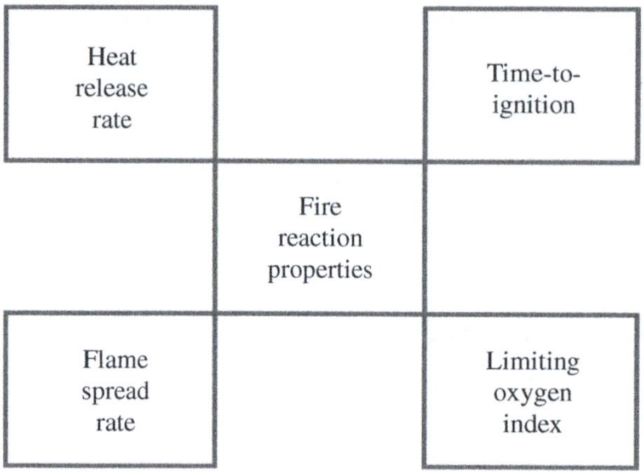

Figure 6.1 Major fire reaction properties that influence fire growth.

oxygen concentration required to support/sustain flaming combustion in the fire surrounding. High LOI value of a material indicates that it is highly suitable for use in fire risk applications.

Apart from the above-mentioned properties, there are two important properties that have a vital role not only on the performance of the material but also on human health and survival in the fire zone; they are *smoke density* and *gas toxicity*. Smoke density depends on the concentration of carbon soot and water vapor within the plume of a flame. During a fire, the release of thick smoke may lead to uncertainty and disturbance among people trying to escape from the fire zone, which prolong their exposure to toxic smokes that may lead to unconsciousness and death. Due to the hyperventilation behavior, inhalation of toxic gases like CO and HCl will cause major issues [1].

Fire retardancy of polymers and its composites is a major concern, since it is essential to decrease the fire risk and ensure a fire safety environment. Most of the flame retardants developed in the past (e.g. Bromine, chlorine, and fluorine are the well-known elements in the halogens chemical group) are efficient, but the usage of some halogen retardants during fire are very harmful to the human health and affects our ecosystem [2]. This compelled researchers to extend their studies for alternate flame retardants. In the recent years, there has been a tremendous increase in the utilization of polymers and their composites in a growing range of applications, which is mainly due to their notable strength to weight ratio, chemically inert property, corrosion resistance, ease of manufacturing, and low cost. Besides these, biobased composites are ecofriendly in nature and familiar due to their sustainability. But the most critical issue with biobased composites is their least resistance to fire.

Both the biobased reinforcement and the polymeric matrix are highly susceptible to flame and combustion. In order to overcome this, flame-retardant compounds are introduced during the fabrication of polymer composites and in textiles in order to inhibit or delay the flame production to prevent the spread of fire. Cellulose (Figure 6.2) is a major component of the plant fiber and Figure 6.3 shows the step by step decomposition process of cellulose during combustion [5].

In order to provide a safe living environment, superior performance of the material, even in a perilous situation, is mandatory since fire is unexpected. Therefore, biobased composites have to clear stringent safety regulations for their use in various sectors. Hence, much

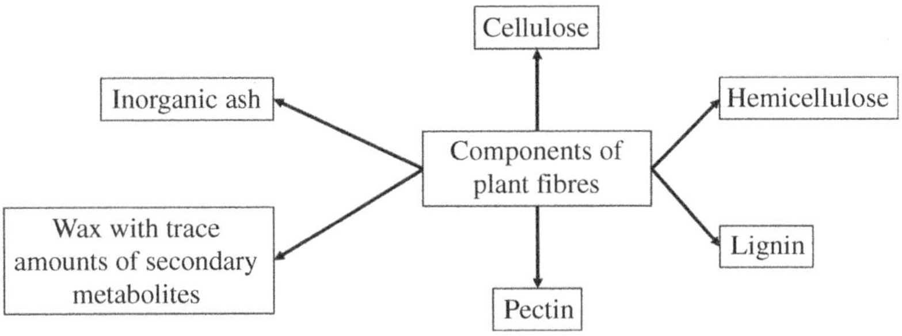

Figure 6.2 Major components of plant fibers. *Source:* Based on Kozłowski and Władyka-Przybylak [3].

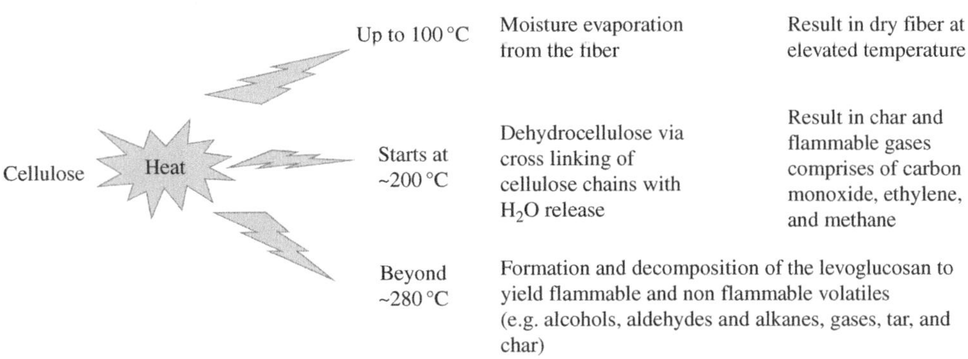

Figure 6.3 Schematic representation of the cellulose degradation process. *Source:* Based on Kim et al. [4].

attention should be paid to the flame retardancy of biocomposites and the enhancement of their fire properties.

Commonly used matrix materials for the manufacturing of composites are thermosets or thermoplastics and biopolymers. In fire proof applications, the most widely used polymer is PP, which belongs to the polyolefins family. Next to PP, the family of polyvinyl plastics are the majorly used materials after polyolefins in the world, which include PVC, PS, ABS; cellulose derivatives, including cellulose nitrate, acetate, butyrate, and propionate; and aromatic compounds, including polycarbonate (PC), phenoxy resins, polysulfone, and polyimide as thermoplastic materials.

Thermoset polymers mainly used in composites are epoxy resins, which have versatile applications and unsaturated polyesters that are predominantly used in structural and transport applications; phenolic resins; melamine – formaldehyde and urea – formaldehyde amino resins, and polyurethane.

6.1.1 Flame Retardants

In general, flame retardants are chemical compounds incorporated in plastic compounds to provide varying degrees of flammability protection. In view of biobased composites, reinforcement of natural fillers slightly reduces the flammability of the biocomposites; this reduction is due to the minimization of a highly flammable polymer quantity in the composites. Yet, this marginal reduction of flammability could not be sufficient to pass through the

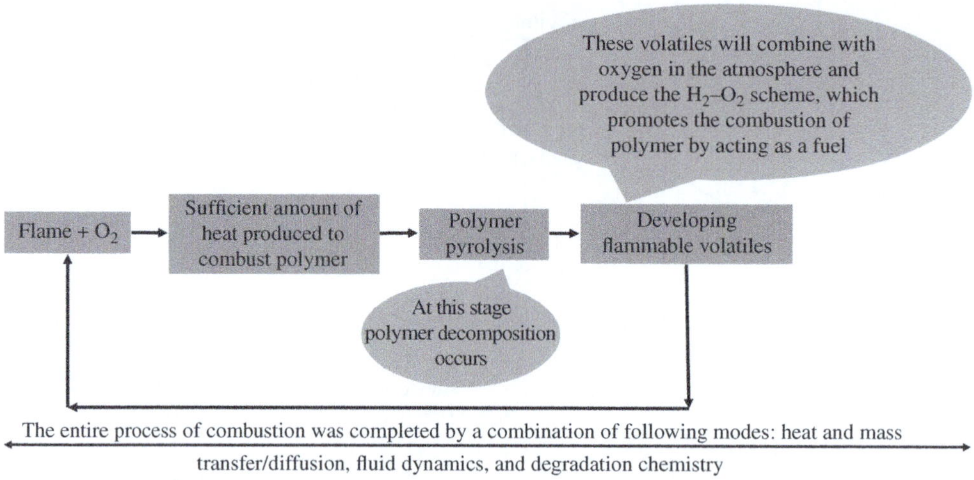

Figure 6.4 Combustion cycle of polymeric materials. *Source:* Based on Dasari et al. [6].

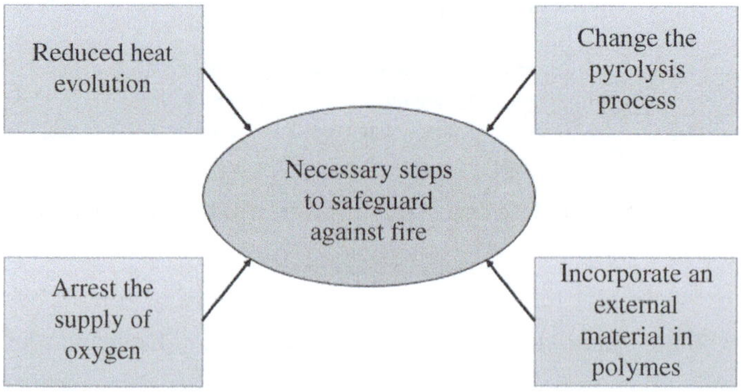

Figure 6.5 Steps to accomplish the reduced flammability in polymer composites.

stringent fire safety rules formed by government authorities. In order to decelerate the fire/combustion process, incorporation of flame retardants in any form is highly required. The combustion of polymeric materials and its composites is a highly complicated process [6], which is illustrated in Figure 6.4.

Figure 6.5 illustrates the necessary steps for the protection of polymer biobased composites from fire. The first one is the reduction of heat evolution, which discourages further combustion of a polymer biocomposite. The second one is the reformation of the pyrolysis process to decrease the quantity of flammable volatile compounds released. The third step is the provision of a barrier between the atmospheric air (oxygen) and the flammable material, which helps in slowing down combustion. The final step is the addition of flame retardants, in any phase, to the parent material to slow down or prevent the start/spread of fire.

In summary, the objectives of any fire retardant are to reduce the intensity of heat generation, cut off oxygen supply, and stop feeding the fuel for combustion.

6.1.2 Types of Flame Retardants

Usually, flame retardants are majorly categorized into two types, i.e. additive and reactive compounds.

Additive flame retardants are mixed with the host polymer matrix during composite manufacturing; they contact physically with the matrix but do not react/bond chemically with it at any stage of composite fabrication. These compounds are utilized for all types of polymers, i.e. both thermosets and thermoplastics. The major chemical elements found in these compounds, which are responsible for flame retardation, are antimony used in the form of antimony oxide; boron in the form of boron oxide; aluminum in the form of alumina trihydrate; and phosphorus, bromine, or chlorine. Besides these, additive compounds rarely contain elements like barium, iron, and sulfur. During flaming of the composites, the added additives in a polymer deliver water vapor [7, 8] or appear as hydrated metal salts that decompose endothermically, resulting in the reduced evolution of flammable volatiles and the minimized HRR of the composite.

Reactive flame retardants are permanently bonded with the host polymer (especially with thermosets) during synthesis and combined into the resin molecular network structure. These fire retardants are mainly based on halogens (either chlorine or bromine), phosphorus, inorganic metals, and melamine. Bromine and chlorine are the common retardants because of their effective action in the pyrolysis zone.

6.2 Types of Biobased Polymer Composites Used in a Flame-Retardant Application

In this section, the biobased composites developed for flame-retardant applications are classified in a different perspective from that of the conventional approach. In the primary category of flame-retardant biocomposites, either the *matrix* material should be a biopolymer (derived from plants, e.g. soybean-oil-based copolymers containing silicon or boron) or the *reinforcement* should be a natural fiber. In the second category, the reinforcement is dedicatedly made for manufacturing a biocomposite or it may be a byproduct that can be utilized as a waste natural filler in polymer composites. In this chapter, we concentrate on the second category, reinforcement based biocomposites.

6.3 Role and Effect of Natural Byproducts on the Flame-Retardant Behavior of a Biocomposite

Natural fiber reinforced biocomposites are gaining importance not only because of its comparable strength and ecofriendly nature, but also due to its availability and low cost as compared with GFRs and CFRs used in the manufacture of composites. In addition, natural waste reinforcement based biocomposites (listed in Figure 6.6) have advantageous features similar to that of natural fibers and are available for almost zero cost. Some of the advantages of utilizing waste natural resources in polymer composites are reduced landfilling, increased sustainability of green environment, and waste minimization. Apart from the above list, some textile-based wastes like recycled wool fibers and fired plant-based wastes like biochar are utilized as reinforcements in the manufacture of polymer composites to increase the flame retardancy of biocomposites.

6.3.1 Flammability of Biochar Reinforced Biocomposites

One of the major drawbacks in using natural fiber reinforcement in polymer composites is its degradability and thermal stability. Therefore, to overcome this issue, instead of using

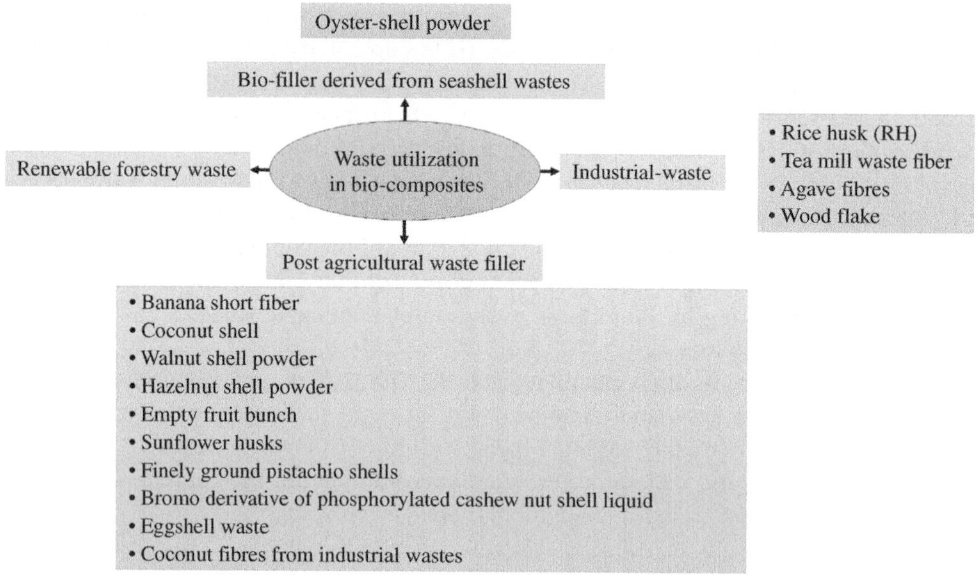

Figure 6.6 Various kinds of wastes used as a reinforcement in the manufacture of biocomposites.

organic wastes directly in the manufacture of biocomposites, biochar or biocarbon derived from various biomasses such as paunch grass, pine wood saw dust, date palm, poultry litter, and sewage sludge are utilized as a reinforcement. Lignocellulosic wastes are valorized by thermochemical conversion to produce a carbonaceous and renewable material called biochar. Biochar can be produced from waste biomass using various thermal decomposition techniques such as pyrolysis, combustion, gasification, and liquefaction. Among the listed processes, pyrolysis is the most commonly used technique that can convert waste biomass into biochar by thermal decomposition. This biochar has better thermal stability than the natural fibers and the pores on its surface provide better physical bonding with matrix. For a biochar, the surface area and pore volume are important factors. Higher surface area along with increased pore volume of biochar might provide good bonding between the filler and the matrix. Biocomposites prepared using various wastes are demonstrated by Das et al. [9]. Mostly, the biochar has a porous honeycomb structure consisting of a high concentration of carbon and aromatized material. These porous honeycomb structures of the biochar filler allow to fill the liquid polymer inside and create a physical bonding which could result in the improvement of mechanical properties of the biocomposites [10]. Biochar is produced from natural wastes through pyrolysis and is recommended as an effective waste management and disposal solution. The long-term availability and quality of raw material for the production of biochar and its economy are the major concerns. The condition of the raw materials and pyrolysis are the influencing factors that control the physical and chemical properties of the end product called "biochar" [11]. The physicochemical properties of biochar are dependent on numerous factors such as raw material type, carbonization processes, pyrolysis temperature, heating rate, and residence time. The biochar/PP biocomposites were fabricated by Das et al. and their mechanical properties, thermal stability, and flammability were tested. The study concluded that both the tensile and flexural modulus of the biocomposites increased with the increase in biochar concentration. Also, the reinforced biocomposites showed increased thermal stability, lower HRR, and less smoke release when compared to pristine PP. The major reason for the enhancement of mechanical properties is good physical bonding between the biochar and the matrix, which is accomplished due to the porous structure of the biochar [12]. The lignocellulosic biowaste from date palm were pyrolysed at different temperatures and the effect of temperature on the physical structure and surface morphology was studied. The biochar/PP composites were prepared and they concluded that the reinforced PP showed better thermal stability and enhanced stiffness. Also, the authors suggested surface functionalization of the filler for further enhancement of mechanical, thermal, and flammability properties [13]. The chemical reaction between the polymer and biochar is insignificant; therefore, the interfacial bonding between the wood filler and the matrix was achieved by the coupling agent added maleated anhydride polypropylene (MAPP). Ikram et al. studied the mechanical and flammability characteristics of wood/pine wood biochar/PP biocomposites and demonstrated the variation in properties of as received PP and maleated anhydride polypropylene. They concluded that the addition of the MAPP coupling agent and wood particles has a significant role in the enhancement of tensile and flexural properties, but the HRR remains unaffected [14]. For the further enhancement of mechanical properties and flammability of the biocomposites, a hybridization technique was followed. The influence of a hybrid form of waste biochar/wool fillers was studied by Das et al. Biocomposites possessing hybrid fillers significantly minimized the peak HRR, smoke release, and showed elevated mass loss rate as compared to pristine PP. The LOI was enhanced because of the hybridization with wool [15].

In summary, the biochar derived from biowastes such as wood dust, grasses [16], etc. using any suitable combustion technique can be effectively used in the fabrication of biocomposites, which significantly reduce the landfilling of agro-wastes and also provide a new platform to the development of new materials.

6.3.2 Commonly Used Agro-wastes to Improve the Flame Retardancy of a Biocomposite

The use of wood-based plastic biocomposites has progressively increased and drawn more attention in various fields such as those of automotive and building materials in the last two decades. Concerns for environmental safety and diminishing petrochemical resources are the major reasons behind the development of biocomposites. But, enhancement of flame retardancy of the natural fiber-based biocomposites is highly important, which also limits its application. Wang et al. prepared the powder form of ammonium polyphosphate (APP) and converted it into a microencapsulated form. Further, they compared its flame-retardant performance with conventional APP in wood flour/PP biocomposites. It was observed that the microencapsulated APP-filled wood flour/PP biocomposites have better fire retardancy and mechanical properties than the APP-filled composites [17]. Similarly, Guan et al. aimed to improve the flame retardancy of wood flour reinforced PP composites but they have modified the APP using an ion exchange reaction with ethanolamine. The presence of hydroamine groups and phosphate acid in the treated composites has significantly improved the flame retardancy of the wood flour biocomposites. Also, because of the improved filler/matrix compatibility, the flexural properties of the treated biocomposites were improved significantly [18]. Large types of additives such as bromine-based, magnesium hydroxide, zinc borate, melamine phosphate, and APP are used to improve the fire retardancy of biocomposites. From the aforementioned list, Stark et al. concluded that

Table 6.1 Various fire-retardant properties of a biocomposite.

Type of biocomposites		Time to ignition (s)	Peak heat release rate (kW/m^2)	Total heat release (MJ/m^2)	References
Biochar/PP		16.3±0.5 ↓	477.22±22.1 ↓	98.31±2.0 ↑	[10]
Rice Husk/PP		17.3±0.6 ↓	678.1±2.7 ↓	101.5±2.5 ↓	[20]
Wood flour/PP	With APP	19.6 ↓	312 ↓	78.9 ↓	[21]
	With MPP	20.4 ↓	352 ↓	98.6 ↓	
	With $Al(OH)_3$	24.8 ↓	467 ↓	99 ↓	

Symbol indicates the increase (↑) or decrease (↓) of the property with respect to pristine resin.

zinc borate exhibited poor performance and worst oxygen index compared to other fire retardants. But, the flexural properties of zinc borate-added wood–polyethylene composites were better than the others [19].

In polymer composites, if the matrix is based on renewable resources and also contain agro-wastes or other fillers, the end composites would be a completely biobased, sustainable composite. But, most of the researches have used PP as a matrix material; this is because of its nondegradability, durability, and low cost. Table 6.1 shows the effect of various waste elements in PP composites under fire environment. The main reason behind the usage of rice husk in the preparation of fire-retardant composites is, it has significant amount of inorganic compounds and nonflammable silica. Aforesaid reasons could reduce the fire spread of a material while using rice husk as a reinforcement in polymer composites [22]. Also, these reasons made an agro-waste, namely rice husk, a commonly used reinforcement in the manufacture of fire-retardant polymer composites next to wood dust or wood flour. Das et al. collected five types of rice husks, prepared rice husk reinforced PP biocomposites, and compared their mechanical properties and fire behavior with pristine PP [20]. Compared to other lignocelluloses (sisal, flax, etc.), rice husk has a significant content of inorganic silica. While burning the rice husk, the presence of silica in the rice husk left a residue, which forms a thin shielding layer at the burning region and acts as a flame barrier mechanism of a biocomposite. Formation of this thin shielding layer would be continuous and quite stronger when the rice husk concentration is maximum (more than 50 wt%). This char layer would stop spreading the flame and act as an insulating layer for the unfired region. The effect of rice husk on flame-resistant properties of high-density polyethylene (HDPE) biocomposites was studied and the shielding layer formation and the detailed mechanism was proposed by Zhao et al. [23]. Phan et al. prepared rice husk-filled polyurethane foam samples with and without loading aluminum diethylphosphinate and aluminum hydroxide. The authors concluded that sole rice husk in polyurethane did not pass the UL94 HB standard test and burned entirely with a low LOI. But, 7 wt% of aluminum diethylphosphinate (OP) and 32 wt% of aluminum hydroxide loaded rice husk/polyurethane composites showed better thermal stability and superior fire retardancy [24]. Effect of ethylene-glycidyl methacrylate compatibilizer in rice husk reinforced polymer blend ecocomposites (contain recycled-HDPE/recycled-polyethylene terephthalate (PET)) was studied and its flame-retardant properties were characterized. The thermal stability of a compatibilizer added polymer blend has showed increase in the thermal stability. Also, the flame-retardant properties were enhanced with increase in rice husk concentration [25]. Apart from the majorly used fillers like rice husk and wood flour, other biobased wastes are also reinforced in polymers to improve the fire retardancy of a polymer composite. Tea mill waste fiber [26], walnut shells powder [27], forestry waste [28], finely milled pistachio shells [29], hazelnut shell [30], seashell powder as biofiller [31], and oyster shell powder [32–34] are used as a biobased reinforcement to improve the fire resistance of the polymer composites.

In summary, the usage of agro-waste materials as attractive reinforcements in composite preparation from the perspective of their contribution is increasing the thermal stability and flammability properties.

6.4 Role and Effect of Biobased Natural Fibers on the Flammability of a Biocomposite

Recently, natural fiber reinforced composites have been extended its role in automobile, structural, and heat insulating applications. This prolonged application is due to the advantages like high strength, low cost, and availability and depletion of petrochemicals. Therefore, improving the flame retardancy of natural fiber reinforced biobased composites is essential. It has been reported in the earlier research that the incorporation of some types of lignocellulosic fillers (such as flax fiber) can reduce the flammability of the polymers [35]. Primarily, the effect of untreated and sole (without any agents) natural fibers were reinforced and their flame-retardant characteristics were studied. Chai et al. compared the HRR of the flax/epoxy with glass fiber (GF)/epoxy composites and concluded that the flax fiber reinforced composites were ignited quickly, released more heat, and deformed significantly when compared to GF reinforced epoxy composites [36]. Borysiak et al. prepared pine wood particles reinforced PP composites and compared the flammability performance of reinforced PP with pristine PP. They concluded that the pine wood composites are quickly ignited than the pristine PP, even though pine wood composites disclosed lower HRR and emitted minimal toxic gas as compared to pristine PP [37].

In order to overcome the quicker ignition and other fire-retardancy properties, some chemical fire retardants were used to treat the natural fibers and test their performances. Suardana et al. used diammonium phosphate (DAP) to enhance the flame retardancy of coir and jute fibers; further, these fibers were reinforced in two different matrixes and fabricated PLA and PP biocomposites. The 5% DAP treated fiber showed best low burning rate in both matrixes and consequently formed more char after burning [38]. Shumao et al. incorporated APP solely into the ramie fiber and into the PLA matrix and prepared ramie/PLA composites and compared their flame-retardancy performance with ramie/PLA (both are APP-treated) composites. Compared to APP-treated ramie/PLA composite, other two composites containing 10.5 wt% APP achieved *V*–0 rating in TGA, the APP can effectively encouraged char residue at elevated temperature, as a result of which flame retardancy was enhanced [39]. Idumah et al. fabricated hybrid kenaf flour/graphene nanoplatelets (GNP) reinforced PP nanocomposites and concluded that the nanocomposites took longer time to ignite and showed very low fire growth index. They also predicted that the addition of GNP's have played an important role on the materials flame retardancy. The GNP char have restricted oxygen contact and acted as a shielding [40]. The surface treatment of fibers using alkaline and silane were not much effective in improving the flame retardancy of the biocomposites. Shukor et al. prepared polylactic acid (PLA) biocomposites in which the reinforcement used was alkali-treated kenaf fiber. In PLA/kenaf biocomposites, they incorporated 15 wt% of polyethylene glycol (PEG) and different concentration of APP. The results showed that the APP was very effective in enhancing LOI due to improved char residue formation at elevated temperatures, but the mechanical properties were strongly affected with the addition of APP. Also, the reduction of LOI was observed with an increase in NaOH concentration during alkali treatment. This was due to the removal of lignin on the fibers' surface, which led to the lack of carbonization agent and consequently reduced char formation [41]. Asim et al. used silane to treat the fibers extracted from pineapple leaf and kenaf, and incorporated these fibers in phenolic hybrid composites. The authors concluded that the treated hybrid composites were not very effective in improving the flammability of PALF/KF hybrid composites [42].

In summary, natural fiber reinforced biocomposites are a good alternate for synthetic fiber reinforced composites, but, some secondary techniques like surface treatment or addition of fire retardants like APP are required to improve the flame-retardancy characteristics of the biocomposites.

6.5 Summary

Natural fibers and its composites are flammable. Hence, in order to utilize these biocomposites in structural and transport fields, flame retardancy against fire needs to be improved for safety reasons and pass fire safety regulations. In general, composites made of thermoset resins need a high concentration of flame-retardant additives to attain satisfactory outcomes in terms of flammability which may significantly reduce their mechanical properties. All the methods discussed in this chapter focused toward reducing the rate of heat release during combustion, delay in time to ignition, and reduction in flame spread. Utilizing agro-waste materials are attractive reinforcements in composite preparation from the perspective of their contribution in increasing the thermal stability and flame retardancy.

References

1 Mouritz, A.P. and Gibson, A.G. (2006). Fire Properties of Polymer Composite Materials. The Netherlands: *Springer*.

2 Lu, S.Y. and Hamerton, I. (2002). Recent developments in the chemistry of halogen-free flame retardant polymers. *Prog. Polym. Sci.* 27 (8): 1661–1712.

3 Kozłowski, R. and Władyka-Przybylak, M. (2008). Flammability and fire resistance of composites reinforced by natural fibers. *Polym. Adv. Technol.* 19 (6): 446–453.

4 Kim, N.K., Dutta, S., and Bhattacharyya, D. (2018). A review of flammability of natural fibre reinforced polymeric composites. *Compos. Sci. Technol.* 162: 64–78.

5 Horrocks, A.R. (1983). An introduction to the burning behaviour of cellulosic fibres. *J. Soc. Dye. Colour.* 99 (7–8): 191–197.

6 Dasari, A., Yu, Z.Z., and Cai, G.P. (2013). Recent developments in the fire retardancy of polymeric materials. *Prog. Polym. Sci.* 38 (9): 1357–1387.

7 Hornsby, P.R. (1996). The application of hydrated mineral fillers as fire retardant and smoke suppressing additives for polymers. *Macromol. Symp.* 108 (1): 203–219.

8 Rothon, R.N. and Hornsby, P.R. (1996). Flame retardant effects of magnesium hydroxide. *Polym. Degrad. Stab.* 54 (2–3): 383–385.

9 Das, O., Sarmah, A.K., and Bhattacharyya, D. (2015). A sustainable and resilient approach through biochar addition in wood polymer composites. *Sci. Total Environ.* 512: 326–336.

10 Das, O., Sarmah, A.K., and Bhattacharyya, D. (2016a). Biocomposites from waste derived biochars: mechanical, thermal, chemical, and morphological properties. *Waste Manag.* 49: 560–570.

11 Srinivasan, P., Sarmah, A.K., and Smernik, R. (2015). A feasibility study of agricultural and sewage biomass as biochar, bioenergy and biocomposite feedstock: production, characterization and potential applications. *Sci. Total Environ.* 512: 495–505.

12 Das, O., Bhattacharyya, D., and Hui, D. (2016b). Mechanical and flammability characterisations of biochar/polypropylene biocomposites. *Compos. Part B: Eng.* 106: 120–128.

13 Elnour, A.Y., Alghyamah, A.A., and Shaikh, H.M. (2019). Effect of pyrolysis temperature on biochar microstructural evolution, physicochemical characteristics, and its influence on biochar/polypropylene composites. *Appl. Sci.* 9 (6): 1149.

14 Ikram, S., Das, O., and Bhattacharyya, D. (2016). A parametric study of mechanical and flammability properties of biochar reinforced polypropylene composites. *Compos. Part A: Appl. Sci. Manu.* 91: 177–188.

15 Das, O., Kim, N.K., Sarmah, A.K., and Bhattacharyya, D. (2017). Development of waste based biochar/wool hybrid biocomposites: Flammability characteristics and mechanical properties. *J. Clean. Prod.* 144: 79–89.

16 Behazin, E., Misra, M., and Mohanty, A.K. (2017). Sustainable biocarbon from pyrolyzed perennial grasses and their effects on impact modified polypropylene biocomposites. *Compos. Part B: Eng.* 118: 116–124.

17 Wang, W., Zhang, S., and Wang, F. (2016). Effect of microencapsulated ammonium polyphosphate on flame retardancy and mechanical properties of wood flour/polypropylene composites. *Polym. Compos.* 37 (3): 666–673.

18 Guan, Y.H., Huang, J.Q., and Yang, J.C. (2015). An effective way to flame-retard biocomposite with ethanolamine modified ammonium polyphosphate and its flame retardant mechanisms. *Ind. Eng. Chem. Res.* 54 (13): 3524–3531.

19 Stark, N.M., White, R.H., and Mueller, S.A. (2010). Evaluation of various fire retardants for use in wood flour–polyethylene composites. *Polym. Degrad. Stab.* 95 (9): 1903–1910.

20 Das, O., Hedenqvist, M.S., and Prakash, C. (2019). Nanoindentation and flammability characterisation of five rice husk biomasses for biocomposites applications. *Compos. Part A: Appl. Sci. Manu.* 125: 105566.

21 Umemura, T., Arao, Y., and Nakamura, S. (2014). Synergy effects of wood flour and fire retardants in flammability of wood–plastic composites. *Energy Procedia* 56: 48–56.

22 Schirp, A. and Barrio, A. (2018). Fire retardancy of polypropylene composites reinforced with rice husks: from oxygen index measurements and cone calorimetry to large-scale single-burning-item tests. *J. Appl. Polym. Sci.* 135 (37): 46654.

23 Zhao, Q., Zhang, B., and Quan, H. (2009). Flame retardancy of rice husk-filled high-density polyethylene ecocomposites. *Compos. Sci. Technol.* 69 (15–16): 2675–2681.

24 Phan, H.T., Nguyen, B.T., and Pham, L.H. (2019). Excellent fireproof characteristics and high thermal stability of rice husk-filled polyurethane with halogen-free flame retardant. *Polymers* 11 (10): 1587.

25 Chen, R.S., Ahmad, S., and Gan, S. (2016). Effect of polymer blend matrix compatibility and fibre reinforcement content on thermal stability and flammability of ecocomposites made from waste materials. *Thermochim. Acta* 640: 52–61.

26 Donmez Cavdar, A., Kalaycioglu, H., and Mengeloğlu, F. (2016). Technological properties of thermoplastic composites filled with fire retardant and tea mill waste fiber. *J. Compos. Mater.* 50 (12): 1627–1634.

27 Mohammed, A.J. (2014). Study the effect of adding powder Walnut shells on the mechanical properties and the flame resistance for low density polyethylene (LDPE). *Int. J. Sci. Technol.* 3 (1): 18–22.

28 Chen, G.G., Hu, Y.J., and Peng, F. (2018). Fabrication of strong nanocomposite films with renewable forestry waste/montmorillonite/reduction of graphene oxide for fire retardant. *Chem. Eng. J.* 337: 436–445.

29 Salasinska, K., Polka, M., and Gloc, M. (2016). Natural fiber composites: the effect of the kind and content of filler on the dimensional and fire stability of polyolefin-based composites. *Polimery* 61: 255–265.

30 Salasinska, K., Barczewski, M., and Borucka, M. (2019). Thermal stability, fire and smoke behaviour of epoxy composites modified with plant waste fillers. *Polymers* 11 (8): 1234.

31 Moustafa, H., Youssef, A.M., and Duquesne, S. (2017). Characterization of bio-filler derived from seashell wastes and its effect on the mechanical, thermal, and flame retardant properties of ABS composites. *Polym. Compos.* 38 (12): 2788–2797.

32 Shah, A.U.R., Prabhakar, M.N., and Saleem, M. (2017). Development of biowaste encapsulated polypropylene composites: thermal, optical, dielectric, flame retardant, mechanical, and morphological properties. *Polym. Compos.* 38 (2): 236–243.

33 Chen, X., Zhang, X., and Zhang, X. (2020). Influence of isopropyl tris (dioctylphosphoryloxy) titanate for flame-retardant TPU based on oyster shell powder. *J. Thermal. Anal. Cal.* 139 (1): 197–206.

34 Chong, M.H., Chun, B.C., and Chung, Y.C. (2006). Fire-retardant plastic material from oyster-shell powder and recycled polyethylene. *J. Appl. Polym. Sci.* 99 (4): 1583–1589.

35 Helwig, M. and Paukszta, D. (2000). Flammability of composites based on polypropylene and flax fibers. *Mole. Crys. Liq. Crys. Sci. Technol. Sect. A.* 354 (1): 373–380.

36 Chai, M.W., Bickerton, S., and Bhattacharyya, D. (2012). Influence of natural fibre reinforcements on the flammability of bio-derived composite materials. *Compos. Part B: Eng.* 43 (7): 2867–2874.

37 Borysiak, S., Paukszta, D., and Helwig, M. (2006). Flammability of wood–polypropylene composites. *Polym. Degrad. Stab.* 91 (12): 3339–3343.

38 Suardana, N.P.G., Ku, M.S., and Lim, J.K. (2011). Effects of diammonium phosphate on the flammability and mechanical properties of bio-composites. *Mater. Des.* 32 (4): 1990–1999.

39 Shumao, L., Jie, R., and Hua, Y. (2010). Influence of ammonium polyphosphate on the flame retardancy and mechanical properties of ramie fiber-reinforced poly(lactic acid) biocomposites. *Polym. Int.* 59 (2): 242–248.

40 Idumah, C.I., Hassan, A., and Bourbigot, S. (2017). Influence of exfoliated graphene nanoplatelets on flame retardancy of kenaf flour polypropylene hybrid nanocomposites. *J. Anal. Appl. Pyrol.* 123: 65–72.

41 Shukor, F., Hassan, A., and Islam, M.S. (2014). Effect of ammonium polyphosphate on flame retardancy, thermal stability and mechanical properties of alkali treated kenaf fiber filled PLA biocomposites. *Mater. Des.* 54: 425–429.

42 Asim, M., Paridah, M.T., and Saba, N. (2018). Thermal, physical properties and flammability of silane treated kenaf/pineapple leaf fibres phenolic hybrid composites. *Compos. Struct.* 202: 1330–1338.

7

Failure Mechanisms of Biobased Composites

Dipen Kumar Rajak[1], Durgesh D. Pagar[2], and Catalin I. Pruncu[3, 4]

[1] *Department of Mechanical Engineering, Sandip Institute of Technology and Research Centre, Nashik, Maharashtra, India*
[2] *Department of Mechanical Engineering, K. K. Wagh Institute of Engineering Education and Research, Nashik, Maharashtra, India*
[3] *Department of Mechanical Engineering, Imperial College London, London, UK*
[4] *Department of Mechanical Engineering, School of Engineering, University of Birmingham, Birmingham, UK*

7.1 Introduction

Global issues of plastic pollution and depleting petroleum resources compel the research community to focus on developing environment-friendly alternatives. Since environmental sustainability is an important growing aspect in the society, biobased composite materials have been gaining a great deal of attention in recent years due to the ecofriendliness of their constituents. Many plastic industries have focused on producing biobased products derived from renewable resources to plummet the greenhouse gases and carbon dioxide (CO_2) emissions, and to replace the petroleum-based products. The term biodegradable and biobased materials are extensively misconceived in the literature [1–3]. Biodegradability is a complete deterioration of physical and chemical properties of a material, while biobased materials are derived from raw materials that are extracted from renewable resources [4]. A biobased polymer composite comprises reinforcement of fibers or particles in a biobased polymer matrix, achieving a set of a combination of complete or partial biodegradability with other spectacular properties on account of nature of its constituent materials [5]. This results in minimizing the global demands for petroleum-based materials for a variety of applications in the major fields such as construction, aerospace, automobile, marine, and biomedical. Petroleum-based synthetic composite materials are high-cost materials, nonrenewable, responsible for CO_2 emission, and consumes high amount of energy while manufacturing [6–8]. These drawbacks are overcome by implementing biobased composites, which own superior properties such as high strength to weight ratio, high stiffness value, durability, biodegradability, availability in abundance, low toxicity, resistance to wear, corrosion, chemicals, and elevated temperatures [9–13].

Most of the lignocellulosic materials are generated from agricultural practices; besides, they are often disposed of by burning, which causes pollution and global warming. These lignocellulosic materials are mainly composed of lignin, cellulose, and hemicellulose, including agricultural residue materials such as sugarcane bagasse, sweet sorghum, grasses, corn cobs, corn stover, rice straw, nut shells, wheat straw, banana waste, woods, and sponge gourd fibers [14, 15]. Lignocellulosic materials possess significant properties that reveal

Biobased Composites: Processing, Characterization, Properties, and Applications, First Edition.
Anish Khan, Sanjay Mavinkere Rangappa, Suchart Siengchin and Abdullah M. Asiri.

potential applications in the biotechnological world producing a variety of biofuels, enzymes, chemicals, pulp, paper, and animal feedstuff, and are also used to fabricate the biobased composite material [16]. Composites formed by infusing wheat straw in modified urea melamine formaldehyde (UMF) resin and rice straw in urea formaldehyde (UF) resin are used to develop medium density fiberboards (MDF) [17, 18].

Despite the diverse advantages offered by biobased composite materials, they also possess some shortcomings of thermal degradation, fiber–matrix debonding, void formation, and fiber breakage during manufacturing processes, which can further prompt to failure of material. Therefore, it is requisite to study and acknowledge the failure mechanism of biobased materials.

7.1.1 Fiber Reinforcements in Biobased Composites

The function of incorporating natural or synthetic fibers into the matrix of a composite is to increase the strength and the modulus of material [19]. In recent days, many engineering and biomedical applications are using natural fiber reinforced composites (NFRCs) due to their wide variety of properties. Natural fibers are abundantly available as they are derived from natural resources such as plants, animals, or minerals, where plant fibers are predominantly explored by researchers for the fabrication of biobased composites. It includes fibers of pineapple, sisal, abaca, hemp, jute, kenaf, wheat, corn, rice, cotton, kapok, coir, flax, ramie, and wood [6, 20]. Nevertheless, some characteristics of natural fibers degenerate the desired properties of composite materials; yet, through suitable material selection and specific fiber treatments an enhancement in overall material properties can be achieved [21].

7.1.2 Fiber Failures

Natural fibers serve a number of benefits over synthetic fibers such as biodegradability, nontoxic nature, low density, high toughness, high specific stiffness and strength, being naturally available in abundance, and being cheap to purchase [22, 23]. Besides the advantages they offer, natural fibers also carry some drawbacks such as the tendency to absorb moisture, nonhomogeneity, and their hydrophilic nature which makes them really difficult to bond with hydrophobic polymeric matrices in composite materials. Also, they degrade above the temperature of 200 °C where most thermoplastics are processed [24]. During loading conditions, fibers primarily undergo a failure mechanism namely, fiber–matrix debonding, fiber pullout, and fracture.

7.1.2.1 Fiber–Matrix Debonding

Moisture uptake reduces the stability and durability of NFRCs and also causes microbial degradation. After it gets saturated by moisture diffusion, long hemp fibers start sticking out of the surface of cellulose acetate as well as the polyhydroxybutyrate (PHB) biopolymer matrix, ultimately resulting in fiber rupture [25]. When composites come in contact with water, hydrophilic natural fibers absorb some amount of water, which ends up in fiber swelling. Stresses developed in the matrix materials around swollen fibers further result in the evolution of microcracks as shown in Figure 7.1a. Water gets diffused in the interfacial gap between the fibers and matrix as shown in Figure 7.1b, disintegrating water soluble substances from the fiber material. After drying NFRCs, shrunk fibers have lost contact with the matrix materials creating a gap between them as shown in Figure 7.1c, resulting in fiber–matrix debonding [26]. Water uptake behavior was assessed by time-dependent ζ-potential measurements which showed that cornhusk, hemp, sisal, and a majority of natural fibers absorb as much as 30 wt% amount of water [27].

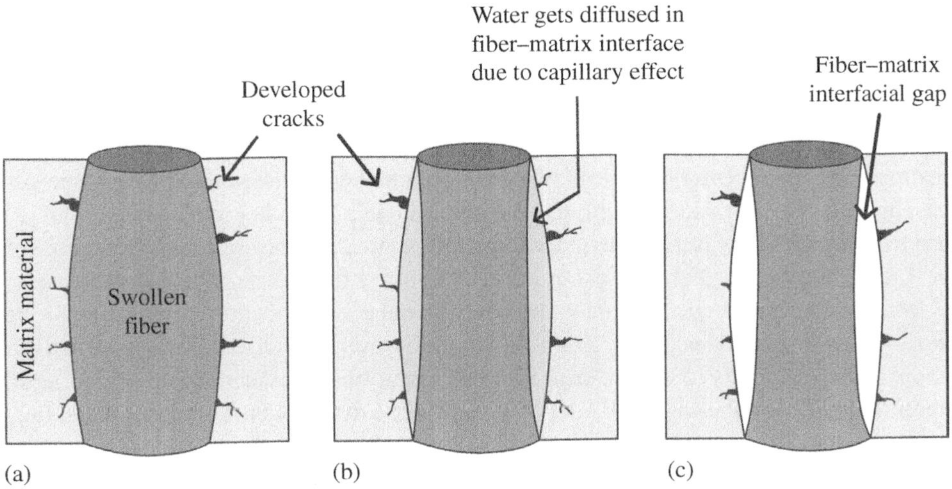

Figure 7.1 Fiber–matrix debonding caused by moisture absorption in fibers. (a) Microcracks development in the matrix due to moisture absorbed by the fiber. (b) Diffused water absorbs water soluble substances from the fiber. (c) Fiber–matrix debonding. *Source:* Based on Moudood et al. [26].

7.1.2.2 Fiber Pullout

Hydrophilic characteristics of natural fibers deteriorate the interfacial bonding between the fibers and hydrophobic polymers. A weak interface or poor compatibility between fibers and matrix materials results in a fiber pullout instead of fracture, which further leads to degradation in mechanical properties of the material, although fiber pullout allows dispersion of energy along the interface serving higher impact strength and ductility values [28]. Short lignocellulose fibers pulled out and vacate the preoccupied space in the feather keratin polymer (FKP) matrix. It was suggested that the use of long fibers or increasing fiber loading would help in minimizing the fiber pullout phenomenon [29]. Fiber pullout criteria, void formation due to fiber pullout, and fiber–matrix debonding can be well understood from Figure 7.2.

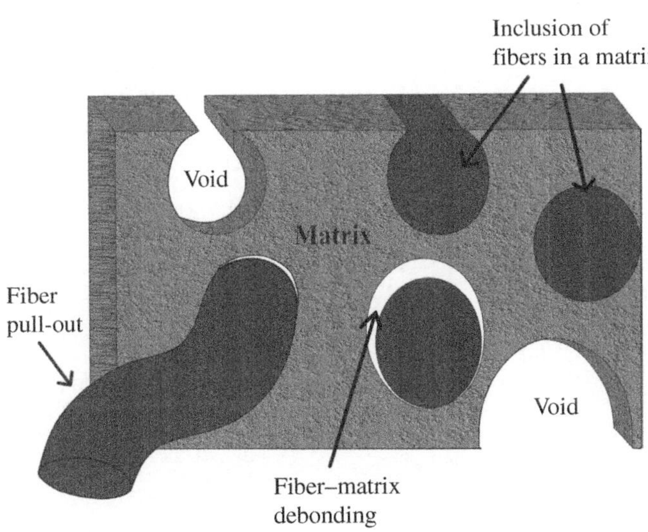

Figure 7.2 Fractured surface analysis showing fiber failures in a fiber reinforced composite.

7.1.2.3 Tear Type Failure

Fiber failure yielded by the acoustic emission (AE) has been depicted in Figure 7.3. Amplitude less than 35 dB shows axial splitting at the boundary layer of the fibers (Figure 7.3a), amplitude ranges between 35 and 60 dB shows generation of transverse microcracks (Figure 7.3b), and amplitude greater than 60 dB shows a tear type fracture in flax fibers (Figure 7.3c). Fiber stacking in the direction perpendicular to the loading conditions offered better resistance to crack propagation. Higher values of fracture toughness were obtained with long fiber reinforcements [31].

Load transfer efficiency is dependent upon the size, shape, aspect ratio, and surface area of fibers. Interfacial stresses between the reinforcement and the matrix material decrease as more load transfer paths get available with more surface areas. Continuous fibers can be arranged in any form such as uni- or multi-directional, woven, braided, or knitted forms, but they hold more probability of surface imperfections prompting decline in the tensile strength.

Therefore, while fabricating NFRCs to overcome these drawbacks of natural fibers, they are pretreated to increase their bonding with the hydrophobic polymers, subsequently improving mechanical and physical properties of NFRCs [32].

7.1.3 Fiber Pretreatments

7.1.3.1 Defibration

Defibration provides more surface area for the matrix material to fill in between the voids of composite structure improving its fiber–matrix bonding. Also, to achieve good strength, natural fibers need to be rich with cellulose content; hence defibration effectively helps to remove noncellulosic components from plant fibers. In traditional retting, natural fibers are confronted with microbial activities for the removal of noncellulosic materials such as pectin, lignin, and hemicelluloses by drowning them into water or by spreading out into the fields after cutting. Reduction in the polygalacturonan (pectin) content from 8 to 3–4 wt% has been observed in field retted hemp fibers [33, 34]. Retting is followed by scotching, where fibers are separated

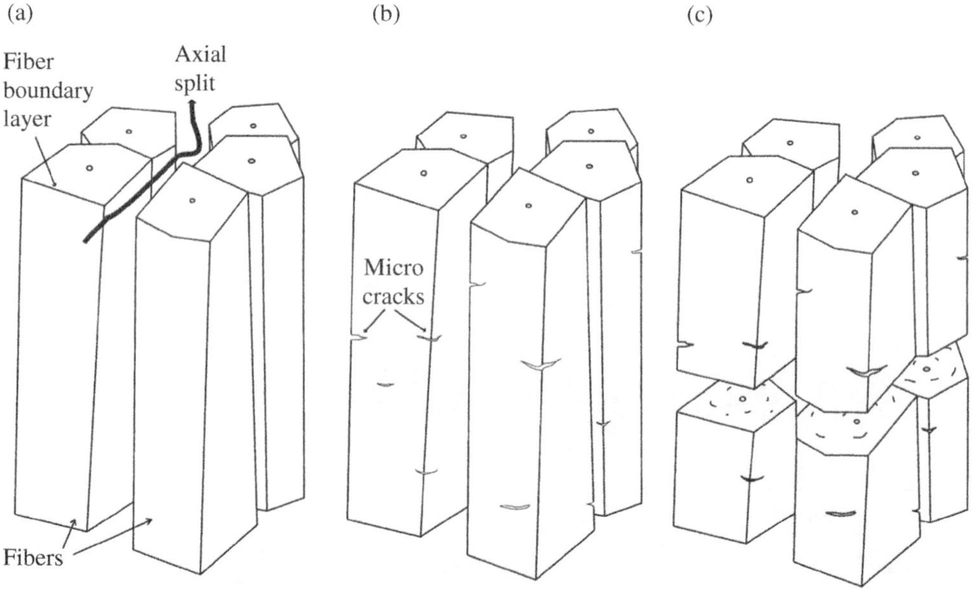

Figure 7.3 Fiber failure caused by acoustic emission. (a) Axial split. (b) Development of microcracks. (c) Tear type failure. *Source:* Based on Romhány et al. [30].

from impurities such as seeds, straw, or stem [35]. A very sharp pressure difference offered by steam explosion process serves a high aspect ratio and the increment of 10–15 wt% cellulose content in the steam exploded hemp fibers [36, 37]. In chemical treatment, chemicals such as ethylene diamine tetra (methylene phosphonic acid) (EDTMPA), ethylenediaminetetraacetic acid (EDTA), sodium hydroxide (NaOH), potassium hydroxide (KOH), and Sodium sulfite (Na_2SO_3) have been applied to remove noncellulosic components [38].

7.1.3.2 Surface Modification

Fiber surface treatments are led by the addition of compatibilizers or chemically treating fibers to improve fiber–matrix interaction. The hydrophilic nature of natural fibers is mainly due to the hydroxyl groups of cellulose which is reduced by exposing them to alkali or acetyl groups resulting in a very clean and smooth fiber surface. This is also achieved by forming ester bonds between hydroxyl groups and carboxylic acid groups in esterification treatment [39, 40]. Hemp fibers treated with alkali and silane solution in a polylactic acid (PLA) composite showed an improvement in tensile properties and hemp fiber–PLA matrix bonding to decrease fiber pull-out with further increase in PLA crystallinity [31]. Reducing fiber pullout by treating a flax fiber reinforced bioresin epoxy composite with acetic anhydride results in a relatively smooth fiber surface with improved fiber–matrix bonding [41]. By implementing plasma treatment, hydrophilic nature of arundo leaf fibers were suppressed by attaining very high degree of continuous adhesion between leaf fibers and biobased epoxy resin with enhanced mechanical properties [42]. Silane-treated flax fiber biobased epoxy resin composite revealed enhancement in fiber/matrix adhesion with reduced water absorption by 20% [43].

7.1.3.3 Coupling Agent

A part of lignin and hemicelluloses were eliminated by the delignification process, which includes treating henequen fibers with sodium hydroxide and anthraquinone. High cellulose content in the henequen fibers allows enhancement in the fiber–matrix interface by introducing 2–4 wt% of maleic anhydride grafted polypropylene (MAPP) as a coupling agent deriving transfer of tensile stresses more effectively. Covalent linkages between the maleic group of MAPP and the hydroxyl groups of cellulose help improve the stress transfer between the constituents of the composite with an increase in tensile strength. It was also observed that the further addition of a coupling agent beyond 4% reduces the tensile strength of the material [44]. Fibers were treated with cellulose nanocrystals (CNC) derived from a date palm tree using methacryloxypropyltrimethoxysilane (MPS) as a coupling agent. An enhancement in the stiffness value of a composite has been observed during uniaxial cyclic loading with minimizing the failure strain. Moreover, successful retention of fiber length was achieved for high performance applications [45].

Adding 6% MAPP as a coupling agent improved interfacial shear strength by more than 45% in a jute strand reinforced composite [46].

7.2 Matrix Materials for Biobased Composites

The matrix in a composite structure determines the shape, surface appearance, durability, and tolerance to the external means while holding the fibrous or particulate reinforcement together which carries majority of the structural loads [5]. Matrix material in a composite structure comprise polymers, metals, ceramics, or sometimes carbon; polymers are the most widely used material due to their high-performance application. Most polymers are derived from petroleum byproducts; however, some biopolymers such as starch, polylactic acid, polyhydroxyalkanoate (PHA) are obtained or developed from natural resources [6, 47].

The term biobased polymer refers to the polymers whose raw materials are derived from renewable sources, while renewable resources are those which get replaced by natural means at a rate comparable or higher than their rate of utilization. For a better understanding, some of the polymers are categorized in Table 7.1, according to the criteria of biodegradability and their source.

Hydrophilic nature of PLA facilitates better interaction with natural fibers than rest of the hydrophobic polymers. Oil-based polymers such as poly ε-caprolactone (PCL), polybutylene succinate (PBS), and polybutylene adipate terephthalate (PBAT) possess ester bonds that gets degraded by the action of certain enzymes secreted by microorganisms; these are therefore categorized as biodegradable polymers. Although derived from natural processes such as glucose fermentation, bio-PE is not a biodegradable polymer. Moreover, natural contents such as cellulose acetate (CA) and ethylene glycol of polysaccharide and bio-PET, respectively, do not make them biodegradable polymers. Yet, the concept of carbon neutrality explains that carbon dioxide generated after burning a biobased plastic gets converted into biomass by the process of photosynthesis; therefore, it barely matters that biobased plastics are biodegradable or not— eventually they are considered as ecofriendly materials [51]. The latest trend of using vegetable oil (VO) in combination with petroleum-based polymer resins brings about an improvement in the brittleness and the fracture toughness of the material. Generally, biobased polymer resins are formulated from VOs such as soybean [52, 53], linseed [54], castor [55], and hemp [56] incorporated with the epoxy resin using a proper hardener. To avoid the toxicity of an amine hardener, researchers have recently developed biobased hardeners derived from renewable resources such as biobased polyamines, biobased resins, water soluble natural acids, terpene-based anhydrides (TPAn), and tannic acid [19].

Table 7.1 Classification of polymers based on biodegradability criteria.

Polymers	Biodegradable	Nonbiodegradable
Biobased (renewable resources)	Poly(lactic acid) (PLA)	Polysaccharide derivatives (with high DS)[a]
	Polyhydroxyalkanoate (PHA)	Polyol–polyurethane
	Polysaccharide derivatives (with low DS)[a]	Bio-polyethylene (bio-PE)
	Poly(amino acid)	Bio-poly(ethylene terephthalate) (bio-PET)
	Starch	
	Polypropylene carbonate (PPC)	
Oil-based (fossil resources)	Poly(ε-caprolactone) (PCL)	Polyethylene (PE)
	Polybutylene succinate (PBS)	Polypropylene (PP)
	Polybutylene adipate terephthalate (PBAT)	Polystyrene (PS)
		Poly(ethylene terephthalate) (PET)

[a] DS, degree of substitution.

Source: Based on Spierling et al. [2]; Faruk et al. [48]; Mohanty et al. [49]; and Chen [50].

7.2.1 Matrix Failure

The transition between the fiber–matrix interface failure and the matrix failure is due to the strain rate. The failure in matrix materials is initiated by means of a plastic strain which is dependent upon strain rate and stress triaxiality, where stress triaxiality is the ratio of mean stress to equivalent stress. Meanwhile, at low strain rates where stress concentration is predominant, crack propagates in the matrix materials more rapidly avoiding failure at interface and fibers as well [57]. In a composite structure, the properties of matrix material determine the failure criteria caused by delamination, crack propagation, water absorption, impact damage, thermal creep, and chemical attack [58]. Presence of contaminations or formation of a resin-rich region caused due to improper mixing of filler materials with matrix affects the structural performance of a composite. Reinforcement of long wood fibers showed poor impregnation due to high aspect ratio, allowing more resin-rich regions. More exposure to matrix-rich regions caused wear failure in composite material due to the matrix cracking phenomenon. Nevertheless, inclusion of hardwood fibers ended up failing on account of matrix cracking, deformation, and micro ploughing while softwood fiber reinforced composite failed due to surface disruption [59]. Preimpregnation and silane treatment prevented fiber surface to fail from fiber–matrix interfacial failure, indeed increasing fiber–matrix interaction leads to matrix failure [60]. Lower ductility of the bioepoxy matrix offered high resistance to impact due to the absence of interfacial gap between hemp fibers and the matrix. However, further increase in impact energy revealed cracking and splitting of matrix on the back surface of the laminate [61].

7.2.2 Matrix Treatment

Impregnation of sunflower oil resulted in lowering the friction coefficient value, consequently reducing specific wear rate of the composite material [59]. Styrene fused methacrylic anhydride modified soybean oil (MMSO), and acetic anhydride modified soybean oil (AMSO) resins reinforced with air-laid flax fiber mats revealed better fiber–matrix adhesion than neat AMSO and MMSO resins to avoid failure [62]. Water, glycerol, octanoic acid, and polyethylene glycol were used as plasticizers for biodegradable corn gluten meal (CGM) matrix. Water has solely brought glass transition temperature (T_g) of CGM from 85° to 17° which was a drastic drop in T_g value compared to other plasticizers [63]. Decline in ultimate tensile strength has been observed on the addition of ground wood pulp due to their hydrophilic character which absorbs water resulting in poor fiber–matrix adhesion. When maleic anhydride grafted polypropylene was added as a coupling agent, the ultimate tensile strength enhanced by 125% with advancement in durability of the composite material, compared to uncoupled composites [64].

7.3 Trends in Biobased Composites

Considering high demand for wood products in the construction industry, alternative biobased composite materials were developed and their potentiality regarding failure mechanism has been tested. Failure surface perpendicular to the axial direction showed fiber rupture in the hemp fiber reinforced cellulose acetate (CA) composite, while failure strains were higher in the weft direction than in the warp. Matrix cracking takes place at lower strain values in hemp fiber reinforced polyhydroxybutyrate composites, although it showed high ductility [65]. Although the use of biobased materials such as vegetable oil-based resins and natural fibers serves the purpose of ecofriendliness, difficulties like poor resin properties and weak fiber–matrix bonding are confronted. Conventionally, inclusion of additives could resolve the problem of resin quality but increase the manufacturing cost of the material. Therefore, to overcome these difficulties a novel biobased composite is developed comprising a biobased resin, epoxidized sucrose soyate (ESS) reinforced with surface-treated flax fibers [66]. An exceptionally novel hybrid composite panel has been developed where fungal mycelium was grown on softwood particles producing mycelium-modified wood and cellulose nano-fiber (CNF) were introduced as a binder material. This complete biodegradable hybrid composite has shown an increase in the rupture and elastic modulus which is then employed for packaging and furniture applications [67]. An alternative to leather used in the apparel and footwear industry, water resistant yet breathable shoes and coats were made of biobased composites developed from renewable resources such as plant oils, natural fiber cloth, and chicken feather fibers to replace [68]. A composite made of powdered walnut and coconut shell incorporated with bioepoxy resins manifested excellent mechanical properties, such as tensile strength of 69.5 MPa, flexural strength (FS) of 14.9 MPa, and impact strength of 21.3 J [69]. Usually metals, plastics, papers, and glass are used for food packaging applications, where glass and metals can be reused but are expensive to manufacture, and the use of paper causes deforestation. Therefore, a biobased composite material was manufactured using sugar palm starch (SPS) bioplastic reinforced with sugar palm fibers (SPF) for food packaging applications. Chemical similarities between both the constituents derived from the same plant provide smooth fracture surfaces with a homogeneous matrix and excellent fiber–matrix adhesion resulting in enhanced mechanical performance and thermal stability [70]. For the applications of edible packaging and bone implants, biobased composite films comprising a polylactic acid (PLA) matrix reinforced with 5 wt% bacterial cellulose nanocrystals (BCNC) have been developed without any surface modification which exhibits a tensile strength of 3.6 MPa [71]. Formaldehyde-based adhesives extracted from petroleum-based resources are replaced by nontoxic, renewable soybean meal-based adhesives, which reveal excellent water-resistant property useful for hydrophobic biobased composites [72]. Biodegradable reinforcements such as coffee silverskin (CSS), an external layer of green coffee bean, was used to develop a biopolyethylene-based composite which shows improvement in the elastic modulus while biopolyethylene (Bio-PE) gets disintegrated into ethylene after dehydration. Here it can be seen that coffee beans are produced in abundance, being the third most consumed drink after water and tea; also, its skin is biodegradable waste [12]. It implies that biobased materials are available in abundance, mostly acquired from waste products, hence being cheap and also serves the purpose of being one of the constituents in the composite material.

7.3.1 Wood Plastic Composites

In recent years, many researchers have put efforts to overcome the global issue of plastic waste dissemination in the environment. Recycled and waste thermoplastics have been utilized to develop wood plastic composites (WPCs), which serve widespread applications in the construction industry, such as decking, roofing, railing, beams, columns windows, and door frames [73]. Furthermore, their advancement in properties such as high stiffness, biodegradability, low cost, light weight, and good mechanical properties reveal their potential applications in automobile and aerospace sectors [74]. Wood, plastics, and additives are the basic components in WPCs where wood dust, wood fibers, or wood flour (WF) have been used as filler materials in the matrix of recycled plastics such as high-density polyethylene (HDPE), polyethylene terephthalate (PET), low-density polyethylene (LDPE), polystyrene (PS), polypropylene (PP), polyvinyl chloride (PVC), and others. WPCs are frequently

Table 7.2 Properties of wood plastic composites [77].

WPC matrix	Yield strength (MPa)	Tensile modulus (GPa)	Elongation (%)	Flexural modulus (GPa)	Notched IZOD impact (J/m)
HDPE	15.5–22.0	1.7–4.4	1.3–15.0	1.4–4.1	16.1–27.7
PP	26.0–32.3	2.7–6.0	1.2–6.2	2.5–4.9	23.3–46.4
PS	22.5–28.9	4.0–5.8	1.5–5.8	3.2	20.3–40.5
PVC	10.0–25.0	1.0–8.0	0.3–1.0	1.0–6.0	

manufactured by thermoplastic shaping techniques such as extrusion, injection, and compression molding [75, 76]. WPC properties acquired by varying weight percentage of wood incorporated with some plastics are presented in Table 7.2.

Polymer blends of recycled polypropylene/polyethylene (r-PP/PE) were obtained from rejected unused disposable diapers (RUDD), which was used to develop a WPC reinforced with kenaf fibers. At 60 wt% fiber content, a FS of 14.21 MPa and a flexural modulus (FM) of 1544 MPa have been observed which were 15 and 85% increased, respectively [78]. However, decrease in impact strength of materials with an increase in the fiber loadings due to extended contact between the fibers has been observed [79]. High wood filler content up to 85 wt% in wood/PP composites developed by extrusion molding showed a bending strength of 50 MPa. Also, to examine weather resistance characteristics with addition of different additives, wood/PP composite has been exposed to weathering condition without any post treatments for a duration of five months. It was observed that acrylic resin coating with addition of hindered amino ester showed best results [80]. Addition of bamboo charcoal (BC) let cover the more surface area of WF with LDPE which helps to improve WF–matrix interface resulting in enhancement of water resistivity and thermal properties [81]. Adding small amounts of thermal fillers such as flake graphite (FG) in polyethylene (PE)-based poplar wood fiber reinforced WPC increased the thermal conductivity of the material [82].

7.3.1.1 Failure in WPC

Increasing wood content has shown negative effects on the tensile strength of WPCs [83]. Microbial attack, fiber debonding, and degradation in fiber–matrix interface of WFC is mainly due to moisture absorption phenomenon caused by the contents of WF such as cellulose, hemicelluloses, lignin, and pectins. This phenomenon can be avoided by treating WPC with emulsions such as paraffin or pickering emulsions, which show enhancement in hardness, water repellency, and corrosion resistance subsequently improving mechanical, friction, and wear properties of the material [84]. Furthermore, the performance is dependent upon their degradation level, the source they are derived from (single source or the combination of recycled and unused plastics), storage, and processing conditions [85]. Samples of WPC formulated from LDPE, WF particles, and recycled plastic blends of PP and PE derived from household and construction solid waste are subjected to accelerated weathering, xenon-arc light, and freeze–thaw cycling. Reduction in FS and FM by 2–26 and 2–30%, respectively, has been observed affecting the durability of WPC material [76].

Composites based on poly(3-hydroxybutyrate-*co*-3-hydroxyvalerate) (PHBV) and WF exposed to natural weathering confronted a gradual drop in mechanical properties such as tensile strength and tensile modulus besides loss in material weight due to moisture-induced fungal attack [86]. Composite samples were manufactured using recycled PP and PE obtained from construction plastic waste. Tensile modulus and rigidity of material has improved due to the inclusion of WF, while incompatibility between WF and plastic bends (PB) resulted in poor strength [87]. Tensile strength of pure PP was observed to be 29.78 MPa while addition of poplar wood fibers from 50 to 90 wt% degrades the tensile strength of the PP-based WPC by 50% [88]. The major cost in WPC is of plastic; using recycled materials eventually cut the material cost, thus reducing the price of the composite [89]. Lifecycle assessment (LCA) of WPCs composed of virgin materials and recycled materials have carried out using end-of-life (EoL) pathways. WPCs composed of secondary materials procured from recycled post-consumer thermoplastics and wood products have appeared to be ecologically and technically superior alternative. Handling the waste materials without diffusing hazardous products into the environment and recycling while maintaining their quality would be the challenge for such WPCs [90].

7.3.2 Hybrid Combination

Hybridization is a sustainable and economical way of utilizing waste materials more efficiently by combining two or more reinforcements in a polymer matrix. Hybridization of composite materials provides flexibility over type, shape, or size of reinforcements to achieve remarkable properties that cannot be attained in composite systems containing only one type of reinforcement [91]. A hybrid combination of petroleum blends and vegetable oil resins composed of epoxidized methyl linseedate (EML) and unsaturated polyester (UPE) were reinforced by nanoclay and hemp fibers to form a hybrid biobased composite. EML addition deteriorated the diffusivity barrier properties of the material which was further improved by the addition of nanoclay. Microcrack formation caused by stress concentration in a brittle polymer matrix was provoked by a stiff nanoclay which was then terminated by the addition of bioresin. In spite of the fact that addition of EML improved the ductility and toughness of the composite material alleviating fiber fractures, it reduces the stiffness giving rise to fiber–matrix interfacial gaps and fiber pullout phenomenon [92]. Agricultural waste residues of coir fibers and coir shell particles of 20 wt% in thermoplastic matrix are used to develop a hybrid composite. Homogeneous dispersion of both biofillers into the PP matrix manifested 50% improvement in Young modulus compared to pure PP [93]. Fiber splitting and fiber–matrix interfacial failure has been observed in a hybrid composite with a different combination of kenaf, bamboo, and coir fiber reinforcements in the PLA matrix. The existing impurities inside untreated bamboo fiber cell walls degrade strength and the stiffness of the material, further failing it during the tensile loading. Also, decohesion of cell walls due to poor wettability of coir fiber with a PLA matrix results in fiber pull out. However, combination of kenaf and coir fibers with a PLA matrix composite revealed highest FM of 15 GPa and combining all three fibers in PLA matrix showed highest tensile strength of 187 MPa among all other combinations [94]. Hybrid composites derived from waste products involving recycled polypropylene (RPP) reinforced with recycled newspaper fiber (RNF) and poplar wood flour (PWF) performed well on account of tensile and FM compared to pure PP. Increasing RNF content in a hybrid composite results in improving the water absorption phenomenon due to the existence of holocellulose in RNF, while increasing PWF content significantly improves thermal stability of the material but reduces the tensile, flexural, and impact modulus [95, 96]. An enhancement in tensile and flexural properties has been observed in hybrid composites incorporating a polyester matrix, reinforced with short banana and sisal fibers. However, there is a negative effect on the impact strength of the composite with increase in banana fiber content [97].

Hybrid composites consisting of liquid methyl methacrylate (MMA), an organic thermoplastic resin reinforced with hybridized plain weave ultrahigh molecular weight polyethylene fibers (UHMWPE) and carbon fibers, is manufactured by vacuum-assisted resin infusion (VARI). The void content in a composite laminate is calculated as shown in Equations (7.1) and (7.2),

$$V = \left(\frac{T_d - M_d}{T_d}\right) \times 100 \tag{7.1}$$

$$T_d = \frac{100}{\left(\frac{R}{D} + \frac{r}{d}\right)} \tag{7.2}$$

where V is the vol% of void content, M_d is the measured density of composite material, T_d is the theoretical composite density, R is wt% of resin in a composite structure, and D is the density of resin, while r is reinforcement wt% and d is density of reinforcement (g/cm^3). The result shows an average void content of 0.75% in hybrid plain weave UHMWPE and carbon fiber composite laminate with composite density of 1.20 g/cm^3 [98].

7.4 Adapted Manufacturing Technologies

The mechanical property that majority of the composites depend upon is the fiber–matrix adhesion, which is directly related to manufacturing conditions. Therefore, to overcome the defects occurring in biobased composites that would consequently be responsible for the material failure, efficient use of manufacturing processes needs to be studied.

7.4.1 Injection Molding

Injection molding is a high-volume closed molding process, which offers the fabrication of complex shapes in short cycle times with a minimal requirement of finishing on molded products. For outdoor structural applications, a biobased composite footbridge of 14 m span across the Dommel River, The Netherlands, has been fabricated using vacuum injection molding process. A vacuum bag lets the bioresin to be sucked into the nonwoven hemp and woven flax fiber mat by means of pressure to produce a laminate structure. LCA carried out to determine the overall environmental impact on the bridge predicted 50 years of service life [99]. Structural stability, high precision fabrication, good shear thinning property, and less viscosity were the requirements for manufacturing thin circuits and flexible implant electrodes. Therefore, biodegradable and biocompatible chitosan–zinc composites comprising 2% chitosan and 0.0375 g/ml of zinc powder have been prepared by the injection molding process for the application of biological sensing phenomena [100]. Well-compounded pellets of PP with chopped jute fibers were injection molded to form a composite, which showed significant improvement in the tensile strength and modulus due to excellent fiber/matrix bonding, good fiber wetting, and uniform fiber distribution. Micropores or voids which cause material failure are diminished by vacuum degassing performed during the fabrication of a composite material [101]. Coffee capsules made of PLA and poly(butylene) succinate co-adipate (PBSA) were fabricated using injection molding to achieve a greater deal of dimensional accuracy and stability [102]. Desired properties in biobased composites are acquired by varying nanoclay addition, mixing times, or injection pressures. Increasing injection pressure from 500 to 900 bars results in an increase in the tensile properties of soy/nanoclay-based composites [103].

7.4.2 Liquid Composite Molding

Improvised liquid composite molding (LCM) is a hybrid version of vacuum-assisted resin transfer molding (VARTM) and resin transfer molding (RTM). A negative (vacuum) pressure was applied to force resin into cellulose nano-fiber (CNF) preforms to avoid air trapping. Stronger fiber–matrix adhesion was attained for adequate load transfer [104]. A solid mold surface is replaced by wrapping nylon bagging sheets around the whole part unveiling the hidden bottom surface of mold to access and visually inspect the resin flow with provision for additional vacuum and injection lines [105]. A composite was fabricated by using biobased epoxy resin and thermoset polyurethane resin reinforced with woven flax fibers. A very smooth surface finish with homogenous thickness was achieved by RTM to avoid the formation of macrovoids [24]. Composite laminates composed of biobased thermoset resin reinforced with hemp fiber are produced by RTM. Parameters such as resin viscosity, fabric permeability, and gelation time allows the resin to completely infuse the reinforcement along with rapid and extensive filling of mold cavity, offering homogeneous filling to eliminate the void defects [106]. Void formation in a matrix structure is caused by formerly entrapped air or moisture during processing or mixing of a matrix with filler materials. Void formation in matrix structure leads to the porosity of composites, which further causes internal corrosion, low compression/tensile strength, and delamination. In manufacturing techniques such as RTM and VARTM, entrapped air during resin flow is the main cause for void formation along with the nucleation of dissolved gases, and generation of gas while curing [107–109]. Void formation is minimized by controlling the stages of its formation, motion, and compression or by optimizing the capillary number. Also, by controlling injection pressure and by a proper application of vacuum the effects of voids are reduced in LCM [110].

7.5 Other Failure Criteria

There is a lot of interfacial area present between the platelets and the matrix of nanocomposite materials. Therefore, to improve the toughness of the biobased brittle ceramic matrix composite, failure mechanisms such as crack bridging and platelet pullout were studied. It was observed that decohesion and platelet sliding takes place below the critical interface value and beyond that slipping of matrix atoms occurs [111]. Mechanical joining of assembly parts by means of rivets or bolts requires the production of deep circular holes; one of the most cost-effective machining processes proposed for composite materials is mechanical drilling. But, due to characteristics of composite materials such as an anisotropic and nonhomogeneous structure with high abrasive nature, a conventional drilling operation causes delamination, fiber–matrix separation, interlaminar cracking, cutting jet lag, and hole tapering [112, 113]. To overcome these failures, ultrasonic drilling (USD) has been significantly applied to improve the quality of drilled holes in conventional FRP composites. Decrease in drilling force (thrust and torque) and burr formation has been observed, consequently reducing the effect of delamination and surface roughness defects by 50% [114]. In the absence of tool–material interactions, a noncontact-type machining process called laser drilling avoids tool wear, machine–tool deflections, drill bits vibration, and delamination caused by the thrust force. As fibers possess a lower coefficient of thermal expansion compared to the matrix material, contraction perpendicular to the fiber direction produces residual tensile stresses after curing. As the load increases, crack propagation across the laminate increases rapidly [115].

Generally, lignocellulosic fibers are brittle in nature and show high dispersion with respect to their inherent tensile strength. Due to variability in their morphology, statistical analysis is required to calculate the mean tensile strength of natural fibers. Mean tensile strength of such natural fibers were calculated using Weibull analysis, where probability of failure is given by Equation (7.3),

$$P_{\mathrm{f}}(\sigma)=1-\mathrm{e}^{\left[-\left(\frac{\sigma}{\eta}\right)\right]^{\beta}}, \tag{7.3}$$

where, β is Weibull modulus, σ is the fiber tensile strength, and η is the fiber characteristic strength.

The objective of such micromechanical studies is to find out the strength contribution of each constituent (i.e. matrix, filler, and interface) to the strength of the composite. On that account, composite strength is given by corresponding stress according to Equation (7.4),

$$\sigma^{R}_{\text{composite}}=\sigma^{R}_{\text{fibers}}V_{\text{fibers}}+\sigma^{0}_{\text{matrix}}\left(1-V_{\text{fibers}}\right), \tag{7.4}$$

where, $\sigma^{R}_{\text{composite}}$ is the composite strength, $\sigma^{R}_{\text{fibers}}$ is the fiber strength, V_{fibers} is the fiber volume fraction, and $\sigma^{0}_{\text{matrix}}$ is the matrix strength at failure.

In compression molded flax fiber reinforced polyester composites, failure strain of polyester matrix is greater than the flax fibers. Therefore, during rupture fibers are likely to break first and the matrix material will not be able to bear further stress, consequently failing the composite. At the beginning of the flax fiber failure where deformation reaches at about 2.5%, the strength was estimated at 500 MPa [31, 116]. Weibull analysis underestimated the strength of the henequen fiber reinforcements as 450 MPa, where the actual tensile strength was 500 MPa [44].

Under tensile loading, the crack craze formation in self-reinforced PLA (SR-PLA) composites is independent of fiber characteristics and fiber–matrix adhesion, but it occurs due to the brittle nature of PLA matrix. As a consequence of crazing, failure fibers did not get completely utilized due to poor load transfer from the matrix to the fibers during the application of low strain loads. It limits the strength of the material and further causes splitting cracks which leads to failure of unidirectional SR-PLA composites [117].

Fracture toughness is the measure of resistance offered by the material to the extension of a crack. A semicircular shape specimen is taken for fracture testing and fracture toughness (K_{IC}) is calculated as presented in Equation (7.5),

$$K_{\text{IC}}=\frac{PS}{BW^{3/2}}f\left(\frac{a}{W}\right), \tag{7.5}$$

where P is the ultimate load, S is the span length, B is the specimen thickness, W is the width, and a is the length of a crack. Results showed that the fracture toughness value of bagasse fiber reinforced composites (FRCs) is highest at 1.108 MPa√m among banana FRC (0.460 MPa√m) and coconut fiber reinforced epoxy resin composite (1.094 MPa√m) [118].

The water uptake of lignocellulosic fibers is evaluated by a time-dependent Zeta-potential measurement. The formation of an electrochemical layer is due to the adsorption of ions at the interface of the solid and electrolyte solution. Equations (7.6) and (7.7) show decaying caused due to rapid swelling of lignocellulosic fibers as a function of time as $\zeta=f(t)$,

$$-\frac{\mathrm{d}\zeta}{\mathrm{d}t}=k\left(\zeta_{t}-\zeta_{\infty}\right) \tag{7.6}$$

$$-\ln\frac{\zeta_t-\zeta_\infty}{\zeta_0-\zeta_\infty}=kt, \tag{7.7}$$

where ζ_t is ζ-potential at time t, ζ_∞ is ζ-potential as a function of $\zeta = f(t)$, ζ_0 is ζ-potential of the dry material instantly measured after the fibers came in contact with the electrolyte solution, and k is the relative constant rate of the swelling. Rapid swelling of fibers leads to increase in surface area of lignocellulose fibers, consequently decreasing the value of ζ-potential due to the disintegration of water-soluble components. For flax fibers and corn husks, the value of ζ-potential decreases with time and for the rest of the lignocellulose fibers ζ-potential increases from ζ_0 asymptotically to a constant but larger value than ζ_∞ [27].

7.6 Conclusion

Biobased composites derived from renewable resources asserted to be the superior alternatives for the majority of petroleum-based products in a variety of fields, alleviating the repercussions of global warming and environmental pollutions. Rapidly developing WPCs reveal their potential applications in construction industries serving better strength and durability than wood products.

Studying the failure mechanism of biobased composite materials manifested that failure caused by the development of interfacial gap between the filler and matrix materials was dominating compared to other failure phenomenon. It was primarily prompted due to hydrophilic characteristic of natural materials which can be overcome by chemically treating filler materials, using a certain type of coupling agents or plasticizers.

Type of filler or matrix materials, percentage amount of filler in matrix, manufacturing type, and the processing conditions are some of the major influencing factors affecting failure mechanism in biobased composite materials.

Self-reinforced polymer matrix composites could be the best future material suggested as it eliminates the greater part of failure defects induced by biobased composites. Exceptional compatibility between filler and matrix materials can be achieved as they are acquired from the same material.

References

1 Abhilash, M. and Thomas, D. (2017). Biopolymers for biocomposites and chemical sensor applications. In: *Biopolymer Composites in Electronics*, 405–435. Elsevier.
2 Spierling, S., Knüpffer, E., Behnsen, H. et al. (2018). Bio-based plastics – a review of environmental, social and economic impact assessments. *Journal of Cleaner Production* 185: 476–491.
3 Katiyar, V. (2017). *Bio-based Plastics for Food Packaging Applications*. Smithers Pira.
4 Babu, R.P., O'connor, K., and Seeram, R. (2013). Current progress on bio-based polymers and their future trends. *Progress in Biomaterials* 2 (1): 8.
5 Chen, G.Q. and Patel, M.K. (2011). Plastics derived from biological sources: present and future: a technical and environmental review. *Chemical Reviews* 112 (4): 2082–2099.
6 Väisänen, T., Das, O., and Tomppo, L. (2017). A review on new bio-based constituents for natural fiber–polymer composites. *Journal of Cleaner Production* 149: 582–596.
7 Rus, A.Z.M. (2010). Polymers from renewable materials. *Science Progress* 93 (3): 285–300.
8 Dicker, M.P., Duckworth, P.F., Baker, A.B. et al. (2014). Green composites: a review of material attributes and complementary applications. *Composites Part A: Applied Science and Manufacturing* 56: 280–289.

9 Gupta, A. and Kumar, A. (2010). Composites materials: addressing the climate change. *Asia Pacific Business Review* 6 (1): 78–89.

10 Zhang, Y., Zhang, J., Chen, M. et al. (2019). A tough, water-resistant, high bond strength adhesive derived from soybean meal and flexible hyper-branched aminated starch. *Polymers* 11 (8): 1352.

11 Ganster, J., Erdmann, J., and Fink, H.-P. (2013). *Bio-Based Composites. Polimery* 6: 423. https://doi.org/10.14314/polimery.2013.423.

12 Dominici, F., García García, D., Fombuena, V. et al. (2019). Bio-polyethylene-based composites reinforced with alkali and palmitoyl chloride-treated coffee silverskin. *Molecules* 24 (17): 3113.

13 Murthy, P.S. and Naidu, M.M. (2012). Production and application of xylanase from *Penicillium sp.* utilizing coffee by-products. *Food and Bioprocess Technology* 5 (2): 657–664.

14 Satyanarayana, K.G., Arizaga, G.G., and Wypych, F. (2009). Biodegradable composites based on lignocellulosic fibers – an overview. *Progress in Polymer Science* 34 (9): 982–1021.

15 Arevalo-Gallegos, A., Ahmad, Z., Asgher, M. et al. (2017). Lignocellulose: a sustainable material to produce value-added products with a zero waste approach – a review. *International Journal of Biological Macromolecules* 99: 308–318.

16 Iqbal, H.M.N., Kyazze, G., and Keshavarz, T. (2013). Advances in the valorization of lignocellulosic materials by biotechnology: an overview. *BioResources* 8 (2): 3157–3176.

17 Halvarsson, S., Edlund, H., and Norgren, M. (2008). Properties of medium-density fibreboard (MDF) based on wheat straw and melamine modified urea formaldehyde (UMF) resin. *Industrial Crops and Products* 28 (1): 37–46.

18 Li, X., Wu, Y., Cai, Z., and Winandy, J.E. (2013). Primary properties of MDF using thermomechanical pulp made from oxalic acid pretreated rice straw particles. *Industrial Crops and Products* 41: 414–418.

19 Mustapha, R., Rahmat, A.R., Abdul Majid, R., and Mustapha, S.N.H. (2019). Vegetable oil-based epoxy resins and their composites with bio-based hardener: a short review. *Polymer–Plastics Technology and Materials* 58 (12): 1311–1326.

20 Faruk, O. and Ain, M.S. (2013). Biofiber reinforced polymer composites for structural applications. In: *Developments in Fiber-Reinforced Polymer (FRP) Composites for Civil Engineering*, 18–53. Woodhead Publishing.

21 John, M.J. and Thomas, S. (2008). Biofibres and biocomposites. *Carbohydrate Polymers* 71 (3): 343–364.

22 Rajak, D.K., Pagar, D.D., Kumar, R., and Pruncu, C.I. (2019). Recent progress of reinforcement materials: a comprehensive overview of composite materials. *Journal of Materials Research and Technology* 8 (6): 6354–6374.

23 Rajak, D.K., Pagar, D.D., Menezes, P.L., and Linul, E. (2019). Fiber-reinforced polymer composites: manufacturing, properties, and applications. *Polymers* 11 (10): 1667.

24 Cuinat-Guerraz, N., Dumont, M.J., and Hubert, P. (2016). Environmental resistance of flax/bio-based epoxy and flax/polyurethane composites manufactured by resin transfer moulding. *Composites Part A: Applied Science and Manufacturing* 88: 140–147.

25 Christian, S.J. and Billington, S.L. (2012). Moisture diffusion and its impact on uniaxial tensile response of biobased composites. *Composites Part B: Engineering* 43 (5): 2303–2312.

26 Moudood, A., Rahman, A., Öchsner, A. et al. (2019). Flax fiber and its composites: an overview of water and moisture absorption impact on their performance. *Journal of Reinforced Plastics and Composites* 38 (7): 323–339.

27 Baltazar-y-Jimenez, A. and Bismarck, A. (2007). Wetting behaviour, moisture up-take and electrokinetic properties of lignocellulosic fibres. *Cellulose* 14 (2): 115–127.

28 Haq, M., Burgueño, R., Mohanty, A.K., and Misra, M. (2008). Hybrid bio-based composites from blends of unsaturated polyester and soybean oil reinforced with nanoclay and natural fibers. *Composites Science and Technology* 68 (15–16): 3344–3351.

29 Barone, J.R. (2009). Lignocellulosic fiber-reinforced keratin polymer composites. *Journal of Polymers and the Environment* 17 (2): 143.

30 Romhány, G., Karger-Kocsis, J., and Czigány, T. (2003). Tensile fracture and failure behavior of thermoplastic starch with unidirectional and cross-ply flax fiber reinforcements. *Macromolecular Materials and Engineering* 288 (9): 699–707.

31 Sawpan, M.A., Pickering, K.L., and Fernyhough, A. (2011). Improvement of mechanical performance of industrial hemp fibre reinforced polylactide biocomposites. *Composites Part A: Applied Science and Manufacturing* 42 (3): 310–319.

32 Kalia, S., Kaith, B.S., and Kaur, I. (2009). Pretreatments of natural fibers and their application as reinforcing material in polymer composites – a review. *Polymer Engineering & Science* 49 (7): 1253–1272.

33 Placet, V., Day, A., and Beaugrand, J. (2017). The influence of unintended field retting on the physicochemical and mechanical properties of industrial hemp bast fibres. *Journal of Materials Science* 52 (10): 5759–5777.

34 Liu, M., Ale, M.T., Kołaczkowski, B. et al. (2017). Comparison of traditional field retting and Phlebia radiata Cel 26 retting of hemp fibres for fibre-reinforced composites. *AMB Express* 7 (1): 58.

35 Thygesen, A., Madsen, B., Bjerre, A.B., and Lilholt, H. (2011). Cellulosic fibers: effect of processing on fiber bundle strength. *Journal of Natural Fibers* 8 (3): 161–175.

36 Liu, M., Thygesen, A., Summerscales, J., and Meyer, A.S. (2017). Targeted pre-treatment of hemp bast fibres for optimal performance in biocomposite materials: a review. *Industrial Crops and Products* 108: 660–683.

37 Pielhop, T., Amgarten, J., von Rohr, P.R., and Studer, M.H. (2016). Steam explosion pretreatment of softwood: the effect of the explosive decompression on enzymatic digestibility. *Biotechnology for Biofuels* 9 (1): 152.

38 Liu, M., Meyer, A.S., Fernando, D. et al. (2016). Effect of pectin and hemicellulose removal from hemp fibres on the mechanical properties of unidirectional hemp/epoxy composites. *Composites Part A: Applied Science and Manufacturing* 90: 724–735.

39 Kabir, M.M., Wang, H., Lau, K.T., and Cardona, F. (2013). Effects of chemical treatments on hemp fibre structure. *Applied Surface Science* 276: 13–23.

40 Ali, A., Shaker, K., Nawab, Y. et al. (2018). Hydrophobic treatment of natural fibers and their composites – A review. *Journal of Industrial Textiles* 47 (8): 2153–2183.

41 Loong, M.L. and Cree, D. (2018). Enhancement of mechanical properties of bio-resin epoxy/flax fiber composites using acetic anhydride. *Journal of Polymers and the Environment* 26 (1): 224–234.

42 Scalici, T., Fiore, V., and Valenza, A. (2016). Effect of plasma treatment on the properties of *Arundo Donax L.* leaf fibres and its bio-based epoxy composites: a preliminary study. *Composites Part B: Engineering* 94: 167–175.

43 Fathi, B., Foruzanmehr, M., Elkoun, S., and Robert, M. (2019). Novel approach for silane treatment of flax fiber to improve the interfacial adhesion in flax/bio epoxy composites. *Journal of Composite Materials* 53 (16): 2229–2238.

44 Tarrés, Q., Vilaseca, F., Herrera-Franco, P.J. et al. (2019). Interface and micromechanical characterization of tensile strength of bio-based composites from polypropylene and henequen strands. *Industrial Crops and Products* 132: 319–326.

45 Hajlane, A., Kaddami, H., and Joffe, R. (2017). Chemical modification of regenerated cellulose fibres by cellulose nano-crystals: towards hierarchical structure for structural composites reinforcement. *Industrial Crops and Products* 100: 41–50.

46 López, J.P., Boufi, S., El Mansouri, N.E. et al. (2012). PP composites based on mechanical pulp, deinked newspaper and jute strands: a comparative study. *Composites Part B: Engineering* 43 (8): 3453–3461.

47 Avella, M., Buzarovska, A., Errico, M.E. et al. (2009). Eco-challenges of bio-based polymer composites. *Materials* 2 (3): 911–925.

48 Faruk, O., Bledzki, A.K., Fink, H.P., and Sain, M. (2012). Biocomposites reinforced with natural fibers: 2000–2010. *Progress in Polymer Science* 37 (11): 1552–1596.

49 Mohanty, A.K., Vivekanandhan, S., Pin, J.M., and Misra, M. (2018). Composites from renewable and sustainable resources: challenges and innovations. *Science* 362 (6414): 536–542.

50 Chen, G.Q. (2010). Sustainable bioplastics for future applications. *Biotechnology Journal* 5 (11): 1117–1117.

51 Iwata, T. (2015). Biodegradable and bio-based polymers: future prospects of eco-friendly plastics. *Angewandte Chemie International Edition* 54 (11): 3210–3215.

52 Gupta, A.P., Ahmad, S., and Dev, A. (2010). Development of novel bio-based soybean oil epoxy resins as a function of hardener stoichiometry. *Polymer–Plastics Technology and Engineering* 49 (7): 657–661.

53 Ratna, D. (2001). Mechanical properties and morphology of epoxidized soyabean-oil-modified epoxy resin. *Polymer International* 50 (2): 179–184.

54 Mosiewicki, M., Borrajo, J., and Aranguren, M.I. (2005). Mechanical properties of woodflour/linseed oil resin composites. *Polymer International* 54 (5): 829–836.

55 Paluvai, N.R., Mohanty, S., and Nayak, S.K. (2015). Epoxidized castor oil toughened diglycidyl ether of bisphenol A epoxy nanocomposites: structure and property relationships. *Polymers for Advanced Technologies* 26 (12): 1575–1586.

56 Manthey, N.W., Cardona, F., Francucci, G., and Aravinthan, T. (2013). Thermo-mechanical properties of epoxidized hemp oil-based bioresins and biocomposites. *Journal of Reinforced Plastics and Composites* 32 (19): 1444–1456.

57 Koyanagi, J. and Sato, M. (2019). Time and temperature dependence of transverse tensile failure of unidirectional carbon fiber-reinforced polymer matrix composites. In: *Creep and Fatigue in Polymer Matrix Composites*, 253–268. Woodhead Publishing.

58 Lotfi, A., Li, H., Dao, D.V., and Prusty, G. (2019). Natural fiber-reinforced composites: a review on material, manufacturing, and machinability. *Journal of Thermoplastic Composite Materials* https://doi.org/10.1177/0892705719844546.

59 Akpan, E.I., Wetzel, B., and Friedrich, K. (2018). A fully biobased tribology material based on acrylic resin and short wood fibres. *Tribology International* 120: 381–390.

60 Herrera-Franco, P.J. and Valadez-Gonzalez, A. (2004). Mechanical properties of continuous natural fibre-reinforced polymer composites. *Composites Part A: Applied Science and Manufacturing* 35 (3): 339–345.

61 Scarponi, C., Sarasini, F., Tirillo, J. et al. (2016). Low-velocity impact behaviour of hemp fibre reinforced bio-based epoxy laminates. *Composites Part B: Engineering* 91: 162–168.

62 Adekunle, K., Åkesson, D., and Skrifvars, M. (2010). Biobased composites prepared by compression molding with a novel thermoset resin from soybean oil and a natural-fiber reinforcement. *Journal of Applied Polymer Science* 116 (3): 1759–1765.

63 Beg, M.D.H., Pickering, K.L., and Weal, S.J. (2005). Corn gluten meal as a biodegradable matrix material in wood fibre reinforced composites. *Materials Science and Engineering: A* 412 (1–2): 7–11.

64 Mendez, J.A., Vilaseca, F., Pelach, M.A. et al. (2007). Evaluation of the reinforcing effect of ground wood pulp in the preparation of polypropylene-based composites coupled with maleic anhydride grafted polypropylene. *Journal of Applied Polymer Science* 105 (6): 3588–3596.

65 Christian, S.J. and Billington, S.L. (2011). Mechanical response of PHB- and cellulose acetate natural fiber-reinforced composites for construction applications. *Composites Part B: Engineering* 42 (7): 1920–1928.

66 Taylor, C., Amiri, A., Paramarta, A. et al. (2017). Development and weatherability of bio-based composites of structural quality using flax fiber and epoxidized sucrose soyate. *Materials & Design* 113: 17–26. https://doi.org/10.1016/j.matdes.2016.10.002.

67 Sun, W., Tajvidi, M., Hunt, C.G. et al. (2019). Fully bio-based hybrid composites made of wood, fungal mycelium and cellulose nanofibrils. *Scientific Reports* 9 (1): 3766.

68 Cao, H., Wool, R.P., Bonanno, P. et al. (2014). Development and evaluation of apparel and footwear made from renewable bio-based materials. *International Journal of Fashion Design, Technology and Education* 7 (1): 21–30.

69 Chandramohan, D. and Kumar, A.J.P. (2017). Experimental data on the properties of natural fiber particle reinforced polymer composite material. *Data in Brief* 13: 460–468.

70 Sahari, J., Sapuan, S.M., Zainudin, E.S., and Maleque, M.A. (2013). Mechanical and thermal properties of environmentally friendly composites derived from sugar palm tree. *Materials & Design* 49: 285–289.

71 Wardhono, E.Y., Kanani, N., Alfirano, A., and Rahmayetti, R. (2019). Development of polylactic acid (PLA) bio-composite films reinforced with bacterial cellulose nanocrystals (BCNC) without any surface modification. *Journal of Dispersion Science and Technology*: 1–8. https://doi.org/10.1080/01932691.2019.1626739.

72 Ferdosian, F., Pan, Z., Gao, G., and Zhao, B. (2017). Bio-based adhesives and evaluation for wood composites application. *Polymers* 9 (2): 70.

73 Ayrilmis, N., Benthien, J.T., Thoemen, H., and White, R.H. (2012). Effects of fire retardants on physical, mechanical, and fire properties of flat-pressed WPCs. *European Journal of Wood and Wood Products* 70: 215–224.

74 Najafi, S.K. (2013). Use of recycled plastics in wood plastic composites – a review. *Waste Management* 33 (9): 1898–1905.

75 Ratanawilai, T. and Taneerat, K. (2018). Alternative polymeric matrices for wood-plastic composites: effects on mechanical properties and resistance to natural weathering. *Construction and Building Materials* 172: 349–357.

76 Turku, I., Kärki, T., and Puurtinen, A. (2018). Durability of wood plastic composites manufactured from recycled plastic. *Heliyon* 4 (3): e00559.

77 Biron, M. (2013). Thermoplastic composites. In: *Thermoplastics and Thermoplastic Composites*, 769–829. William Andrew.

78 Taufiq, M.J., Mansor, M.R., and Mustafa, Z. (2018). Characterisation of wood plastic composite manufactured from kenaf fibre reinforced recycled-unused plastic blend. *Composite Structures* 189: 510–515.

79 Samariha, A., Bastani, A., Nemati, M. et al. (2013). Investigation of the mechanical properties of bagasse flour/polypropylene composites. *Mechanics of Composite Materials* 49 (4): 447–454.

80 Takatani, M. and Okamoto, T. (2008). Wood/plastic composite of high filler content. *Molecular Crystals and Liquid Crystals* 483 (1): 326–338.

81 Li, X., Lei, B., Lin, Z. et al. (2014). The utilization of bamboo charcoal enhances wood plastic composites with excellent mechanical and thermal properties. *Materials & Design* 53: 419–424.

82 Zhang, X., Hao, X., Hao, J., and Wang, Q. (2018). Heat transfer and mechanical properties of wood–plastic composites filled with flake graphite. *Thermochimica Acta* 664: 26–31.

83 Chan, C.M., Vandi, L.J., Pratt, S. et al. (2018). Composites of wood and biodegradable thermoplastics: a review. *Polymer Reviews* 58 (3): 444–494.

84 Jiang, L., He, C., Fu, J., and Wang, L. (2018). Serviceability analysis of wood–plastic composites impregnated with paraffin-based pickering emulsions in simulated sea water–acid rain conditions. *Polymer Testing* 70: 73–80.

85 Teuber, L., Osburg, V.S., Toporowski, W. et al. (2016). Wood polymer composites and their contribution to cascading utilisation. *Journal of Cleaner Production* 110: 9–15.

86 Chan, C.M., Pratt, S., Halley, P. et al. (2019). Mechanical and physical stability of polyhydroxyalkanoate (PHA)-based wood plastic composites (WPCs) under natural weathering. *Polymer Testing* 73: 214–221.

87 Turku, I., Keskisaari, A., Kärki, T. et al. (2017). Characterization of wood plastic composites manufactured from recycled plastic blends. *Composite Structures* 161: 469–476.

88 Zhang, H. (2014). Effect of a novel coupling agent, alkyl ketene dimer, on the mechanical properties of wood–plastic composites. *Materials & Design* 59: 130–134.

89 Keskisaari, A. and Kärki, T. (2018). The use of waste materials in wood-plastic composites and their impact on the profitability of the product. *Resources, Conservation and Recycling* 134: 257–261.

90 Sommerhuber, P.F., Wenker, J.L., Rüter, S., and Krause, A. (2017). Life cycle assessment of wood–plastic composites: analysing alternative materials and identifying an environmental sound end-of-life option. *Resources, Conservation and Recycling* 117: 235–248.

91 Kumar, R., Singh, T., and Singh, H. (2015). Solid waste-based hybrid natural fiber polymeric composites. *Journal of Reinforced Plastics and Composites* 34 (23): 1979–1985.

92 Haq, M., Burgueño, R., Mohanty, A.K., and Misra, M. (2014). Hybrid bio-based composites from UPE/EML blends, natural fibers, and nanoclay. *Macromolecular Materials and Engineering* 299 (11): 1306–1315.

93 Essabir, H., Bensalah, M.O., Rodrigue, D. et al. (2016). Structural, mechanical and thermal properties of bio-based hybrid composites from waste coir residues: Fibers and shell particles. *Mechanics of Materials* 93: 134–144.

94 Yusoff, R.B., Takagi, H., and Nakagaito, A.N. (2016). Tensile and flexural properties of polylactic acid-based hybrid green composites reinforced by kenaf, bamboo and coir fibers. *Industrial Crops and Products* 94: 562–573.

95 Ashori, A. (2010). Hybrid composites from waste materials. *Journal of Polymers and the Environment* 18 (1): 65–70.

96 Ashori, A. and Sheshmani, S. (2010). Hybrid composites made from recycled materials: moisture absorption and thickness swelling behavior. *Bioresource Technology* 101 (12): 4717–4720.

97 Idicula, M., Neelakantan, N.R., Oommen, Z. et al. (2005). A study of the mechanical properties of randomly oriented short banana and sisal hybrid fiber reinforced polyester composites. *Journal of Applied Polymer Science* 96 (5): 1699–1709.

98 Kazemi, M.E., Shanmugam, L., Lu, D. et al. (2019). Mechanical properties and failure modes of hybrid fiber reinforced polymer composites with a novel liquid thermoplastic resin, Elium®. *Composites Part A: Applied Science and Manufacturing* 125: 105523.

99 Blok, R., Smits, J., Gkaidatzis, R., and Teuffel, P. (2019). Bio-based composite footbridge: design, production and in situ monitoring. *Structural Engineering International* 29 (3): 453–465.

100 Peng, H., Wang, K., and Huang, Z. (2019). An injection molding method to prepare chitosan–zinc composite material for novel biodegradable flexible implant devices. *Materials and Manufacturing Processes* 34 (3): 256–261.

101 Pailoor, S., Murthy, H.N., Hadimani, P., and Sreenivasa, T.N. (2019). Effect of chopped/continuous fiber, coupling agent and fiber ratio on the mechanical properties of injection-molded jute/polypropylene composites. *Journal of Natural Fibers* 16 (1): 126–136.

102 Barletta, M., Aversa, C., Puopolo, M., and Donninelli, A. (2019). Effect of micro-lamellar talc on dimensional accuracy and stability in injection molding of PLA/PBSA blends. *Polymer–Plastics Technology and Materials* 58 (7): 776–788.

103 Bourny, V., Perez-Puyana, V., Felix, M. et al. (2017). Evaluation of the injection moulding conditions in soy/nanoclay based composites. *European Polymer Journal* 95: 539–546.

104 Barari, B., Ellingham, T.K., Qamhia, I. et al. (2016). Mechanical characterization of scalable cellulose nano-fiber based composites made using liquid composite molding process. *Composites Part B: Engineering* 84: 277–284.

105 Dweib, M.A., Hu, B., Shenton, H.W. III, and Wool, R.P. (2006). Bio-based composite roof structure: manufacturing and processing issues. *Composite Structures* 74 (4): 379–388.

106 Di Landro, L. and Janszen, G. (2014). Composites with hemp reinforcement and bio-based epoxy matrix. *Composites Part B: Engineering* 67: 220–226.

107 Lundstrom, T.S., Gebart, B.R., and Lundemo, C.Y. (1993). Void formation in RTM. *Journal of Reinforced Plastics and Composites* 12 (12): 1339–1349.

108 Lundström, T.S. and Gebart, B.R. (1994). Influence from process parameters on void formation in resin transfer molding. *Polymer Composites* 15 (1): 25–33.

109 Afendi, M., Banks, W.M., and Kirkwood, D. (2005). Bubble free resin for infusion process. *Composites Part A: Applied Science and Manufacturing* 36 (6): 739–746.

110 Mehdikhani, M., Gorbatikh, L., Verpoest, I., and Lomov, S.V. (2019). Voids in fiber-reinforced polymer composites: a review on their formation, characteristics, and effects on mechanical performance. *Journal of Composite Materials* 53 (12): 1579–1669.

111 Mathiazhagan, S. and Anup, S. (2019). Effect of interface strength on the mechanical behaviour of bio-inspired composites: a molecular dynamics study. *Mechanics of Materials* 132: 93–100.

112 Abrao, A.M., Faria, P.E., Rubio, J.C. et al. (2007). Drilling of fiber reinforced plastics: a review. *Journal of Materials Processing Technology* 186 (1–3): 1–7.

113 Panchagnula, K.K. and Palaniyandi, K. (2018). Drilling on fiber reinforced polymer/nanopolymer composite laminates: a review. *Journal of Materials Research and Technology* 7 (2): 180–189.

114 Khashaba, U.A. (2013). Drilling of polymer matrix composites: a review. *Journal of Composite Materials* 47 (15): 1817–1832.

115 Ogin, S.L., Brøndsted, P., and Zangenberg, J. (2016). Composite materials: constituents, architecture, and generic damage. In: *Modeling Damage, Fatigue and Failure of Composite Materials*, 3–23. Woodhead Publishing.

116 Charlet, K., Jernot, J.P., Gomina, M. et al. (2010). Mechanical properties of flax fibers and of the derived unidirectional composites. *Journal of Composite Materials* 44 (24): 2887–2896.

117 Goutianos, S., Van der Schueren, L., and Beauson, J. (2019). Failure mechanisms in unidirectional self-reinforced biobased composites based on high stiffness PLA fibres. *Composites Part A: Applied Science and Manufacturing* 117: 169–179.

118 Gope, P.C., Singh, V.K., and Rao, D.K. (2015). Mode I fracture toughness of bio-fiber and bio-shell particle reinforced epoxy bio-composites. *Journal of Reinforced Plastics and Composites* 34 (13): 1075–1089.

8

Recent Advances and Technologies of Biobased Composites

E. Biswas[1], S. Hawkins[1], K. Monroe[1], T. F. Garrison[2], and R. L. Quirino[1]

[1] *Chemistry and Biochemistry Department, Georgia Southern University, Statesboro, GA, USA*
[2] *Chemistry Department, King Fahd University of Petroleum and Minerals, Dhahran, Saudi Arabia*

8.1 Introduction

In recent years, there has been great activity in the research on biobased composites [1–7]. With constant fluctuation in crude oil price worldwide, the market for all petroleum-based products became highly susceptible to changes on a short-time scale. Despite different emphases on environmental concerns and oil supply over the years by different countries and governmental bodies, there has been a continuous interest in supporting the development of materials made from biorenewable resources. A growing awareness of environmental issues has resulted in diverse policies and initiatives by many governmental bodies to promote more sustainable practices. Within this context, biobased composites represent a strategic alternative to the established petroleum-based polymer and composites industry.

Polymer composites are relatively easy to prepare and allow for the incorporation of existing technologies into systems that combine the benefits of other materials. Biobased composites have been engineered to offer a range of economically competitive materials exhibiting tailored properties for specific applications, such as increased strength [8], light weight [9], and self-healing abilities [10]. More recently, the development of multifunctional, biobased materials has gained considerable attention. The importance and prevalence of biobased materials is expected to grow in the coming decades, driven by a need to address dwindling crude oil reserves, accumulation of plastic waste, and increasing levels of CO_2 in our short carbon cycle.

In this chapter, the terms "biobased composites" and "biocomposites" are used interchangeably to denote all reinforced polymer–matrix composites in which at least one of the components is from biorenewable compounds. Conversely, the term "conventional" is employed in reference to common, non-biobased systems. The focus of this chapter encompasses biobased polymer matrices with nonrenewable inorganic and synthetic organic reinforcements, conventional polymer matrices reinforced with biobased materials, and fully biobased systems (biobased reinforcements with biobased matrices). However, this chapter does not consider examples of conventional (not biobased) composites applied in biological systems.

Since the intention is to provide a brief overview of the recent advances in the research related to biobased composites, this chapter will emphasize the technical literature reporting

Biobased Composites: Processing, Characterization, Properties, and Applications, First Edition.
Anish Khan, Sanjay Mavinkere Rangappa, Suchart Siengchin and Abdullah M. Asiri.

new technologies and advances made over the course of the last decade. For contextualization purposes, discussion of seminal works is included where needed. The chapter is organized into sections that cover the recent technological advances in matrices, reinforcements, and processing/fabrication techniques of biobased polymer composites. Concluding remarks highlight potential directions for future research on biobased composites based on current trends and needs.

8.2 Recent Advances on Biobased Matrices

In a polymer composite, the matrix acts as a binder that holds reinforcement materials together and distribute the load when stress is applied to the composite. Conventional matrices, such as epoxy resins, petroleum-based polyesters, polyurethanes, and polyolefins, are typically used as structural materials in construction, automotive, biomedical, protective coatings, and food packaging industries. Biobased matrices, such as carbohydrate-derived systems, plant oil-based polymers, biobased polyesters, natural rubber, and proteins, show promise as alternative matrix materials for these applications. In biobased polymer composites reinforced with lignocellulosic materials, the matrix is more likely to be susceptible to environmental degradation due to moisture adsorption, thermal damage, and chemical exposure than in conventional composite materials. Recent advances in biobased matrices have focused on improving the performance and interfacial adhesion with reinforcing materials. As a result, stronger and more competitive materials that address several of the known issues of biobased composites have been obtained.

8.2.1 Carbohydrate-Based Matrices

The benefits of conventional reinforcements in a biobased starch matrix have been demonstrated with a number of reinforcing materials [11, 12]. Different amounts of graphene oxide have been incorporated into a pea starch matrix to produce biocomposite films with significantly improved properties in comparison to nonreinforced films. Graphene oxide was observed to form strong intermolecular interactions with the starch matrix, contributing to the compatibility between matrix and reinforcement. The addition of graphene oxide was shown to decrease UV transmittance and moisture uptake in comparison to the nonreinforced pea starch film. Furthermore, thermogravimetric analysis (TGA) demonstrated that the addition of graphene oxide improved the thermal stability of the biocomposite film. Likewise, the film's Young modulus and tensile strength also increased with the addition of graphene, while elongation at break decreased with reinforcement loading [11].

A number of other conventional materials have been suggested as reinforcement in starch matrices, including talc, which was incorporated to thermocompressed cornstarch-based films to improve the material's properties [12]. Indeed, corn starch thermoplastic films have been reinforced with various loadings of talc nanoparticles, with 3wt% resulting in an increase in yield stress and Young modulus. It was also observed that an increase in talc loading resulted in a decrease in water vapor and oxygen permeability. Furthermore, the addition of talc nanoparticles provided alternative methods for opening bags and packaging made from the material. Peeling bags were made possible and were proposed in replacement of tearing, demonstrating additional benefits for the use of talc as reinforcement [12]. A similarly structured matrix, cellulose, may also be reinforced with conventional materials such as silver nanoparticles [13]. In fact, cellulose is one of the most abundant natural polymers. It is a major component of all plant species, and due to its widespread availability and the refinement of extraction methods long employed and perfected in the paper industry, it has become a cheap resource. Cellulose also exhibits a very homogeneous structure, with consistent properties, making it an interesting biopolymer for the preparation of biobased composites, acting as either the matrix, or the reinforcement, as discussed later in the text. In a recent report, a cellulose matrix was reinforced with silver nanoparticles through a hydrothermal method using a microcrystalline cellulose solution, silver nitrate, and aluminum chloride hexahydrate. Silver nanoparticles were well dispersed over the surface and within the cellulose matrix, resulting in a composite with excellent antimicrobial properties toward both gram-negative and gram-positive bacteria [13]. The use of a polar matrix enhanced the even dispersion of the nanoparticles and represents a new direction for the fabrication of inorganic/organic hybrid materials.

8.2.2 Plant Oil-Based Matrices

Plant oils are another common biobased resource used for matrix applications. Their widespread availability combined with easy extraction methods (mechanical pressing/solvent extraction) ensure virtually unending supply and low production cost. From a chemical perspective, all vegetable oils share a very similar and simple structure, which allows for straightforward and well-defined chemical transformations. Yet, the set of three carboxyl groups and carbon–carbon double bonds (as depicted in Figure 8.1) allows for various strategies for the development of novel biobased polymers, which are often proposed as matrices in biocomposites for various applications, including construction and furniture. Virgin cooking oil monomers have been modified through acid-catalyzed ring-opening to form bioepoxies or polyols, which served as the major matrix component in composites with various loadings of acid-treated graphite. It has been demonstrated that increase in treated graphite loading yielded composites with improved Young modulus and tensile strength, and decreased elongation at break. The observed improvement in mechanical properties has been attributed to the dispersion of graphite and its interfacial adhesion with the oil-based matrix [14]. In addition, plant oil-based alkyd resins have been attractive due to their easy synthesis and low cost. In order to improve mechanical strength and thermal stability, expanded graphite was added to reinforce a blend of alkyd and epoxy resins prepared from Jatropha curcas oil. Transmission electron microscopy was used to evaluate the formation and homogeneity of the composite, showing an even dispersion of expanded graphite in the Jatropha oil matrix. Furthermore, the composite with the greatest expanded graphite loading exhibited the greatest flame retardancy, thermal stability, and tensile strength [15]. In both cases, the increase in stiffness of biocomposites prepared with vegetable oils in comparison to

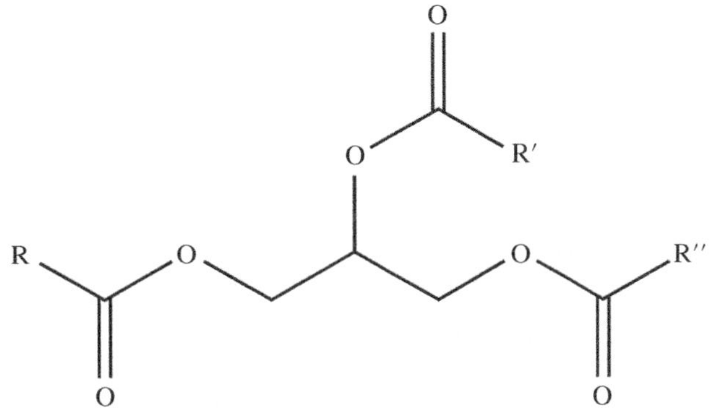

Figure 8.1 Chemical structure of vegetable oils. R, R′, and R″ are linear carbon chains that vary in length and the number of unsaturations according to the vegetable oil.

other matrices is attributed to a better graphite distribution, which is the result of fiber–matrix interactions and compatibility.

Epoxidized soybean oil (ESO) has been researched heavily as a matrix for the preparation of biobased composites [16]. In one instance, ESO has been crosslinked with polycaprolactone (PCL) to form a partially biobased thermosetting polymer network through an acid cure for application as a shape memory material [16]. Mechanical testing reveals an increase in overall maximum stress from 1.0 to 12.3 MPa with an addition of 75 wt% of PCL to ESO. In addition, shape memory recovery was observed using heat deformation [16].

ESO-based matrices have been prepared with maleate half-esters and blended with a resol resin containing *p*-tertiary butyl phenol (*p*-TBP) and *p*-nonylphenol (*p*-NP), resulting in tough thermosetting materials [16]. DMA of samples cured at 150 °C revealed a storage modulus at 30 °C of 40 MPa for samples containing *p*-TBP and 10 MPa for samples containing *p*-NP [16]. When the cure temperature was increased to 190 °C, these values increased to 1.1 GPa for ESO-*p*-TBP and 180 MPa for ESO-*p*-NP [16]. Two glass transition temperatures (T_gs) were found for each of the samples. ESO-*p*-TBP presented T_gs at 33 and 25 °C, while ESO-*p*-NP exhibited T_gs at 65 and 45 °C [16]. Lastly, maximum tensile strength was recorded at 4 MPa for ESO-*p*-TBP and at 13 MPa for ESO-*p*-NP [16]. It is apparent, with TGA, that all samples studied had a good thermal resistance, as indicated by the temperature at 5 wt% degradation (T_{95}) [16]. ESO-*p*-TBP cured at 190 °C had the highest thermal resistance, with $T_{95} = 237$ °C, while ESO-*p*-TBP cured at 150 °C had the lowest thermal resistance, with $T_{95} = 164$ °C [16].

8.2.3 Biobased Polyester Matrices

Natural polyesters like polylactic acid (PLA) and polyhydroxyalkanoates (PHAs) have been heavily studied lately as biobased matrices for composite applications. Figure 8.2 provides the generic chemical structures of PLA and PHAs. The need for a chemically and economically reliable biobased plastic with a potential to replace commodity polyolefins, such as polyethylene (PE), lead to the development of PLA. Despite its known limitations, such as low thermal stability, PLA has been used for applications as a biobased matrix in biocomposites. For example, a PLA matrix was reinforced by various percentages of hydrophobic fumed silica nanoparticles through the solution casting method to create a composite with improved mechanical properties. The composites were observed under a SEM, revealing good dispersion of silica nanoparticles throughout the matrix. The resulting composites, bearing a silica load of 0.2 wt%, exhibited improved tensile and barrier properties in comparison to the pure polymer while the properties of composites with higher silica content did not significantly improve in comparison to the neat polymer [17].

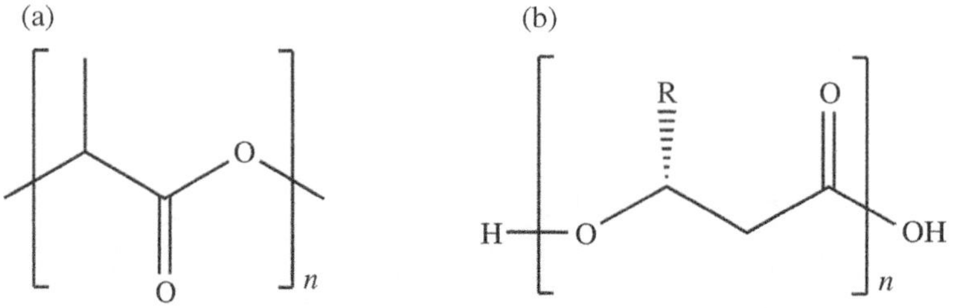

Figure 8.2 Chemical structure of (a) PLA and (b) PHAs. R represents alkyl chains of variable length.

The known limitations of PLA have created a need for the development of alternative systems, such as PHAs. PHAs are a type of plastic that can be biologically produced through various bacteria by breaking down other materials, such as waste oils and spent coffee grounds, and can be used to replace common petroleum-based plastics, with the advantage of being fully biodegradable. Due to their natural origin and the ease of genetic modification of the producing organisms to maximize production yield, PHAs have recently become an interesting bioplastic alternative. The addition of PHA to PLA can improve heat tolerance, as well as mechanical strength of pure PLA. However, it is noteworthy that pure PHA exhibits low melt strength and poor "adaptability processing," significantly limiting its widespread commercial application [18].

The addition of reinforcing agents to PHA at the nanoscale can also lead to a variety of composite materials [19]. The reinforcing effect of calcium phosphate nanoparticles in PHA composites has been investigated, revealing an improvement in compressive strength, bioactivity, and cell proliferation rate in bone tissue engineering (BTE) for patients suffering from bone disease. In addition, due to the biodegradable nature of PHA, its use in a BTE composite is ideal since the structure works as a temporary scaffold [20]. PHA nanocomposites have also been found to have the potential to be used for applications, such as a drug delivery system for humans [20], as well as artistry and clothing [21].

Chemical modification of PHA has been employed to improve interfacial adhesion between matrix and biobased reinforcements, such as palm fiber. More specifically, the grafting of maleic anhydride onto PHA chains has led to stronger fiber–matrix interactions in palm fiber composites in which the fibers were treated with a coupling agent. The resulting PHA-g-MA/TPF composites exhibited highly homogeneous dispersions of palm fiber in the PHA matrix. The increased compatibility due to maleic anhydride grafting and treatment of the palm fibers resulted in a significantly higher tensile strength and water resistance compared to composites without maleic anhydride grafting or fiber treatment [22]. In a similar study, maleic anhydride grafting of PHA and treatment of tea plant fibers resulted in PHA-g-MA/*t*-TPF composites that also exhibited similar increases in compatibility, tensile strength, and water resistance in comparison to pure PHA and PHA/PF composites [23].

8.2.4 Natural Rubber

Natural rubber has been long known and exhibits unique elastomeric properties. Its application as a biobased matrix is of great interest. The reinforcement of natural rubber with functionalized carbon nanotubes results in a great improvement of mechanical properties. Natural rubber was indeed reinforced with acid-treated multiwalled carbon nanotubes (MWCNTs). The MWCNTs were first functionalized with a blend of acids and poly(diallyldimethylammonium chloride), and then added to latex particles. The resulting composites exhibited a homogenous dispersion of carbon nanotubes and strong adhesion between the natural rubber matrix and reinforcement, a property that was largely responsible for significant improvement of mechanical properties. The composite was found to demonstrate significantly increased tensile strength, higher glass transition temperature, and greater thermal stability when compared to the pure natural rubber polymer [24].

8.2.5 Collagen

One of the most abundant animal proteins, collagen, has also been used as an effective matrix in biobased composites, exhibiting excellent properties. The isolation of collagen from animal tissues can be achieved in high yields with a series of simple steps [25]. A matrix with high crosslink density can be obtained by the use of polyethylenimine as a crosslinking agent, resulting in a significant improvement of the mechanical properties. When collagen hydrogels were reinforced with polyethylenimine-functionalized single-walled carbon nanotubes (PEI-SWNT), Raman spectroscopy demonstrated that the SWNT network was not crosslinked. The mechanical properties of the collagen/PEI-SWNT composites were greatly improved. Indeed, an improvement in Young modulus of 4.5–9 times was reported for the composites in comparison to PEI-crosslinked collagen hydrogels or collagen/SWNT composites [26].

8.3 Recent Advances on Biobased Reinforcements

It is well understood that reinforcements are critical for the enhancement of biobased materials' properties, with great advances achieved in the use of biobased reinforcements and hybrid composites. Currently, wood fibers and pulp are among the most promising biobased reinforcements, which have been recently combined with different plant-based oils and resins in industrial, agricultural, and medical applications. Unlike petrochemical and commonly used conventional materials, biocomposites often lack consistent properties, such as thermal stability and durability, compromising their widespread industrial use. In the past decade, there has been great improvement in the development of biobased materials with more competitive properties as detailed in the next few sections. With the need to reduce the global usage of petrochemicals, advancements in technologies that make use of bioderived and biorenewable resources have become critical to attain sustainability.

8.3.1 Biobased Fiber Reinforcements

The addition of flax and basalt fibers to a poly malic–lactic acid (PMLA) matrix resulted in better properties than a PMLA composite prepared with flax reinforcement alone [27]. The natural flax fiber has a density of 1.45 g/cm^3 and hardness comparable to glass fibers. The flax fibers were impregnated with malic-lactic acid prior to polymerization. Composites with loads varying from 40 to 60 wt% were prepared and their properties, namely tensile strength, flexural strength, and water absorption, were assessed. It was observed that both flexural modulus and strength decreased with an increase in reinforcement loading. Indeed, the flexural modulus decreased from 10 to 7 GPa while the strength decreased from 86 to 50 MPa as the loading increased from 40 to 60 wt%. It was concluded that the ideal flax fiber load was 40 wt%. For comparison purposes, prior research have shown that an ESO resin reinforced with a nonwoven flax, an acrylated ESO resin reinforced with flax, and an acrylated modified soybean oil resin reinforced with flax, gave flexural modulus and strength of 0.66–0.94 GPa and 125–129 MPa, 1.2 GPa and 27 MPa, and 6.9 GPa and 98 MPa, respectively [16]. The chemical structures of ESO and a modified acrylated ESO are provided in Figure 8.3. The flexural results for ESO-based and PMLA-based composites are comparable. It should be noted that tensile results were similar to those obtained for biocomposites reinforced with inorganic fibers and also indicate 40 wt% as the ideal fiber load.

A PMLA matrix has also been reinforced with a blend of biobased flax fibers and mineral-based basalt fibers to make a hybrid composite material [27]. The addition of the mineral basalt fibers improved tensile and flexural modulus, and interfacial shear strength and strain [27]. The hybrid composites had a performance comparable to that of glass fiber composites [27]. For flax/basalt loadings of 40, 50, and 60 wt%, the flexural modulus obtained

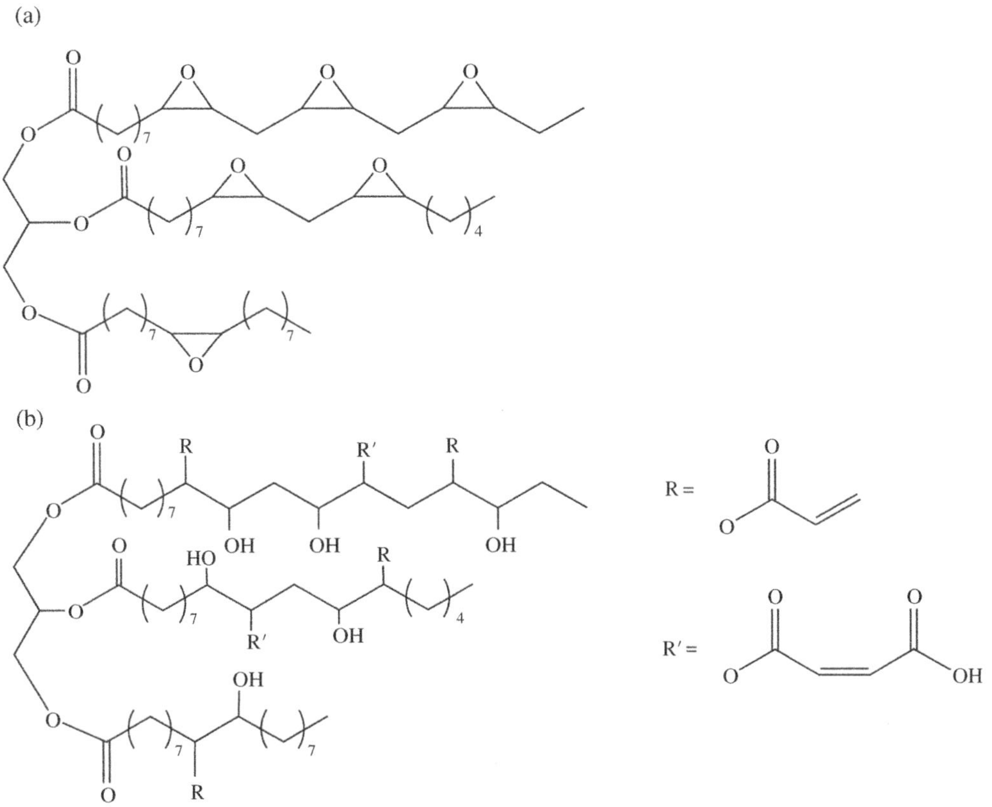

Figure 8.3 (a) Epoxidized soybean oil (ESO). (b) Acrylated epoxidized soybean oil modified with maleic anhydride.

was 10.8, 11.1, and 14.0 GPa, respectively, while the flexural strength was 59, 66, and 75 MPa, respectively, with strains varying between 3.3 and 1% [27]. Significant water absorption was observed for composites with increasing loads of fiber reinforcement, with overall optimum results obtained for 40 wt% fiber loading [27].

Natural hemp fiber was used as reinforcement in a biobased epoxy laminate at a 40 vol% load [28]. Due to the hydrophilic nature of the biocomposite, a water absorption test revealed a 11% increase in weight over 30 days. Despite the issues associated to water absorption, good mechanical properties were obtained, as indicated by a maximum tensile stress of 63 MPa, elongation at break of 1.8%, and Young modulus of 5.9 GPa [28].

Alternatively, traditional paper processing techniques can also be applied to nontraditional sources of natural fibers, such as the fibers from yellow nutsedge (*Cyperus esculentus*), to make low-cost nonwoven fabric materials for biobased polymer composites. The rolls of the nonwoven fibers made in this manner have low densities, are porous, and are easy to handle [29]. Nonwoven fabrics may also be prepared by blending kenaf [30] and flax fibers [31, 32] with binder fibers to improve cohesion. Polymers used as binder fibers include polypropylene and poly(vinyl alcohol).

Despite the environmental advantages biobased matrices with biobased reinforcements have, their mechanical properties are often inferior to many synthetic materials. To lessen the environmental impact of synthetic materials and to move toward environmentally friendly composites, biobased reinforcements have often been used in conventional matrices. Most commonly, plant fibers, such as hemp, tree bark, and plant leaf have been used as reinforcement in synthetic matrices.

Similarly, other biobased fibers can be used as reinforcement for conventional polymers, such as HDPE. It has been shown that HDPE matrices reinforced with tree bark fibers of varying coarse size, load, and type of bark (black spruce or aspen) produce excellent biocomposites. For example, composites reinforced with black spruce exhibited superior strength, and displayed a more brittle behavior than those reinforced with aspen bark, while composites prepared with either type of bark exhibited inferior strength in comparison to composites reinforced with wood flour [33].

A number of plant leaf fibers have been used as an effective reinforcement for conventional matrices. Lignocellulosic materials have been traditionally used as reinforcement in petroleum-based polymers, such as polystyrene, polypropylene, and other polymers with low melting temperatures. Indeed, polymers with low melting temperatures are often required for the use of plant leaf fibers as reinforcement due to their low thermal stability. However, pineapple leaf fibers exhibit a low lignin content and therefore can be used as reinforcement for a wider range of materials, including nylon. 20 wt% of alkaline-treated, silane-treated, and untreated pineapple fibers were added to a nylon matrix and synchrotron wide angle X-ray scattering showed that both the pineapple leaf fibers and the nylon crystallites exhibited high levels of orientation, attributed to the polar nature of both pineapple leaf fibers and the nylon matrix. These high levels of orientation contributed to the significant improvement in the mechanical properties of the composites. The composites exhibited superior tensile and flexural strength in comparison to the pure matrix, with the greatest improvements observed in the alkaline-treated and silane-treated pineapple leaf fiber composites. A similar high degree of orientation and improved mechanical properties have also been observed in curaua/nylon and kenaf/nylon composites. The improved performance of the composites may also be explained by the polar nature of both the nylon matrix and the biobased reinforcement [34].

8.3.2 Wood Biochar-Based Reinforcements

Though wood flour has been studied as an excellent reinforcement for both biobased and conventional matrices, other wood products have also been evaluated as reinforcement in the preparation of biobased composites. A relatively new method of utilizing wood-derived materials as reinforcement is through the addition of wood biochar to polypropylene with maleated anhydride polypropylene as a compatibilizer between the matrix and reinforcement. The pyrolysis of waste wood produces highly carbonated wood biochar with high surface area, making it an excellent reinforcement for wood/polypropylene composites. Composites containing 12 and 18 wt% biochar exhibited the greatest improvement in ductility and thermal stability, while composites containing 24 wt% biochar exhibited the greatest improvement in tensile strength [35]. Biobased composites can also be obtained by the reinforcement of a biobased matrix with conventional materials that can be specifically designed to impart desired properties, such as set glass transition temperatures, hydrophobicity, hydrophilicity, etc.

8.3.3 Biobased Nanocomposite Reinforcements

8.3.3.1 Cellulose Nanocomposites

As a semi-crystalline polymer (Figure 8.4), natural cellulose is formed of both amorphous and crystalline domains. Cellulose nanocrystals (CNC) are a common nanostructure

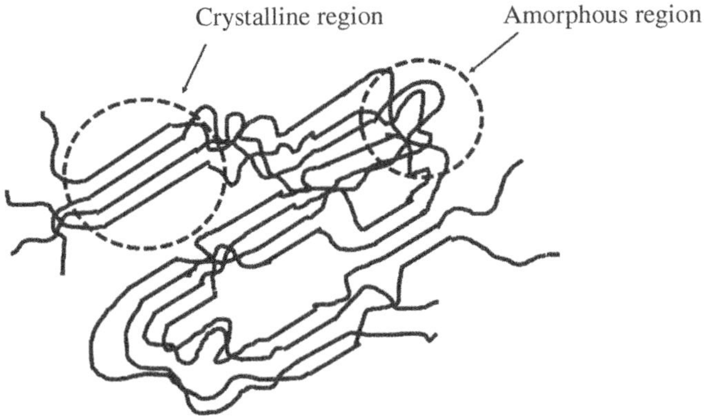

Figure 8.4 Crystalline and amorphous regions of lignin.

obtained from the partial hydrolysis of cellulose. Upon mild hydrolysis conditions, amorphous cellulose gets hydrolyzed, leaving the crystalline domains, often denoted as cellulose nanowhiskers or nanocrystals. The addition of CNC to a variety of resins leads to biocomposites. For example, CNC-reinforced furan-based resin was used as a more environmentally safe coating for aluminum surfaces than phenolic coatings. In such systems, CNC helped improve the mechanical properties [36]. Likewise, the addition of 0.1 wt% of nanocellulose to carbonized polyacrylonitrile-based carbon fibers resulted in a 19% increase in tensile modulus and a 27% increase in tensile strength [37].

In other applications, the addition of cellulose nanorods to common plastics has been used as a strategy for the preparation of light, flexible, and transparent materials for use in electronics. Adding a network of cellulose nanorods to said plastics has been shown to highly improve thermal stability [38]. Cellulose nanofibers have also been used as reinforcement for PLA to overcome some of the shortcomings of pure PLA, such as low melt strength. However, there are still significant issues with using nanocellulose as reinforcement for PLA. As cellulose has high hydrophilicity, PLA is not chemically compatible with it, therefore compromising dispersion of the nanoparticles [39].

8.3.3.2 Other Nanocomposites

Another approach is preparing nanostructures from lignin for use as reinforcement in biocomposites [40]. Other biobased nanocomposites include PLA reinforced with Eucalyptus sawdust nanoparticles. The use of the nanoparticles resulted in an improvement of mechanical properties, as well as properties related to degradation [41].

8.4 Recent Advances on Biobased Composite Processing

In addition to the recent developments in the use of biobased fibers and matrices for making polymer composites, there has been significant advances in technology for the processing of biobased composites. A wide range of processing techniques are currently being used for making biobased polymer composites, including automated spray lay-up [42], wet lay-up [31, 32], vacuum resin infusion [43, 44], pultrusion [45], extrusion/injection molding [46], and resin transfer molding [47]. These well-established fabrication techniques continue to see widespread use with continued refinements and improvements in their application to natural fibers. Furthermore, advanced processing techniques like additive manufacturing, are nowadays being applied to natural fibers. The following section will briefly highlight the latest developments in the fabrication processes.

8.4.1 Extrusion and Injection Molding Techniques

Conventional melt processing of natural fiber reinforcements with thermoplastic resins using an extruder, followed by injection molding are important industrial processing techniques for biobased composites [46]. Major advantages include continuous and large-scale production capabilities. Although there are a number of different types of extruders, the most commonly used, industrially, are twin-screw extruders [48, 49]. Recent models of flow behavior in injection molding of natural fiber-based composites show that nonbranched fibers have better flowability and homogeneity compared to branched fibers [50]. Natural fibers are prone to damage from the high temperatures and stresses experienced during the compounding process [51, 52]. Recently, there have been considerable research efforts on the characterization, analysis, and modeling of damage to lignocellulosic fibers by twin-screw extrusion [48, 49, 53]. Different breakage behaviors are observed depending on the botanical source of the fibers [53]. Furthermore, changes to the dimensions of the lignocellulosic fibers are also influenced by the fiber sources [48] and the processing conditions [49]. A separate study showed that increased screw speed improves the dispersion of natural fibers, but also increases barrel temperatures resulting in thermal degradation of the fibers [54]. Limitations of extrusion/injection molding include the size of the natural fibers (restricted to 2 mm or less), as well as an upper limit for total fiber reinforcement of 50 vol% [31]. Furthermore, preprocessing treatments of fiber reinforcement are generally required in order to meet the particle size and/or density requirements [55].

8.4.2 Wet Lay-Up Techniques

Composite preparation by wet lay-up exhibit several advantages in comparison to other techniques, including cost, high fiber loadings, ease of use, and a broad range of compatible materials. Wet lay-up is being applied to biobased fiber reinforcements, such as nonwoven yellow nutsedge-based fiber rolls [29] and nonwoven flax [31, 32], to obtain higher fiber loadings than the ones that can be achieved with other techniques. Target applications of the wet-laid composites from nonwoven fabrics include thermal and acoustic insulation for the construction [31, 32] or automotive industries [29].

8.4.3 3D Printing of Biobased Composites

Additive manufacturing, also known as 3D printing, is now being used to produce biobased polymer composites. Fused deposition modeling (FDM) is a form of additive manufacturing that has garnered the most attention for use with biobased materials. FDM is sometimes referred to as fused filament fabrication (FFF), and may be used to 3D print filaments in layers without the use of molds [56, 57]. Advantages of FDM include low cost, operating speed, and design flexibility. A very recent study reported the optimization and control of process parameters for using a material extrusion technique similar to FDM for the manufacture of large-scale objects [58].

PLA is commonly used as a thermoplastic resin for 3D printing with FDM [59]. Recently, there have been efforts in the direction of blending PLA with other biobased materials to improve the properties of the printed thermoplastic filaments. In one study, PLA was blended with up to 5 wt% of inexpensive Kraft lignin (sulfate lignin), resulting in thermoplastic filaments with a similar Young modulus to neat PLA, but also decreased tensile strength and percent elongation [60]. Wu and Tsou studied the grafting of acrylic acid onto PLA and blending with rice husks chemically treated to improve compatibility between the filler and PLA in the printed materials [59]. Their findings showed that the modifications improved both the interfacial adhesion and the tensile properties of the filaments.

Besides PLA, other biobased thermoplastic resins are being used for 3D printing with FDM. Biobased PE, produced commercially by Braskem (Sao Paulo, Brazil), was blended with thermomechanical pulp fibers and then printed using FDM to produce composites with better tensile properties than the neat bio-PE [61]. Despite the weaker tensile properties of FDM-printed biocomposites in comparison to compression molded composites, there may be other potential advantages. For example, a different study investigated the potential for using Woodfill Fine, a commercially available material from ColorFabb containing a blend of PLA, poly(hydroxyalkanoate) (PHA), and recycled wood fibers, to 3D print filaments with a moisture gradient-induced bending response [62]. The controlled 3D printing of hygroscopic filaments may enable an actuated bending response to moisture for targeted applications. The hygromorphic biobased composites mimic the behavior of naturally occurring bilayer microstructural actuators found in pine cones that are used for seed dispersal [62].

The FDM technique has been enhanced by the printing of continuous natural fibers with PLA. Jute yarn was one of the first continuous natural fibers to be 3D printed via FFD with a PLA matrix using in-nozzle impregnation [57]. Another recent study has reported the use of FFD to 3D print continuous flax fibers as reinforcement with a PLA matrix as a potential material for structural applications. The resulting biocomposites had comparable tensile properties as similar 3D printed composites made from continuous glass fibers and polyamide [56].

8.5 Conclusion

The advancements made in the field of biobased composites in the last decade, as shown in this chapter, suggest that a great variety of materials can be designed and prepared from biorenewable resources with interesting properties that grant them potential to replace current petroleum-based polymers and composites. Some of the key impediments for the more widespread use of biocomposites at an industrial scale are the variability in material properties, and the variation in the local availability of biorenewable resources. These hurdles, along with other economic aspects, can be overcome in the near future as the biocomposites industry matures and specializes in key applications. After all, many of these hurdles have been experienced by the crude oil industry before its widespread establishment in the early twentieth century. The somewhat recent concept of biorefineries is interesting due to the diversification of products obtained by a single resource, but the idea is limited because it tries to mimic the petroleum stream. A more promising approach for making biocomposites more competitive might involve the development of specialized materials, with target applications and consistent properties.

In the realm of biobased polymer composites, PLA- and ESO-based resins have been consolidated as promising matrices, while a great variety of natural fibers and other materials have been studied as reinforcements. Due to its consistent structure and properties, cellulose shows great promise as a reinforcing agent. Advancements in cellulose processing have led to cellulose nanoparticles, such as cellulose nanowhiskers and nanocrystals, which have helped improve final biocomposite properties. The importance and prevalence of biobased materials is expected to continue growing over the next few decades, benefitting from government policies to incentivize the development of more sustainable practices.

References

1 Wool, R.P. and Sun, X.S. (2005). *Bio-Based Polymers and Composites*. Boston: Elsevier Academic Press.

2 Avella, M., Buzarovska, A., Errico, M. et al. (2009). Eco-challenges of bio-based polymer composites. *Materials* 2 (3): 911–925.

3 Quirino, R.L., Garrison, T.F., and Kessler, M.R. (2014). Matrices from vegetable oils, cashew nut shell liquid, and other relevant systems for biocomposite applications. *Green Chem.* 16 (4): 1700–1715.

4 Zhang, C., Quirino, R.L., and Sun, J. (2018). Biobased polymers and composites. *Int. J. Polym. Sci.* 2018: 1–1.

5 Quirino, R.L. and Kessler, M. (2015). Vegetable oil-based resins and composites. In: *Sustainable Composites and Advanced Materials*, 51–86. Lancaster: DEStech Publications.

6 Johns, A. and Quirino, R.L. (2017). Free-radical and ionic polymerisation of regular and conjugated soybean oil for sustainable bioplastics and biocomposites. In: *Soy-Based Bioplastics*, 309–321. Akron: Smithers Rapra Publisher.

7 Murawski, A., Diaz, R., Inglesby, S. et al. (2019). Synthesis of bio-based polymer composites: fabrication, fillers, properties and challenges. In: *Polymer Nanocomposites in Biomedical Engineering*, 29–55. Basel: Springer Nature Switzerland AG.

8 Quirino, R.L., Woodford, J., and Larock, R.C. (2012). Soybean and linseed oil-based composites reinforced with wood flour and wood fibers. *J. Appl. Polym. Sci.* 124: 1520–1528.

9 Stevulova, N., Kidalova, L., Cigasova, J. et al. (2013). Lightweight composites containing hemp hurds. *Proc. Eng.* 65: 69–74.

10 Hondred, P.R., Salat, L., Mangler, J., and Kessler, M.R. (2014). Tung oil-based thermosetting polymers for self-healing applications. *J. Appl. Polym. Sci.* 131 (12): 40406.

11 Li, R., Liu, C., and Ma, J. (2011). Studies on the properties of graphene oxide-reinforced starch biocomposites. *Carbohydr. Polym.* 84: 631–637.

12 López, O., Castillo, L., García, M. et al. (2015). Food packaging bags based on thermoplastic corn starch reinforced with talc nanoparticles. *Food Hydrocoll.* 43: 18–24.

13 Dong, Y.Y., Fu, L.H., Liu, S. et al. (2015). Silver-reinforced cellulose hybrids with enhanced antibacterial activity: synthesis, characterization, and mechanism. *RSC Adv.* 5 (118): 97359–97366.

14 Rus, A.Z.M., Abdullah, N.M., Abdullah, M.F.L., and Idris, M.I.F. (2015). Graphite/bio-based epoxy composites: the mechanical properties interface. *Appl. Mech. Mater.* 799–800: 115–119.

15 Gogoi, P., Boruah, M., Bora, C., and Dolui, S.K. (2014). Jatropha curcas oil based alkyd/epoxy resin/expanded graphite (EG) reinforced bio-composite: evaluation of the thermal, mechanical and flame retardancy properties. *Prog. Org. Coat.* 77 (1): 87–93.

16 Baroncini, E.A., Yadav, S.K., Palmese, G.R., and Stanzione, J.F. (2016). Recent advances in bio-based epoxy resins and bio-based epoxy curing agents. *J. Appl. Polym. Sci.* 133 (45): 44103.

17 Pilic, B., Radusin, T., Ristic, I. et al. (2016). Hydrophobic silica nanoparticles as reinforcing filler for poly(lactic acid) polymer matrix. *Hem. Ind.* 70 (1): 73–80.

18 Xu, Y., Li, W., and Liang, K. (2016). High heat-resistant PLA/PHA composite and preparation method thereof. Chinese Patent CN106,467,657-A.

19 Nielsen, C.L. (2018). Investigations of polyhydroxyalkanoate secretion and production using sustainable carbon sources. Master thesis. Utah State University.

20 Cichoń, E., Haraźna, K., Skibiński, S. et al. (2019). Novel bioresorbable tricalcium phosphate/polyhydroxyoctanoate (TCP/PHO) composites as scaffolds for bone tissue engineering applications. *J. Mech. Behav. Biomed. Mater.* 98: 235–245.

21 Feng, X., Di, S., Chen, Z., Chen, M., Pan, Z. (2016). A kind of PHA/PLA composite fibers imitate spun silk silk shell fabric and preparation method thereof. Chinese Patent CN107,541,842-A.

22 Wu, C., Liao, H., and Cai, Y. (2017). Characterisation biodegradability and application of palm fibre-reinforced polyhydroxyalkanoate composites. *Polym. Degrad. Stab.* 140: 55–63.

23 Wu, C. (2013). Preparation, characterization and biodegradability of crosslinked tea plant-fibre-reinforced polyhydroxyalkanoate composites. *Polym. Degrad. Stab.* 98 (8): 1473–1480.

24 Peng, Z., Feng, C., Luo, Y. et al. (2010). Self-assembled natural rubber/multi-walled carbon nanotube composites using latex compounding techniques. *Carbon* 48 (15): 4497–4503.

25 Scholz, A., Lewis, R.L., Bachan, M. et al. (2017). Biocomposites from the reinforcement of a tung oil-based thermosetting resin with collagen. *Mat. Chem. Frontier* 1 (1): 1795–1803.

26 Homenick, C., Sheardown, H., and Adronov, A. (2010). Reinforcement of collagen with covalently-functionalized single-walled carbon nanotube crosslinkers. *J. Mater. Chem.* 20: 2887–2894.

27 Bakare, F.O., Ramamoorthy, S.K., Åkesson, D., and Skrifvars, M. (2016). Thermomechanical properties of bio-based composites made from a lactic acid thermoset resin and flax and flax/basalt fibre reinforcements. *Comp. Part A: Appl. Sci. Manufact.* 83: 176–184.

28 Landro, L.D. and Janszen, G. (2014). Composites with hemp reinforcement and bio-based epoxy matrix. *Comp. Part B: Eng.* 67: 220–226.

29 Carbonell, A., Boronat, T., Fages, E. et al. (2015). Wet-laid technique with *Cyperus esculentus*: development, manufacturing and characterization of a new composite. *Mater. Des.* 86: 887–893.

30 Hao, A., Zhao, H., and Chen, J.Y. (2013). Kenaf/polypropylene nonwoven composites: the influence of manufacturing conditions on mechanical, thermal, and acoustical performance. *Comp. Part B: Eng.* 54: 44–51.

31 Fages, E., Gironés, S., Sánchez-Nacher, L. et al. (2012). Use of wet-laid techniques to form flax-polypropylene nonwovens as base substrates for eco-friendly composites by using hot-press molding. *Polym. Compos.* 33: 253–261.

32. Fages, E., Cano, M., Gironés, S. et al. (2013). The use of wet-laid techniques to obtain flax nonwovens with different thermoplastic binding fibers for technical insulation applications. *Text. Res. J.* 83: 426–437.

33 Yemele, M.C.N., Koubaa, A., Cloutier, A. et al. (2010). Effect of bark fiber content and size on the mechanical properties of bark/HDPE composites. *Comp. Part A: Appl. Sci. Manufact.* 41 (1): 131–137.

34 Nopparut, A. and Amornsakchai, T. (2016). Influence of pineapple leaf fiber and it's surface treatment on molecular orientation in, and mechanical properties of, injection molded nylon composites. *Polym. Test.* 52: 141–149.

35 Das, O., Sarmah, A.K., and Bhattacharyya, D. (2015). A novel approach in organic waste utilization through biochar addition in wood/polypropylene composites. *Waste Manag.* 38: 132–140.

36 Asaro, L., Seoane, I.T., Fasce, L.A. et al. (2019). Development of low environmental impact protective coatings based on a furan resin and cellulose nanocrystals. *Prog. Org. Coat.* 133: 229–236.

37 Jiang, E., Maghe, M., Zohdi, N. et al. (2019). Influence of different nanocellulose additives on processing and performance of PAN-based carbon fibers. *ACS Omega* 4 (6): 9720–9730.

38 Biswas, S.K., Sano, H., Yang, X.P. et al. (2019). Highly thermal-resilient AgNW transparent electrode and optical device on thermomechanically superstable cellulose nanorod-reinforced nanocomposites. *Adv. Opt. Mater.* 7 (15): 201900532.

39 Vatansever, E., Arslan, D., and Nofar, M. (2019). Polylactide cellulose-based nanocomposites. *Int. J. Biol. Macromol.* 137: 912–938.

40 Addepally, U., Gandham, V., Palety, K.K., and Kanakaraju, Y. (2019). Lignin-based carbon nanomaterials – the future scope. *Mater. Perform. Charact.* 8 (3): 20180153.

41 Silva, C.G., Campini, P.A.L., Rocha, D.B., and Rosa, D.S. (2019). The influence of treated eucalyptus microfibers on the properties of PLA biocomposites. *Compos. Sci. Technol.* 179: 54–62.

42 Hanafee, Z.M., Khalina, A., Norkhairunnisa, M. et al. (2019). The effect of different linear robot travel speed on mass flowrate of pineapple leaf fiber (PALF) automated spray up composite. *Comp. Part B: Eng.* 156: 220–228.

43 Shah, D.U., Schubel, P.J., Clifford, M.J., and Licence, P. (2014). Mechanical property characterization of aligned plant yarn reinforced thermoset matrix composites manufactured via vacuum infusion. *Polym.Plast. Technol. Eng.* 53: 239–253.

44 Afolabi, L.O., Megat-Yusoff, P.S.M., Ariff, Z.M., and Hamizol, M.S. (2019). Fabrication of *Pandanus tectorius* (screw-pine) natural fiber using vacuum resin infusion for polymer composite application. *J. Mater. Res. Technol.* 8: 3102–3113.

45 Yang, Y., Murakami, M., and Hamada, H. (2012). Molding method, thermal and mechanical properties of jute/PLA injection molding. *J. Polym. Environ.* 20: 1124–1133.

46 Marinho, J.F., Braga, N.F., Krohn, A. et al. (2015). Melt processing of polymer biocomposites. *Polímeros* 25: 133–136.

47 Ashworth, S., Rongong, J., Wilson, P., and Meredith, J. (2016). Mechanical and damping properties of resin transfer moulded jute–carbon hybrid composites. *Compos. Part B: Eng.* 105: 60–66.

48 Giuseppe, E.D., Castellani, R., Budtova, T., and Vergnes, B. (2017). Lignocellulosic fiber breakage in a molten polymer. Part 2. Quantitative analysis of the breakage mechanisms during compounding. *Comp. Part A: Appl. Sci. Manufact.* 95: 31–39.

49 Berzin, F., Beaugrand, J., Dobosz, S. et al. (2017). Lignocellulosic fiber breakage in a molten polymer. Part 3. Modeling of the dimensional change of the fibers during compounding by twin screw extrusion. *Comp. Part A: Appl. Sci. Manufact.* 101: 422–431.

50 Elsabbagh, A., Ramzy, A., Steuernagel, L., and Ziegmann, G. (2019). Models of flow behaviour and fibre distribution of injected moulded polypropylene reinforced with natural fibre composites. *Comp. Part B: Eng.* 162: 198–205.

51 Berzin, F., Amornsakchai, T., Lemaitre, A. et al. (2018). Processing and properties of pineapple leaf fibers–polypropylene composites prepared by twin-screw extrusion. *Polym. Compos.* 39: 4115–4122.

52 Subasinghe, A.D.L., Das, R., and Bhattacharyya, D. (2015). Fiber dispersion during compounding/injection molding of PP/kenaf composites: flammability and mechanical properties. *Mater. Des.* 86: 500–507.

53 Castellani, R., Giuseppe, E.D., Beaugrand, J. et al. (2016). Lignocellulosic fiber breakage in a molten polymer. Part 1. Qualitative analysis using rheo-optical observations. *Comp. Part A: Appl. Sci. Manufact.* 91: 229–237.
54 Gunning, M.A., Geever, L.M., Killion, J.A. et al. (2014). Melt processing of bioplastic composites via twin screw extrusion and injection molding. *Polym. Plast. Technol. Eng.* 53: 379–386.
55 Burgstaller, C. (2014). A comparison of processing and performance for lignocellulosic reinforced polypropylene for injection moulding applications. *Comp. Part B: Eng.* 67: 192–198.
56 Duigou, A.L., Barbé, A., Guillou, E., and Castro, M. (2019). 3D printing of continuous flax fibre reinforced biocomposites for structural applications. *Mater. Des.* 180: 107884.
57 Matsuzaki, R., Ueda, M., Namiki, M. et al. (2016). Three-dimensional printing of continuous-fiber composites by in-nozzle impregnation. *Sci. Rep.* 6: 23058.
58 Vijay, Y., Sanandiya, N.D., Dritsas, S., and Fernandez, J.G. (2019). Control of process settings for large-scale additive manufacturing with sustainable natural composites. *J. Mech. Des.* 141 (8): 081701.
59 Wu, C.-S. and Tsou, C.-H. (2019). Fabrication, characterization, and application of biocomposites from poly(lactic acid) with renewable rice husk as reinforcement. *J. Polym. Res.* 26 (2): 1–1.
60 Gkartzou, E., Koumoulos, E.P., and Charitidis, C.A. (2017). Production and 3D printing processing of bio-based thermoplastic filament. *Manuf. Rev.* 4: 1.
61 Tarrés, Q., Melbø, J.K., Delgado-Aguilar, M. et al. (2018). Bio-polyethylene reinforced with thermomechanical pulp fibers: mechanical and micromechanical characterization and its application in 3D-printing by fused deposition modelling. *Comp. Part B: Eng.* 153: 70–77.
62 Duigou, A.L., Castro, M., Bevan, R., and Martin, N. (2016). 3D printing of wood fibre biocomposites: from mechanical to actuation functionality. *Mater. Des.* 96: 106–114.

9

Biocomposites for Energy Storage

M. Ramesh[1], J. Maniraj[1], and L. Rajesh Kumar[2]

[1] *Department of Mechanical Engineering, KIT-Kalaignarkarunanidhi Institute of Technology, Coimbatore, Tamil Nadu, India*
[2] *Department of Mechanical Engineering, KPR Institute of Engineering and Technology, Coimbatore, Tamil Nadu, India*

9.1 Introduction

Biocomposites did not exist 50 years ago and the word biocomposite was never uttered in the beginning of twentieth century. In the last two decades, biocomposites are widely used in medical and engineering fields. There were no medical device manufacturers (except limbs, fracture fixation equipment, glass eyes, and dental related devices), no formalized regulatory approval processes, no knowledge of biocompatibility, and certainly no awareness on biocomposites. However, these composites have been used in history, usually with bad or mixed results. The new approach is to use biocomposites that display properties such as biocompatibility, biodegradability, mobility, disposability, fluorescence electromagnetic, encapsulation, and dielectric properties [1].

The structure–property relationship of these biocomposites has been studied on a large scale basis. These composites have become very active areas of research due to the reasons for this fascinating material to grow so fast are complex. Anyway one can name probably the most significant parts of biocomposites that make this field very interesting. As a matter of importance, the need to do interdisciplinary work while tending to a particular issue around there has confronted a few scientists; to have productive work from areas other than their unique field. Many challenges while manufacturing or describing biocomposites lie in the fundamental physicochemical or mechanical properties. In the older days, a wide range of characteristic materials were used as biocomposites, such as wood; paste; elastics; and tissues of living organisms, as well as conventional elements such as glass, iron, zinc, and gold. There were incredibly different reactions from the host to these products. It has been acknowledged that there are substantial variations between the nonliving and living materials. A large scope of materials incorporating each of the traditional materials, for example, metallic elements, ceramics, composites, and polymers have been researched as biocomposites [2]. The biocomposites characterized into a few kinds, for example, biolatent and bioactive, biostable and biodegradable, and so on. Biodegradable or reusable materials are consolidated into the comprehensive tissue or can even completely break down upon an undefined time frame. Metallic materials are generally inactive, ceramic materials may be latent, and dynamic and polymeric materials may be inactive [3].

Biobased Composites: Processing, Characterization, Properties, and Applications, First Edition.
Anish Khan, Sanjay Mavinkere Rangappa, Suchart Siengchin and Abdullah M. Asiri.

Biocomposites are expected to be characterized by a nonlethal, noncancerous, inert toward chemical reactions, unchanging and able to endure the rehydrated power through its entire life.

Biocomposites are under the constant consideration of study. Such composites are used in almost every field of mechanical applications. Edible oils such as soybean and rape seed oil have become an enticing commodity that can be converted into biobased polyols [4]. Despite the biobased matrix, the properties of these composites, including vegetable oil-based polyurethanes, are affected by various factors such as the filler structure and size or preparation conditions and techniques. Standard fillers consisting of cellulose, hemicellulose, and lignin are commonly used for the manufacture of biobased composites. The critical features of these composites are accessibility, naturally amicable, ease, great mechanical properties, and biodegradability. In any case, there are few shortcomings related with poor similarity and interlaminar attachment between the fibers and matrices. The low protection from dampness retention made the utilization of common fiber composites less alluring. So as to dispose of these drawbacks, fibers are treated with reasonable chemical or physical modifications [5–7]. The learning about the thermal decay procedure of regular fibers could foresee fiber-based composites procedure and gauge the impact on the strength of the fibers. The thermal strength of the fibers and the composites may be examined utilizing thermogravimetric examination. The active display of the deterioration and furthermore the learning about actuation energy are significant for an appropriate expectation of the conduct of the composites under various conditions and the basic energy expected to begin a decay response [6].

9.2 Fundamental Concepts

9.2.1 Background

There are various types of composites available for different applications. Biocomposites contain an energizing field that has been essentially and consistently created over the previous years and includes parts of drug, chemical, and materials sciences. These composites have been utilized for surrogates for joints, plates reinforced over bones, bonding of bones, synthetic ligament, teeth inserts, vein prostheses, heart valves, contact focal points, and breast embeds [8]. Henceforth, biocomposites may be required for upgrading the recovery of common tissue, thus advancing the rebuilding of basic, metabolic, and biosynthetic tissue, just as biomechanical exhibitions [9]. The structure of new, economical, biobased materials plays a significant role in enhancing the living environments of people and their betterment in view of the increasing need of implants [10]. Considering this, fundamentally the procedures utilized for biocomposites creation should be reasonable, quick, and to be completed easily. A few techniques have been used for the planning of new bioactive, biocompatible materials [11–20].

Biocomposites have been developed, which can tie deep down through arrangement of a layer over the surface comprising hydroxyl apatite [21]. The sol–gel types are currently utilized for supplying bioactive veneers, fine particles, and attached layers, which emphasize an influence at the molecular level of protein/cell attachment and natural behavior and can be linked as sensors and transplants [22–24]. There are a few works at the utilization of this procedure for creating biocomposites, for example, bioactive glass, permeable bioactive glass, and nano-bioactive glass, among others [25–29]. The conditions of this method were suitable for the planning of the aforementioned materials that provides a mixture of inorganic and natural composites on a nanoscale basis [30]. Bone could be readily considered as an inorganic and natural composite consisting mainly of fibers of hydroxyl apatite and collagen in nanoscale. The advantages of both natural and inorganic properties were associated with synthetic materials obtained by the sol–gel method. In order to create the SiO_2 nanoparticles, a few types of organo-useful alkoxysilane raw materials were assessed. For example, the sol–gel provides focal points with the probability of acquiring homogeneous hybrid materials at low temperatures, thus taking into account consolidation of an assortment of mixes [31–37].

The sol–gel method depends on metal or silicon alkoxide hydrolysis and build-up and is used for acquiring an easy-to-manufacture assortment of high-purity inorganic oxides or hybrid inorganic natural materials [38]. This technique can be used to combine functional silica with the exact size and shape of the molecule [39–46]. In addition to a few areas of application, drug delivery system (DDS) was used all the more as of late biocomposites. For such situations, as they promote worldly and focus on drug discharge, ordinary and biodegradable polymers rise as potential materials. Applications include tissue design, recovery expedients, sensors, diagnostic devices, but DDSs [47]. A perfect DDS ought to have the capacity to lead a naturally dynamic particle at the ideal rate and for the ideal span to the ideal target, in order to keep up the medication smooth within the body at ideal helpful fixations with least fluctuation [8, 46]. Utilization of DDSs defeats the issues identified with ordinary organization routes, for example, oral and intravenous administration. Materials built with hydroxyl apatite as fundamental constituents, silica, traditional materials, clays, engineered polymeric materials, and lipids are just a few examples for the biocomposites used to transport complex particles all around the body. Some lipid nanoparticles, dendrimers, liposomes, micelles, nano- and microsized polymeric particles, nanoparticles of metal, and nanoemulsion are typically used as DDSs. Systems including biodegradable polymers and silica are given distinct consideration.

The drugs are consolidated in a polymer matrix in polymer DDSs. Because biodegradable polymers are degraded to nontoxic substances, after implantation they do not need to be expelled. This makes them an attractive candidate for DDS applications. The rate of drug discharge from the polymer matrices is dependent on a few parameters, such as the existence of the polymer matrix, matrix structure, drug properties, starting drug stacking, and interaction between drug matrices. Therefore, the drugs can be discharged safely through the biodisintegration of the matrix materials [46]. Similar to silica, inorganic materials may give the essential properties for the application of nanoparticles as DDS, especially nondanger, biosimilarity, high soundness, and a hydrophilic and permeable structure. The rate of drug discharge from the structures of silica could be restricted by changing molecule measurements and the permeable structure [47–50].

The sol–gel innovation is applied in the planning of inorganic clay and glass materials. This strategy was initially utilized on silica gels in the middle 1800s [48]. During the initial stages, a homogenous solution of water and solvent was formed from silicate which was derived from tetraethyl-ortho silicate using the sol–gel method. As of late, novel chemicals showed up, which ended up in the blending of newer hybrid materials and new oxides of inorganic nature utilizing this technique. One more procedure termed by way of nonhydrolytic sol–gel created by Acosta et al. [49]. The author utilized the buildup response of a metal or a semi-metal halide and alkoxide for obtaining oxides of those elements. The hydrolytic sol–gel and its counterpart forms are examined by a few scientists for its components [37, 51–53]. The sol–gel course is outstanding due to its straightforwardness and higher costs. This is the normally utilized strategy to combine the nanoparticles including the concurrent hydrolysis and buildup response of salt or alkoxides. Their significance and points of interest of nanoparticles have been experimentally illustrated and they possess few modern solicitations which includes catalysis, pharmaceuticals, colors, photonic gadgets, and midst else [42, 54–56].

9.3 Selection Parameters for Biocomposites

A biocomposite utilized for implant should have some critical properties for its long haul use in the body without dismissal. The plan and choice of biocomposites rely upon various properties which are clarified in this section.

9.3.1 Host Response and Biocompatibility

Host reaction is known to be the reaction of the host entity to the embedded device or material [57]. Researchers have studied the terms biocomposite and biocompatibility to demonstrate the characteristics of biobased materials [58]. Composites that are biocompatible are known as biocomposites, and biocompatibility is an enlightening word that specifies the ability of a material to adopt in a specific application encompassing within the host reaction [59]. Honestly, it suggests amicability of any biocomposite by way of living things. Biocompatibility could be described as the capacity of any biomaterial to live among the human tissues and not posing any hazard to the living tissues. Not only is it just related to danger, but also to every unfavorable impact in a natural system [60, 61]. The material should be devoid of unfavorably influencing the body and complete host atmosphere of connection (ionic arrangement of plasma, bone, intra- and extracellular liquids, soft tissues) [62]. It alludes to a lot of characteristics on which the utilization of the material depends steadily in a living thing. It has to be against pyrogenic, toxic, allergenic, incendiary, and blood compatibility.

9.3.2 Biofunctionality

Biousefulness has a particular role in mechanical and physical aspects. The material is expected to fulfill its enterprise necessities in amenity [62]:

- Load carrying and stress bearing
- Articulation to permit movement
- Control of blood and liquid flow
- Space filling
- Electrical upgrades
- Light transmission
- Sound transmission

9.3.3 Functional Tissue Structure and Pathobiology

Biocomposite materials are embedded inside human organs and tissues by converting them into medical elements and devices. In this manner, the prime ideologies administering the edifice of typical cells, tissue and organ systems, the method of contemplation of the structure and capacity of ordinary and anomalous tissues, and the vital contrivances of illness make the basic contemplation for the humans [62].

9.3.4 Toxicology

Biocomposites ought not to be poisonous, except if it is explicitly designed to cater those necessities (an instance could be a DDS in the form of a smart bomb to target and abolish cancerous cells). Toxicology of biocomposite materials manages the constituents which relocate out of other biobased materials. Most normal workability of the biocomposites is that it is not supposed to emit no matter what from within its mass unless it is explicitly intended for doing so [62].

9.3.5 Design and Manufacturability

Biocomposites ought to be machinable and extrudable. Finite element method is an excellent method utilized in the structure of implantations. Present day nontraditional practices are important for ensuring the eminence required toward the orthopedic instruments manufactured by utilizing biocomposites.

9.3.6 Mechanical Properties

Load bearing capacity of the materials which are used to fabricate the embed would pose an impact on the fissure of the synthetic organs. When it is without sufficient quality, it may fail or rupture. At a point when the interfacial part of the implant of the bone initiate to undergo fracture, construction of a lenient tissue with fibers could relax the bone-embed interface and make them bear relatively higher loads [60]. But this in fact causes more agony to the patient and if it exceeds a certain limit, the implant has to be supplanted with an update system [63]. Probably the most essential properties of biocomposites that ought to be deliberately examined and dissected are elastic modulus, rigidity, flexural quality, hardness, yield strength, fatigue strength, corrosion, and creep characteristics. Meanwhile, physical properties should also be considered while choosing materials for selected applications. For instance, the articular cup of the hip joint has been chosen dependent on the high lubricity and the dialysis membrane has chosen with a predefined porousness. Refraction and clarity in vision are the major prerequisites of the intraocular lens.

For significant applications, like joint substitution, the yield strength of the human bones has to be higher which could be coupled with lower modulus [64, 65]. The magnitude of bone modulus varies within a range relying on the kind of bone used and its estimated direction [66]. Higher difference in Young modulus values of bone and its contained implant could be a factor for higher concentration of stress at the interface and to be a specific burden protecting from common bone which possibly debilitates the bone and segregating away from the interface, thus ending up in implant failure [60, 67]. Hence for selecting a biocomposite, value of modulus should also be given prime focus.

9.3.7 Corrosion Resistance

Singh and Dahotre [68] investigated inhibition toward corrosion as an imperative concern in the choice of metal based biocomposites on the grounds that the metallic implants corrode because of the body liquid which is unavoidable. They discharge disagreeable metal particles that are not biocompatible. The human life might be diminished by the process of corrosion. Okazaki and Gotoh [69] communicated that the dissolved metal particles can either accumulate nearer to tissues at a close proximity to implants or they might be elated to different portions of the body.

9.3.8 Wear and Fatigue Resistance

The low wear resistance or high friction coefficient results in implant slackening [70]. Wear fragments are observed to be naturally active and make an extreme fiery reaction thus causing a demolition of strong bone which backs up the real embed. Soreness between the implant and the bone, which results in corrosion as it emits metal particles that are not biocompatible, is also of high concern. Mechanical stacking may also additionally result in wear, corrosion, and fatigue properties [63]. Its fatigue strength could be derived from the cyclic loads applied to the material. Fracture at the interface due to fatigue may pose threats like extreme implant loosening and stress protecting as observed frequently in the case of hip prostheses [71]. Fatigue properties emphatically rely upon the microstructures. The microstructures of metallic biocomposites modify as indicated by the preparation and heat treatment utilized [57].

9.4 Biocomposites for Energy Storage

In the recent decades, the utilization of energy has been increasing because of fast innovative and industrial advancements all through the world. Around 85% of the worldwide energy request has been provided by nonrenewable energy sources [72–76]. These petroleum products have been vigorously used in power generation and mechanical assembling because of their intrinsic energy density and plenitude; be that as it may, present day society is subject to the procurement and exchange of these energy sources [77–80]. In the meantime, the broad utilization of nonrenewable energy sources has prompted extensive releases of greenhouse gas substances, for example chlorofluorocarbons, methane, ozone, carbon dioxide, and nitrous oxide into the environment. Such ozone-depleting substances denote the discomposure of carbon balance in the climate that exists over a long period of time.

As per the data from the US earth system research laboratory, carbon dioxide level in the environment takes a growth of 2.1 ppm/year during recent years and 1.9 ppm/year in the past 10 years. The long haul variation of environmental CO_2 focus demonstrates a relentless increment of CO_2 in the environment as given in Figure 9.1. From the figure it is seen that

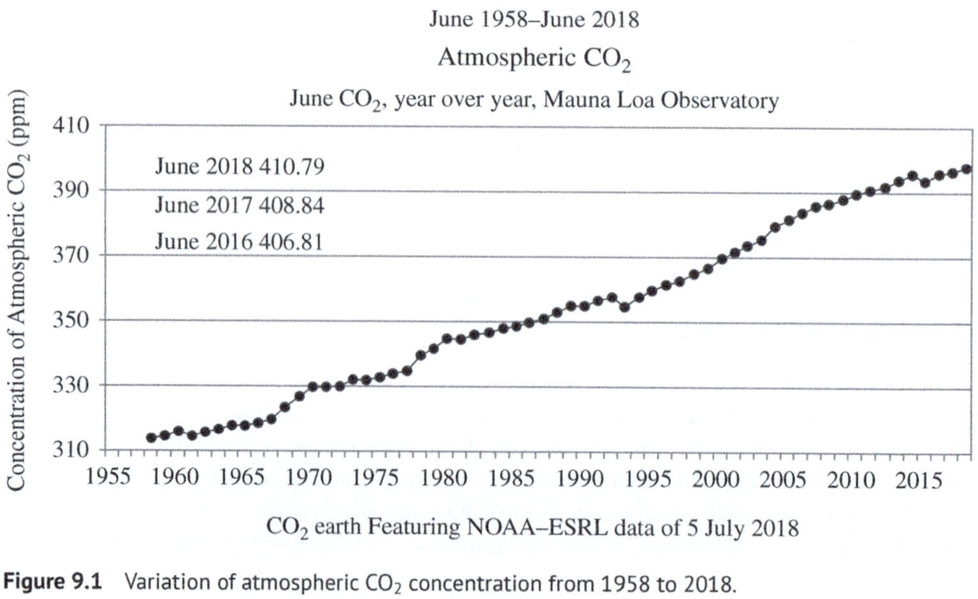

Figure 9.1 Variation of atmospheric CO_2 concentration from 1958 to 2018.

the all-out CO_2 focus in the environment has been expanded from 280 ppm during 1995 to 399 ppm during 2015 [81]. This notable increment in CO_2 level influences the changes in environment along with the increase in actual surface temperature of the earth [81–83]. In this way, progressively stringent guidelines are critically expected to advance CO_2 capture and storage (CCS) and for controlling the environmental contagion to strongly govern the changes in environment. In a perfect world, CO_2 can be changed over into valuable or degradable items, for example, energy fuels. The governmental board on environmental change anticipated that CO_2 discharges could be lowered between 80 and 90% if worldwide power use productive CCS innovations coupled with conceivable correlative advances to improve energy resources [74, 80–83]. A rundown of strategies and materials that are accessible is presented in Figure 9.2 [84].

Currently different types of CCS innovations are created to address the CO_2 outflows based on different conditions like pre- and postcombustion. Technology pertaining to oxy-fuel combustion is illustrated in Figure 9.3 [78]. Amidst current CCS advancements, a great arrangement of consideration was given to postcombustion pursued by oxy-fuel and pre-combustion all world around. These technologies have likewise been pleasantly depicted and refreshed completely by several researchers [78–85] (http://co2now.org/). Most of these endeavors showed that new material is a definitive answer for it to ensure low cost, greatest CO_2 catch, easy detachment, and large storage of energy. Over years, the current necessities were tended to utilize the traditional materials along with the equipment, for example, catalysis, retention, and adsorption of the membrane processes. Though few systems pulled in extensive consideration, just a couple of cases fulfilled genuine industrial set by the research

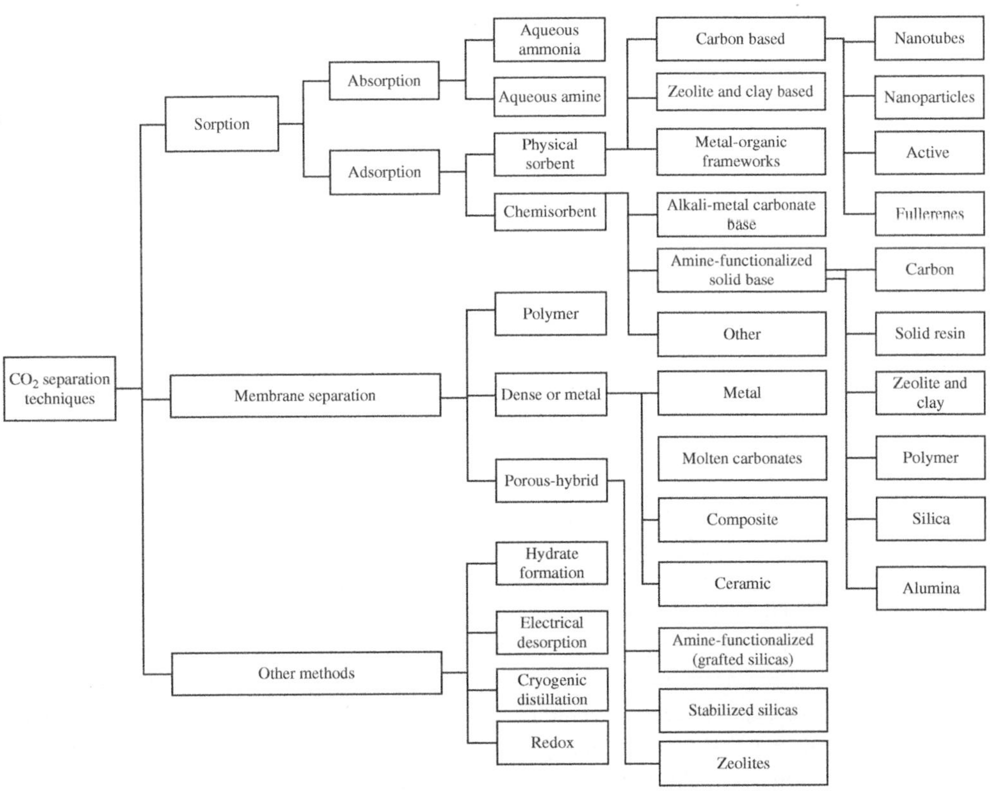

Figure 9.2 Technologies/methods and materials available for CCS. *Source:* Shakerian et al. [84]. © 2015, Elsevier.

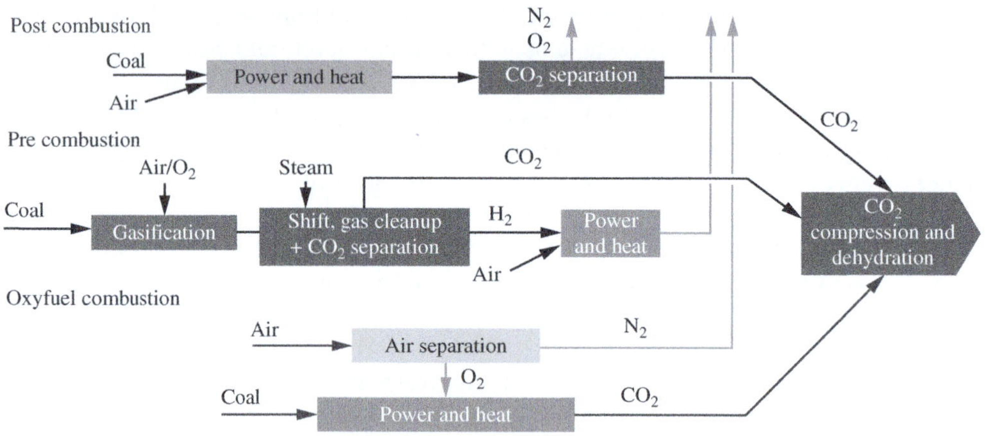

Figure 9.3 Typical CCS technologies for coal-power plants. *Source:* Based on Seoane et al. [78].

laboratory [86, 87]. Regardless, the development of CO_2-based complex molecular elements by means of substance connections and ensuing recovery of CO_2 has been the real confinements for energy and environmental applications.

9.5 Bioinspired Composite Materials

Biocomposite materials possess profoundly methodical designs, intricate morphology, and chemical configurations equipped for assorted applications. Commonly characterized with rigid and indulgent stages, they are organized in various leveled designs with typical proportions spreading over from nanoscale to large scale more alike conventional material structures. Strikingly, such frothy elements exhibit rare arrangements of toughness along with strength, while being hard to replicate artificially [88, 89]. Likewise, these materials are multifunctional and may render multiple design-oriented elucidations that suit well for a predetermined constraints and needs. Such fundamental qualities persuaded the scientists to clone these materials at top priority. As of late, utilization of characteristics of the materials was acquainted with a clarification of the multidimensional space of biocomposites and its execution (Figure 9.4) [89]. In present day science, chemical constituents, design, and structure are the basic criteria that should be figured out in order to make novel materials, for example, composites, polymers, ceramics, and alloys.

Currently, biocomposites are connected to mimic their mechanical highlights for specific uses. Likewise, the broad control of science and microstructure is additionally required for producing newer biocomposites [90–92]. For instance, the elements in bone such as calcium carbonate could act as a catalyst for the oxidation movement during CO_2 transformation. The enzyme catalytic reaction of biocomposites could be largely useful in terms of energy during CO_2 consumption. The importance of biocomposites has been assessed for CCS applications at its best by many researchers. Ideally, the viability of biocomposites are considered broadly at the time of crucial requirement for effective utilization of energy and green innovation of CCS. In addition, on account of low partial pressure of CO_2, the immense quantity of flue gas, and dissipation of H_2O, significant energy utilization is necessary at the recovery phase. Henceforth, catching of CO_2 discharged at elevated temperatures stays difficult for the improvement of bioinspired adsorbents just as in traditional absorbents.

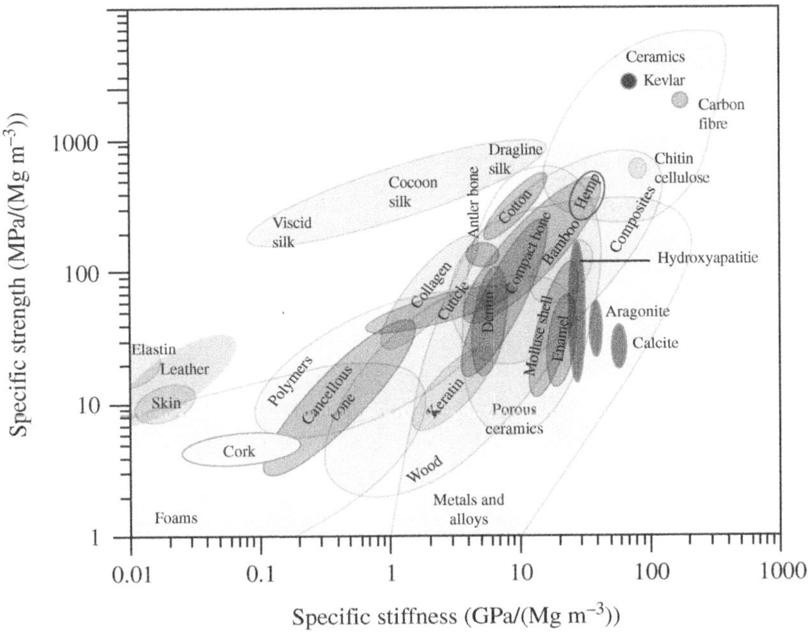

Figure 9.4 Material properties of bioinspired materials. *Source:* Wegst et al. [89]. © 2014, Springer Nature.

9.6 Bioinspired Composites for Energy Storage

Alongside the utilizations of biocomposites to CCS, the most recent improvement of their uses in energy harvesting and storage have been examined. The materials are utilized for electrode making in Li-ion batteries and are termed to be biobatteries; they are sufficiently productive to upgrade through electron exchange, posing enormous electrochemical movement during their high explicit limit and recycling period. For such kind of utilizations, bioparticles or biocomposites were chosen due to their highlights applications [93–103]. The designated materials are used as a cradle of nanocomposites or utilized as bionanocomposites for upgrading the electrochemical capacity during the storing of energy. Few ongoing biocomposites for energy storage applications are listed in Table 9.1. Lu et al. [97] used elastin-like polypeptides encompassing metal-tied motifs protein hydrogel format for a newer bioinspired synthesis of N and F codoped with graphite nanocrystals containing carbon foams. Ionic fluid aids to save the permeable construction of protein matrix, filling in as a fluorine foundation for carbon doping and for the amalgamation of nanocrystals. Subsequent to strengthening at 600 °C in argon, three-dimensional N, F codoped permeable graphitized carbon-supported nanocrystals for batteries which is displayed in Figure 9.5. Likewise, muscle-inspired alluaudite spindle was blended utilizing top-down combination technique as a high potential cathode for sodium particle batteries [98].

The muscle enthused design permitted anode material additionally ideal for 3D conductive system and permeable composite to accomplish quick particle transference and sodium intercalation over single stage response component amid sodium intercalation for its dynamic property (Figure 9.6) [98]. This composite showed attractive electrochemical performance with high working potential and a high-rate ability; alongside these batteries are prepared to accomplish a limit maintenance of 93% after 100 cycles at a temperature of 5 °C

Table 9.1 Bioinspired composite material for energy storage applications.

S. no.	Biocomposites	Bioinspired composite materials	Performance applications/capacity	References
1.	ELK16-FLAG	3-D MnF_2 nanocrystals	—	[98]
2.	Muscle-inspired Alluaudite	$Na_{2+2x}Fe_{2x}(SO_4)_3$/SWNTs	3.8 V	[99]
3.	p8 protein	Hybrid virus-Au–Co_3O_4	890 mAhg[a]	[100]
4.	M13 virus	M 13/$FePO_4$-SWNTs	130 mAhg[a] at 10 °C	[101]
5.	Diphenylalanine peptide	Diphenylalanine peptide : Co_3O_4 nanoparticles (3 : 1)	Solar/fuel cells	[102]
6.	Ferritin	Ferritin : silicon dioxide: iron oxide (2 : 1 : 1)	QD solar cell	[103]
7.	*Desulfovibrio desulfuricans*	Pd NPs	Fuel cells	[104]

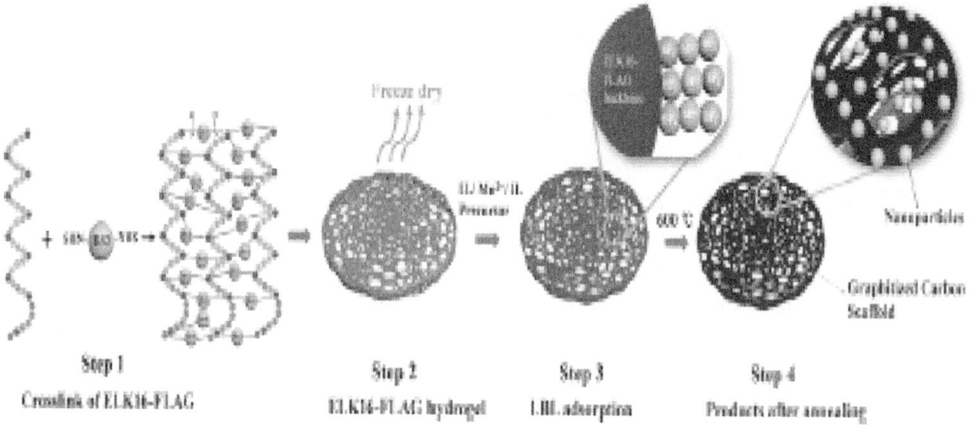

Figure 9.5 N, F codoped 3D porous carbon supported nanocrystals. *Source:* Lu et al. [97]. © 2016, Royal Society of Chemistry.

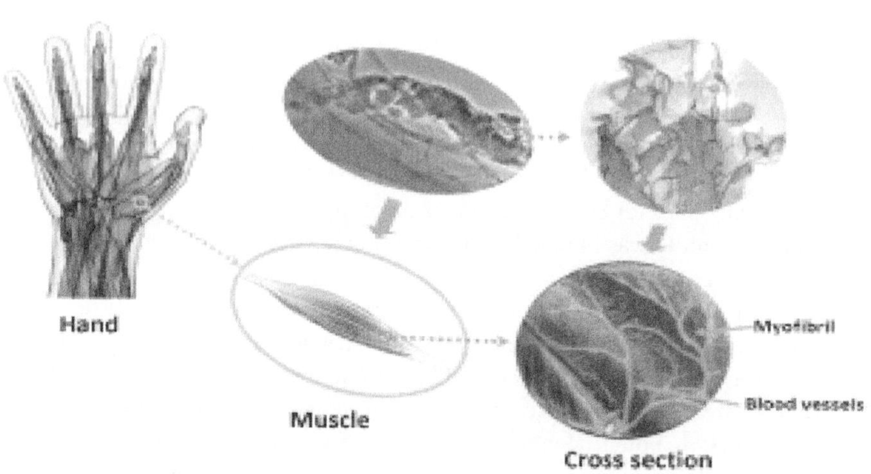

Figure 9.6 Muscle-inspired spindle composite morphology [98].

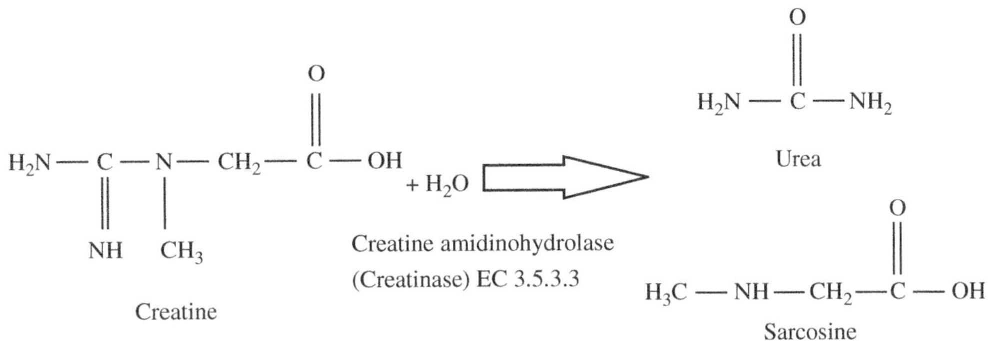

Figure 9.7 Creatine amidinohydrolase catalyzed hydrolysis of creatine.

for predominant electrochemical working [98]. Strikingly, such sorts of electrode material possess numerous points of interest over ordinary materials [98–103]. Nonetheless, regular key difficulties, for example, advancement of ideal structures for improved execution in huge scale devices and their expenses are left to confine their business usage [94–103].

9.7 Enzyme-Based Materials

Clinical analyzers require biosensors to be utilized over and over at the time of interaction with biofluids like urine and blood [104, 105]. They regularly involve a biosensor to work constantly at 37 °C for over seven days [106]. For some biosensors, the constancy of the catalyst stratum constrains the reasonable epoch of a sensor. Enzyme stabilization was given much priority by means of multipoint covalent hold in polyurethanes [107–109]. The noteworthy stabilization that can be granted by means of assimilation into polyurethanes which empower the layered enzyme development make it fit for various applications in amperometric biosensors. Recognition of creatinine intensities in biofluids is an inexorably imperative medical objective. This data is utilized for the assessment of kidney functions and muscle impairment. Specific researches are being carried out to design such sensors for reading the creatinine levels. A biosensor was built upon those changes over creatinine to amperometrically quantifiable hydrogen peroxide [110]. Because of the multifaceted nature, the advancement of the creatinine biosensor was restricted to be moderate [111]. Various strategies are accounted for the restriction of chemicals employed in creatinine biosensors and this process is the important factor in the assurance of the working and storage capacity (Figure 9.7). Creatine amidinohydrolase (creatinase) is a homodimer with subunit molecular weights of 45 kDa. The two dynamic sites of the protein are at the interface of the monomers being shared by each and it is just dynamic. The enzyme has a low utilitarian strength. Added substances, for example, diminishing agents, polyols, and proteins have appeared for expanding the dependability of enzymes [112]. The meager intrinsic stability of creatinase has incited the utilization of protein and was designed for enhancing the dependability [113].

9.8 Biosensing/Bioimaging Applications

The advancement of new kinds of biodetecting/bioimaging has been pulled in enthusiasm among the scientific and research network, in view of the extraordinary effect that these instrumental devices as biosensors have had lately [1]. Dry-granulate ethanol preparation is one among the best techniques for the transformation of starch to bioethanol and is the procedure, for the most part, utilized for creation of bioethanol from sorghum [114]. Biofuel came to spotlight to turn away green-house vaporous outflows and to substitute petroleum product. Biofuel is a nondangerous, sustainable, ecofriendly fuel created from different feed stocks [115]. Densification builds the mass density, improves transportation and coordination, diminishes dust generation, as well as the working costs [116]. For any biotherapeutic use of material, it ought to be biocompatible in nature. The innovation test evaluates the practical cells of any material and the suitability of the cells is connected with the biocompatibility of that material [117]. The inorganic compound favored for the stabilization studies is the mineral segment of bone, a type of calcium orthophosphate known as hydroxyapatite [118]. Nanomolecules at various volume portions can be added to a specific dimension to accomplish required properties dependent on explicit applications [119].

The decarbonization of the energy segment can be made conceivable by incorporating sustainable energy sources with different thermal energy storage systems which have round-trip efficiency of above 96% [120]. A series of aliphatic poly (1,3-propylenedicarboxylate) diols were synthesized and characterized through reacting bio-PDO with various aliphatic di-carboxylic acids [121, 122]. The dissolving and solidifying of both the biobased material and the conventional composite material were taken as beginning temperature purposes of the individual heating and cooling peaks. These were acquired utilizing a program incorporated with the differential examining calorimetry by illustration a line at the purpose of the greatest slant of the main edge of the came about stage change peak and extrapolating pattern on a similar side as the main edge of the peak [123].

Biomimetics can stay away from the broad industrialization and asset extraction done by human. Based on improving new items, the biomimetics can go past simply applying regular properties [124]. Some compostable materials are additionally biodegradable in soil as well as water, yet in a few cases do not biodegrade in soil as well as water [125]. Cyclodextrins are a perfect host system with qualities like that of the enzyme model. Their capacity to frame incorporation complexes has been misused in different regions of research for drug carrier systems, food industry, cosmetics, and more [126]. In spite of the profitable contributions to the mineral bioflocculation and bioseparation science, varieties in the settling behavior and bioflocculation of kaolin and silica in various pH ranges have not been completely explored and contrasted with one another. Such an examination is essential as the two minerals may coincide together [127].

The utilization of enzymes enables us to perform enzymatic hydrolysis, consequently encouraging the accessibility of cellulose as a carbon source for succinate creation [128]. Inferable from the perplexing interactions and broad guideline in metabolism and the restricted learning of this guideline, the development of a cell industrial facility for the most part includes a few rounds of the design–construction–evaluation–optimization cycle [129]. Chemical affecting elements incorporate biomass ratio, solvents concentration; physical components allude to temperature, weight, dissolvable to biomass proportion, impetus focus, heating rate, and habitation time [130]. Natural materials have almost flawless structures and capacities following a great many years' advancement in which the electron exchange is a standout among the most essential and expounded biological procedures [131]. Silica nanoparticles can be utilized as a grid to embodiment of colors. This procedure can limit the photo-degradation of the color and improve its strength in both in vitro and in vivo. These particles can be connected to make papers with amazing properties [132]. Natural materials are principally established by paraffinic and biobased mixes and inorganic materials incorporate salt hydrates and metallic mixes [133].

The worldwide construction division represents 35% of the oil-based polymer composites showcase; however, these composites are not completely degradable because of a protection from microbial decay. Polymers from biodetermined sources, for example, polyhydroxyalkanoates, have the advantage of having a short residency in anaerobic conditions and the consideration of natural fibers has been appeared to speed up degradation [134]. The structure and morphology of cellulose has turned into the primary subject in different looks into around the world. Cellulose is a natural compound which comprises linear polysaccharide of D-glucopyranose units associated with each other with β-(1-4)-glycosidic joins [135]. The effects of biogenic CO_2 are regularly disregarded in a traditional lifecycle investigation [136]. Cellulose is additionally the most copious type of living earthbound biomass and has different applications in current industry [137]. Nanocellulose has a vast surface area and a negative charge of crystalline, can be bound to its surface, with the potential for high payloads and ideal control of dosing [138]. The major chemical of the phosphorus manures depends on the creation of sulfuric acid, the generation of which is related to a large source of sulfur-dioxide and acid mist outflows [139].

The sharp edges of graphene and its subsidiaries are entering bacterial cell membranes destroying the integrity of the cells. Thus, it is critical to investigate a suitable conductive anode material that can bolster high development of biocatalyst microscopic organisms for upgrading the effectiveness [140]. The sorption isotherm of the examined hemp concrete stands in the scope of hygroscopic material [141]. Worldwide production of vegetable oils has encountered a positive development parallel with horticultural progression. Every single vegetable oil has genuinely indistinguishable sub-atomic structure, which are made out of triglyceride particles [142]. The utilization of various protection cover in each wall orientation does not speak to a critical decrease in the worldwide expense of the solutions. From each of the materials analyzed such as mineral wool, polystyrene, and polyurethane, the last one offered the best execution with respect to financial expense, while mineral wool offered lower ecological effect and an increasingly adjusted tradeoff between the two parameters [143].

9.9 Conclusion

Researches in the area of biocomposites have been driven by the search for materials with high biocompatibility and emulation of the artificial synapse. Over recent years, a new type of energy storage system has emerged with biocomposites. Among different biocomposite materials, carbon nanotubes, collagen, chitosan, graphene, metallic nanoparticles, and quantum dots have been used for energy storage. This chapter covered recent advancements in the development of biocomposites for energy storage. The implementation of biocompatible and biodegradable energy storage would be a significant step toward next-generation green composites. From the literature it is observed that biocomposites have notable advantages when compared to traditional materials, for example, high sensitivity, explicitness, movability, cost-viability, conceivable outcomes for scaling down and large scale manufacturing among others.

Notwithstanding, the choice of biocomposite for energy storage system is a critical issue. In addition, these materials have applications for quick, explicit, delicate, reasonable, in-field, online, and/or real-time detection. Some applications of biocomposites prove consistent advancement in the clinical expository divisions. These rising uses of potential high effect in our day-to-day life are discussed in this chapter. We trust that this chapter can give a few experiences to the innovative and motivated composites having engaging benefits of high proficiency for their energy storage and ecological applications, and set forward the improvement of this rising fields of potential high effect in our everyday life.

References

1 Perez, J.A.C., Hernandez, J.E.S., Hussain, S.M. et al. (2018). Bioinspired biomaterials and enzyme-based biosensors for point of-care applications with reference to cancer and bio-imaging. *Biocatal. Agric. Biotechnol.* https://doi.org/10.1016/j.bcab.2018.11.015.

2 Hench, L.L. (1991). Bioceramics: from concept to clinic. *J. Am. Ceram. Soc. Hench* 74: 1487–1510.

3 Bhat, S.V. (2002). Overview of biomaterials. In: *Biomaterials*. Dordrecht: Springer https://doi.org/10.1007/978-94-010-0328-5_1.

4 Liu, W., Xu, K., Wang, C. et al. (2016). Carbon nanofibers reinforced soy polyol-based polyurethane nanocomposites. Thermal and mechanical characterization. *J. Therm. Anal. Calorim.* 123: 2459–2468.

5 Datta, J. and Kopczynska, P. (2015). Effect of kenaffibre modification on morphology and mechanical properties of thermoplastic polyurethane material. *Ind. Crop. Prod.* 74: 566–576.

6 Borsoi, C., Zimmernnam, M.V.G., and Zattera, A.J. (2016). Thermal degradation behavior of cellulose nanofibers and nanowhiskers. *J. Therm. Anal. Calorim.* 126: 1867–1878.

7 Yao, F., Wu, Q., Lei, Y. et al. (2008). Thermal decomposition kinetics of natural fibers: activation energy with dynamic thermogravimetric analysis. *Polym. Degrad. Stab.* 93: 90–98.

8 Nair, L.S. and Laurencin, C.T. (2006). Polymers as biomaterials for tissue engineering and controlled drug delivery. *Adv. Biochem. Eng./Biotechnol.* 102: 47–90.

9 Hench, L.L. (1998). Biomaterials: a forecast for the future. *Biomaterials* 19 (16): 1419–1423.

10 Leeuwenburgh, S.C.G., Malda, J., Rouwkema, J., and Kirkpatrick, C.J. (2008). Trends in biomaterials research: an analysis of the scientific programme of the World Biomaterials Congress 2008. *Biomaterials* 29: 3047–3052.

11 Bonzani, I.C., Adhikari, R., Houshyar, S. et al. (2007). Synthesis of two-component injectable polyurethanes for bone tissue engineering. *Biomaterials* 28 (3): 423–433.

12 Balani, K., Anderson, R., Laha, T. et al. (2007). Plasma-sprayed carbon nanotube reinforced hydroxyapatite coatings and their interaction with human osteoblasts in vitro. *Biomaterials* 28 (3): 618–624.

13 Chu, T.M.G., Warden, S.J., Turner, C.H., and Stewart, R.L. (2007). Segmental bone regeneration using a load-bearing biodegradable carrier of bone morphogenetic protein-2. *Biomaterials* 28 (3): 459–467.

14 Eglin, D., Maalheem, S., Livage, J., and Coradin, T. (2006). In vitro apatite forming ability of type I collagen hydrogels containinrg bioactive glass and silica sol–gel particles. *J. Mater. Sci. Mater. Med.* 17 (2): 161–167.

15 Skelton, K.L., Glenn, J.V., Clarke, S.A. et al. (2007). Effect of ternary phosphate-based glass compositions on osteoblast and osteoblast-like proliferation, differentiation and death in vitro. *Acta Biomater.* 3 (4): 563–572.

16 Misra, S.K., Mohn, D., Brunner, T.J. et al. (2008). Comparison of nanoscale and microscale bioactive glass on the properties of P(3HB)/Bioglass® composites. *Biomaterials* 29 (12): 1750–1761.

17 Lee, H.J., Choi, H.W., Kim, K.J., and Lee, S.C. (2006). Modification of hydroxyapatite nanosurfaces for enhanced colloidal stability and improved interfacial adhesion in nanocomposites. *Chem. Mater.* 18: 5111–5118.

18 Ramila, A., Padilla, S., Munoz, B., and Regí, M.V. (2002). A new hydroxyapatite/glass biphasic material: in vitro bioactivity. *Chem. Mater.* 14: 2439–2443.

19 Dogan, O. and Oner, M. (2006). Biomimetic mineralization of hydroxyapatite crystals on the copolymers of vinylphosphonic acid and 4-vinilyimidazole. *Langmuir* 22: 9671–9675.

20 Iwatsubo, T., Sumaru, K., Kanamori, T. et al. (2006). Construction of a new artificial biomineralization system. *Biomacromolecules* 7: 95–100.

21 Clupper, D.C. and Hench, L.L. (2003). Crystallization kinetics of tape cast bioactive glass 45S5. *J. Non-Cryst. Solids* 318 (1–2): 43–48.

22 Hench, L.L. (1997). Sol–gel materials for bioceramicapplications. *Curr. Opin. Solid State Mater. Sci.* 2 (5): 604–610.

23 Pickup, D.M., Speight, R.J., Knowles, J.C. et al. (2008). Sol–gel synthesis and structural characterisation of binary TiO_2–P_2O_5 glasses. *Mater. Res. Bull.* 43 (2): 333–342.

24 Carla, D., Pickup, D.M., Knowles, J.C. et al. (2007). Structural study of sol–gel and melt-quenched phosphate-based glasses. *J. Non-Cryst. Solids* 353 (18–21): 1759–1765.

25 Saravanakumar, B., Rajkumar, M., and Rajendran, V. (2011). Synthesis and characterization of nanobioactive glass for biomedical applications. *Mater. Lett.* 65: 31–34.

26 Lei, B., Chen, X., Wang, Y. et al. (2010). Fabrication of porous bioactive glass particles by one step sintering. *Mater. Lett.* 64: 2293–2295.

27 Chen, Q.-Z., Li, Y., Jin, L.-Y. et al. (2010). A new sol–gel process for producing Na_2O-containing bioactive glass ceramics. *Acta Biomater.* 6: 4143–4153.

28 Kokubo, T., Kim, H.-M., and Kawashita, M. (2003). Novel bioactive materials with different mechanical properties. *Biomaterials* 24: 2161–2175.

29 Olmo, N., Martín, A.I., Salinas, A.J. et al. (2003). Bioactive sol–gel glasses with and without a hydroxycarbonate apatite layer as substrates for osteoblast cell adhesion and proliferation. *Biomaterials* 24: 3383–3393.

30 Vrancken, K.C., Possemiers, K., Voort, P.V.D., and Vansant, E.F. (1995). Surface modification of silica gels with aminoorganosilanes. *Colloids Surf. A Physicochem. Eng. Asp.* 98 (3): 235–241.

31 Mark, J.E., Lee, C.Y.C., and Bianconi, P.A. (1995). *Hybrid Organic–Inorganic Composites. (ACS Symposium Series 586)*. Washington, DC: American Chemical Society. ISBN: 9780841231481.

32 Cerveau, G., Corriu, R.J.P., Lepeytre, C., and Mutin, P.H. (1998). Influence of the nature of the organic precursor on the textural and chemical properties of silsesquioxane materials. *J. Mater. Chem.* 12 (8): 2707–2714.

33 Corriu, R.A. (1998). New trend in metal-alkoxide chemistry: the elaboration of monophasic organic-inorganic hybrid materials. *Polyhedron* 17 (5-6): 925–934.

34 Corriu, R.J.P. and Leclerq, D. (1996). Recent developments of molecular chemistry for sol–gel processes. *Angew. Chem. Int. Engl.* 35 (13–14): 1420–1436.

35 Shea, K.J., Loy, D.A., and Webster, O. (1992). Arylsilsesquioxane gels and related materials. New hybrids of organic and inorganic networks. *J. Am. Chem. Soc.* 114 (17): 6700–6710.

36 Jackson, C.L., Bauer, B.J., Nakatami, A.I., and Barnes, J. (1996). Synthesis of hybrid organic-inorganic materials from interpenetrating polymer network chemistry. *Chem. Mater.* 8 (3): 727–733.

37 Brinker, C.J. and Scherer, G.W. (1990). *Sol–Gel Science, The Physics and Chemistry of Sol–Gel Processing*. San Diego: Academic Press. ISBN: 0121349705.

38 Nassar, E.J., Ciuffi, K.J., Ribeiro, S.J.L., and Messaddeq, Y. (2003). Europium incorporated in the silica matrix obtained by sol–gel methodology: luminescent materials. *Mater. Res.* 6 (4): 557–562.

39 Beari, F., Brand, M., Jenkner, P. et al. (2001). Organ of functional alkoxysilanes in dilute aqueous solution: new accounts on the dynamic structural mutability. *J. Organomet. Chem.* 625 (2): 208–216.

40 Nassar, E.J., Neri, C.R., Calefi, P.S., and Serra, O.A. (1999). Functionalized silica synthesized by sol–gel process. *J. Non-Cryst. Solids* 247 (1–3): 124–128.

41 Papacídero, A.T., Rocha, L.A., Caetano, B.L. et al. (2006). Preparation and characterization of spherical silica-porphyrin catalysts obtained by the sol–gel methodology. *Colloids Surf.* 275 (1–3): 27–35.

42 Nassar, E.J., Nassor, E.C.O., Avila, L.R. et al. (2007). Spherical hybrid silica particles modified by methacrylate groups. *J. Sol–Gel Sci. Technol.* 43 (1): 21–26.

43 Ricci, G.P., Rocha, Z.N., Nakagaki, S. et al. (2011). Iron–alumina materials prepared by the non-hydrolytic sol–gel route: synthesis, characterization and application in hydrocarbons oxidation using hydrogen peroxide as oxidant. *Appl. Catal. A Gen.* 389 (1–2): 147–154.

44 Matos, M.G., Pereira, P.F.S., Calefi, P.S. et al. (2009). Preparation of a $GdCaAl_3O_7$ matrix by the non-hydrolytic sol–gel route. *J. Lumin.* 129: 1120–1124.

45 Langer, R. (2000). Biomaterials in drug delivery and tissue engineering: one laboratory's experience. *Acc. Chem. Res.* 33 (2): 94–101.

46 Kumar, D.S., Banji, D., Madhavi, B. et al. (2019). Nanostructured porous silicon – a novel biomaterials for drug delivery. *Int. J. Pharm. Pharm. Sci.* 1 (2): 8–16.

47 Arruebo, M., Galan, M., Navascues, N. et al. (2006). Development of magnetic nanostructured silica-based materials as potential vectors for drug-delivery applications. *Chem. Mater.* 18 (7): 1911–1919.

48 Vallet-Regí, M., Balas, F., and Arcos, D. (2007). Mesoporous materials for drug delivery. *Angew. Chem. Int. Ed.* 46: 7548–7558.

49 Acosta, S., Corriu, R.J.P., Leclerq, D. et al. (1994). Preparation of alumina gels by a non-hydrolytic sol–gel processing method. *J. Non-Cryst. Solids* 170 (3): 234–242.

50 Yang, J., Lee, J., Kang, J. et al. (2008). Hollow silica nanocontainers as drug delivery vehicles. *Langmuir* 24 (7): 3417–3421.

51 Hench, L.L. and West, J.K. (1990). The sol–gel process. *Chem. Rev.* 90 (1): 33–72.

52 Bandeira, L.C., Ciuffi, K.J., Calefi, P.S., and Nassar, E.J. (2010). Silica matrix doped with calcium and phosphate by sol–gel. *Adv. Biosci. Biotechnol.* 1 (3): 200–207.

53 de Campos, B.M., Bandeira, L.C., Calefi, P.S. et al. (2010). Protective coating materials on rapid prototyping by sol–gel. *Virtual Phys. Prototyp.* https://doi.org/10.1080/17452759.2010.491938.

54 Pereira, P.F.S., Matos, M.G., Avila, L.R. et al. (2010). Red, green and blue (RGD) emission doped $Y_3Al_5O_{12}$ (YAG) phosphors prepared by non-hydrolytic sol–gel route. *J. Lumin.* 130: 488–493.

55 Cestari, A., Avila, L.R., Nassor, E.C.O. et al. (2009). Characterization of glass ionomer dental cements prepared by a non-hydrolytic sol–gel route. *Mater. Res.* 12 (2): 139–143.

56 Nassar, E.J., Pereira, P.F.S., Ciuffi, K.J., and Calefi, P.S. (2008). *Photoluminescence Research Progress, Recent Development of Luminescent Materials Prepared by the Sol–Gel Process*, 265–285. NY, USA: Nova Science Publishers, Inc.

57 Boretos, J.W. and Eden, M. (1984). *Contemporary Biomaterials, Material and Host Response, Clinical Applications, New Technology and Legal Aspects*, 232–233. Park Ridge: Noyes Publications.

58 Williams, D.F. (1988). Consensus and definitions in biomaterials. In: *Advances in Biomaterials*, 11–16. Elsevier.

59 Black, J. and Hastings, G.W. (1988). *Handbook of Biomaterials Properties*. London: Chapman and Hall.

60 Geetha, M., Singh, A.K., Asokamani, R., and Gogia, A.K. (2009). Ti based biomaterials, the ultimate choice for orthopaedic implants – a review. *Prog. Mater. Sci.* 54: 397–425.

61 Navarro, M., Michiardi, A., Castaño, O., and Planell, J.A. (2008). Biomaterials in orthopaedics. *J. R. Soc. Interface* 5 (27): 1137–1158.

62 Zenz, C.R. (2008). Conference on biomaterials WS2008, for genomics and bioinformatics. In: *A Textbook on Biomaterials*, 9–10.

63 Alvarado, J., Maldonado, R., Marxuach, J., and Otero, R. (2003). Biomechanics of hip and knee prostheses. In: *Applications of Engineering Mechanics in Medicine*. GED-University of Puerto Rico Mayaguez.

64 Long, M. and Rack, H.J. (1998). Titanium alloys in total joint replacement – a materials science perspective. *Biomaterials* 19 (18): 1621–1639.

65 Wang, K. (1996). The use of titanium for medical applications in the USA. *Mater. Sci. Eng. A Struct. Mater. Prop. Microstruct. Process.* 213 (1–2): 134–137.

66 Katz, J.L. (1980). Anisotropy of Young's modulus of bone. *Nature* 283 (5742): 106–107.

67 Au, A.G., James Raso, V., Liggins, A.B., and Amirfazli, A. (2007). Contribution of loading conditions and material properties to stress shielding near the tibial component of total knee replacements. *J. Biomech.* 40 (6): 1410–1416.

68 Singh, R. and Dahotre, N.B. (2007). Corrosion degradation and prevention by surface modification of biometallic materials. *J. Mater. Sci. Mater. Med.* 18 (5): 725–751.

69 Okazaki, Y. and Gotoh, E. (2005). Comparison of metal release from various metallic biomaterials in vitro. *Biomaterials* 26 (1): 11–21.

70 Ramsden, J.J., Allen, D.M., Stephenson, D.J. et al. (2007). The design and manufacture of biomedical surfaces. *CIRP Ann. Manuf. Technol.* 56 (2): 687–711.

71 Teoh, S.H. (2000). Fatigue of biomaterials: a review. *Int. J. Fatigue* 22 (10): 825–837.

72 Rackley, S. (2009). *Carbon Capture and Storage*. Gulf Professional Publishing.

73 Alessandro, D.M., Smit, B., and Long, J.R. (2010). Carbon dioxide capture: prospects for new materials. *Angew. Chem. Int. Ed.* 49: 6058–6082.

74 Li, J.R., Ma, Y., McCarthy, M.C. et al. (2011). Carbon dioxide capture-related gas adsorption and separation in metal–organic frameworks. *Coord. Chem. Rev.* 255: 1791–1823.

75 Leung, D.Y., Caramanna, G., and Maroto-Valer, M.M. (2014). An overview of current status of carbon dioxide capture and storage technologies. *Renew. Sust. Energ. Rev.* 39: 426 443.

76 Jentsch, M., Trost, T., and Sterner, M. (2014). Optimal use of power-to-gas energy storage systems in an 85% renewable energy scenario. *Energy Procedia* 46: 254–261.

77 Hester, R.E. and Harrison, R.M. (eds.) (2010). *Carbon Capture: Sequestration and Storage. Ecosystem Services*. Cambridge: Royal Society of Chemistry.

78 Seoane, B., Coronas, J., Gascon, I. et al. (2015). Metal–organic framework based mixed matrix membranes: a solution for highly efficient CO_2 capture? *Chem. Soc. Rev.* 44: 2421–2454.

79 Yan, J. (2015). Carbon capture and storage (CCS). *Appl. Energy* 148: A1–A6.

80 EPRI (2009). *The Fourth Foundation Report in Strategic Analysis of the Global Status of Carbon Capture and Storage*. Global CCS Institute and Worley Parsons.

81 Choi, S., Drese, J.H., and Jones, C.W. (2009). Adsorbent materials for carbon dioxide capture from large anthropogenic point sources. *ChemSusChem* 2: 796–854.

82 Sumida, K., Rogow, D.L., Mason, J.A. et al. (2011). Carbon dioxide capture in metal–organic frameworks. *Chem. Rev.* 112: 724–781.

83 IPCC (2005). *IPCC Special Report on Carbon Dioxide Capture and Storage*. Cambridge University Press.

84 Shakerian, F., Kim, K.H., Szulejko, J.E., and Park, J.W. (2015). A comparative review between amines and ammonia as sorptive media for post-combustion CO_2 capture. *Appl. Energy* 148: 10–22.

85 Roosa, S.A. and Jhaveri, A.G. (2009). *Carbon Reduction: Policies, Strategies, and Technologies*. The Fairmont Press Inc.

86 Ebner, A.D. and Ritter, J.A. (2009). State-of-the-art adsorption and membrane separation processes for carbon dioxide production from carbon dioxide emitting industries. *Sep. Sci. Technol.* 44: 1273–1421.

87 USDO (2009). *Energy, DOE/NETL's Carbon Capture R&D Program for Existing Coal Fired Power Plants Energy*. USDO.

88 Quan, S., Li, S., Wang, Z. et al. (2015). A bio-inspired CO_2-philic network membrane for enhanced sustainable gas separation. *J. Mater. Chem. A* 3: 13758–13766.

89 Wegst, U.G., Bai, H., Saiz, E. et al. (2014). Bioinspired structural materials. *Nat. Mater.* 14: 23–26.

90 Lau, C.H., Li, P., Li, F.Y. et al. (2013). Reverse-selective polymeric membranes for gas separations. *Prog. Polym. Sci.* 38: 740–766.

91 Su, J.C., Ong, R.C., Wang, P. et al. (2013). Advanced FO membranes from newly synthesized CAP polymer for wastewater reclamation through an integrated FO-MD hybrid system. *AICHE J.* 59 (4): 1245–1254.

92 Mangindaan, D.W., Woon, N.M., Shi, G.M., and Chung, T.S. (2015). P84 polyimide membranes modified by a tripodal amine for enhanced pervaporation dehydration of acetone. *Chem. Eng. Sci.* 122: 14–23.

93 Bhattacharya, P., Du, D., and Lin, Y. (2014). Bioinspired nanoscale materials for biomedical and energy applications. *J. R. Soc. Interface* 11: 20131067.

94 Zhou, H., Fan, T., and Zhang, D. (2011). Biotemplated materials for sustainable energy and environment: current status and challenges. *ChemSusChem* 4: 1344–1387.

95 Lee, Y.J. and Belcher, A.M. (2011). Nanostructure design of amorphous $FePO_4$ facilitated by a virus for 3V lithium ion battery cathodes. *J. Mater. Chem.* 21: 1033–1039.

96 Kannan, A.M., Renugopalakrishnan, V., Filipek, S. et al. (2009). Bio-batteries and bio-fuel cells: leveraging on electronic charge transfer proteins. *J. Nanosci. Nanotechnol.* 2009 (9): 1665–1678.

97 Lu, Y., Zhou, Y.P., Yan, Q.Y., and Fong, E. (2016). Bio-inspired synthesis of N, F co-doped 3D graphitized carbon foams containing manganese fluoride nanocrystals for lithium ion batteries. *J. Mater. Chem. A* 4: 2691–2698.

98 Yu, M., Tiantian, Y., Sen, Z., and Chao, D. (2016). Top-down synthesis of muscle inspired alluaudite $Na^{2+}2xFe_2x(SO_4)_3$/SWNT spindle as a high-rate and high potential cathode for sodium–ion batteries. *J. Mater. Chem. A* 4: 1624–1631.

99 Nam, K.T., Kim, D.-W., Yoo, P.J. et al. (2006). Virus enabled synthesis and assembly of nano wires for lithium ion battery electrodes. *Science* 312: 885–888.

100 Lee, Y.J., Yi, H., Kim, W.-J. et al. (2009). Fabricating genetically engineered high-power lithium–ion batteries using multiple virus genes. *Science* 324: 1051–1055.

101 Ryu, J., Kim, S.W., Kang, K., and Park, C.B. (2009). Synthesis of diphenylalanine/cobalt oxide hybrid nanowires and their application to energy storage. *ACS Nano.* 4: 159–164.

102 Huang, C.H., Igarashi, M., Horita, S. et al. (2010). Novel Si nanodisk fabricated by biotemplate and defect-free neutral beam etching for solar cell application. *Japan J. Appl. Phys.* 49: 04DL16.

103 Yong, P., Mikheenko, I.P., Deplanche, K. et al. (2010). Biorefining of precious metals from wastes: an answer to manufacturing of cheap nanocatalysts for fuel cells and power generation via an integrated biorefinery? *Biotechnol. Lett.* 32: 1821–1828.

104 D'Orazio, P. (2003). Biosensors in clinical chemistry. *Clin. Chim. Acta* 334: 41–69.

105 Malhotra, B.D. and Chaubey, A. (2003). Biosensors for clinical diagnostics industry. *Sens. Actuators B* 91: 117–127.

106 Schaffar, B.P.H. (2002). Thick film biosensors for metabolites in undiluted whole blood and plasma samples. *Anal. Bioanal. Chem.* 372: 254–260.

107 Le Jeune, K.E. and Russell, A.J. (1996). Covalent binding of a nerve agent hydrolyzing enzyme within polyurethane foams. *Biotechnol. Bioeng.* 51: 450–457.

108 Le Jeune, K.E., Mesiano, A.J., Bower, S.B. et al. (1997). Dramatically stabilized phosphotriesterase–polymers for nerve agent degradation. *Biotechnol. Bioeng.* 54: 105–114.

109 Drevon, G.F., Danielmeier, K., Federspiel, W. et al. (2002). High-activity enzyme-polyurethane coatings. *Biotechnol. Bioeng.* 79: 785–794.

110 Tsuchida, T. and Yoda, K. (1983). Multi-enzyme membrane electrodes for determination of creatinine and creatine in serum. *Clin. Chem.* 29: 51–55.

111 Killard, A.J. and Smyth, M.R. (2000). Creatinine biosensors: principles and designs. *Trends Biotechnol.* 18: 433–437.

112 Schumann, J., Mollering, H., and Jaenicke, R. (1993). Intrinsic stability and extrinsic stabilization of creatinase from *Pseudomonas putida*. *Biol. Chem.* 374: 427–434.

113 Schumann, J., Bohm, G., Schumacher, G. et al. (1993). Stabilization of creatinase from *Pseudomonas putida* by random mutagenesis. *Protein Sci.* 2: 1612–1620.

114 Taylor, J., Zhang, K., and Wang, D. (2019). Industrial and nonfood applications. In: Sorghum and Millets: Chemistry, Technology and Nutritional Attributes, 393–420. Woodhead Publishing.

115 Mathimani, T. and Pugazhendhi, A. (2019). Utilization of algae for biofuel, bio-products and bio-remediation. *Biocatal. Agric. Biotechnol.* 17: 326–330.

116 Gilvari, H., de Jong, W., and Schott, D.L. (2019). Quality parameters relevant for densification of bio-materials: measuring methods and affecting factors. *Biomass Bioenergy* 120: 117–134.

117 Kumar, C., Gaur, A., Tiwari, S. et al. (2019). Bio-waste polymer hybrid as induced piezoelectric material with high energy harvesting efficiency. *Compos. Commun.* 11: 56–61.

118 Selvaraju, S., Ramalingam, S., and Rao, J.R. (2018). Inorganic apatite nanomaterial: modified surface phenomena and its role in developing collagen based polymeric bio-composite (Coll-PLGA/HAp) for biological applications. *Colloids Surf. B Biointerf.* 172: 734–742.

119 Alomair, M., Alomair, Y., Tasnim, S. et al. (2018). Analyses of bio-based nano-PCM filled concentric cylindrical energy storage system in vertical orientation. *J. Energy Storage* 20: 380–394.

120 Nazir, H., Batool, M., Francisco, J. et al. (2019). Recent developments in phase change materials for energy storage applications. *J. Heat Mass Transf.* 129: 491–523.

121 Sadowska, J.P., Borowicz, M., Czupryński, B. et al. (2018). Oenotherabiennis seed oil as an alternative raw material for production of bio-polyol for rigid polyurethane–polyisocyanurate foams. *Ind. Crop. Prod.* 126: 208–217.

122 Ruan, M., Luan, H., Wang, G., and Shen, M. (2019). Bio-polyolssynthesizedfrombio-based1, 3-propanedio land applications on polyurethane reactive hotmelt adhesives. *Ind. Crop. Prod.* 128: 436–444.

123 Boussaba, L., Makhlouf, S., Foufa, A. et al. (2018). The palm kernel vegetable fat: a low-cost biobased phase change material for thermal energy storage in buildings. *J. Build Eng.* https://doi.org/10.1016/j.jobe.2018.10.022.

124 Prianka, T.R., Subhan, N., Reza, H.M. et al. (2018). Recent exploration of bio-mimetic nanomaterial for potential biomedical applications. *Mater. Sci. Eng. C Mater. Biol. Appl.* 93: 1104–1115.

125 Briassoulis, D. and Mistriotis, A. (2018). Key parameters in testing biodegradation of bio-based materials in soil. *Chemosphere* 207: 18–26.

126 Cai, H., Huang, Y.L., and Li, D. (2019). Biological metal–organic frameworks on structures, host–guest chemistry and bio-applications. *Coord. Chem. Rev.* 378: 207–221.

127 Ghashoghchi, R.A., Hosseini, M.R., and Ahmadi, A. (2017). Effects of microbial cells and their associated extracellular polymeric substances on the bio-flocculation of kaolin and quartz. *Appl. Clay Sci.* 138: 81–88.

128 García, S.G., Argiz, L., Miguez, P., and Gullon, B. (2018). Exploring the production of bio-succinic acid from apple pomace using an environmental approach. *Chem. Eng. J.* 350: 982–991.

129 Gu, Y., Xu, X., Wu, Y. et al. (2018). Metabolic engineering advances and prospects of bacillus subtilis cellular factories: from rational design to industrial applications. *Metab. Eng.* 50: 109–121.

130 Jiang, W., Kumar, A., and Adamopoulos, S. (2018). Liquefaction of lignocellulosic materials and its applications in wood adhesives-A review. *Ind. Crop. Prod.* 124: 325–342.

131 Lv, Z., Zhou, Y., Han, S.T., and Roy, V.A.L. (2018). From biomaterial-based data storage to bio-inspired artificial synapse. *Mater. Today* 21: 537–552.

132 Mallakpour, S. and Naghdi, M. (2018). Polymer/SiO_2 nanocomposites: production and applications. *Prog. Mater. Sci.* 97: 409–447.

133 Mathis, D., Blanchet, P., Landry, V., and Lagiere, P. (2019). Thermal characterization of bio-based phase changing materials in decorative wood based panels for thermal energy storage. *Green Energy Environ.* 4 (1): 56–65.

134 Miller, S.A. (2018). Natural fiber textile reinforced bio-based composites: mechanical properties, creep, and environmental impacts. *J. Clean. Prod.* 198: 612–623.

135 Mohamed, M.A., Mutalib, M.A., Hir, Z.A.M. et al. (2017). An overview on cellulose-based material in tailoring bio-hybrid nanostructured photocatalysts for water treatment and renewable energy applications. *Int. J. Biol. Macromol.* 103: 1232–1256.

136 Pittau, F., Krause, F., Lumia, G., and Habert, G. (2018). Fast-growing bio-based materials as an opportunity for storing carbon in exterior walls. *Build. Environ.* 129: 117–129.

137 Razavi, M. Bio-based nanostructured materials. *Nanobiomaterials* https://doi.org/10.1016/B978-0-08-100716-7.00002-7.

138 Rex, E., Rosander, E., Royne, F. et al. (2017). A systems perspective on chemical production from mixed food waste: the case of bio-succinate in Sweden. *Resour. Conserv. Recycl.* 125: 86–97.

139 Santos, F.M. and Pires, J.C.M. (2018). Nutrient recovery from wastewaters by microalgae and its potential application as bio-char. *Bioresour. Technol.* 267: 725–731.

140 Sarma, M.K., Quadir, M.G.A., Bhaduri, R. et al. (2018). Composite polymer coated magnetic nanoparticles based anode enhances dye degradation and power production in microbial fuel cells. *Biosens. Bioelectron.* 119: 94–102.

141 Seng, B., Lorente, S., and Magniont, C. (2017). Scale analysis of heat and moisture transfer through bio-based materials-application to hemp concrete. *Energy Build.* 155: 546–558.

142 Syahir, A.Z., Zulkifli, N.W.M., Masjuki, H.H. et al. (2017). A review on bio-based lubricants and their applications. *J. Clean. Prod.* 168: 997–1016.

143 Rivas, A.T., Palumbo, M., Haddad, A. et al. (2018). Multi-objective optimisation of bio-based thermal insulation materials in building envelopes considering condensation risk. *Appl. Energy* 224: 602–614.

10

Analysis of the Physical and Mechanical Properties of A Biobased Composite with Sisal Powder

Kátia Moreira[1], Thiago Santos[1], Caroliny Santos[1], Rubens Fonseca[1], Moises Melo[2], and Marcos Aquino[1]

[1] *Department of Textile Engineering, Textile Engineering Laboratory, Federal University of Rio Grande do Norte, Natal, RN, Brazil*
[2] *Department of Textile Engineering, Laboratory of Characterization of the Textile Materials, Federal University of Rio Grande do Norte, Natal, RN, Brazil*

10.1 Introduction

Natural fiber polymer composites are being used in a broad variety of applications, because they represent an economical and at the same time an ecological alternative compared to conventional materials derived from petroleum. But, large amounts of organic waste and waste from industrial and agricultural processes are underutilized as low value added materials [1]. Numerous reasons have generated the interest and expectation of scientists and engineers of materials to work with vegetable fibers in polymeric composites replacing the artificial fibers. The source of material is one such reason because plant fibers are available worldwide and require little energy for their production [2, 3]. Some polymer composites reinforced with vegetable fibers are technologically viable and interesting substitutes for wood and similar materials [4].

10.2 Biobased Composites

Among the various natural fibers, sisal fiber stands out as a promising reinforcement in polymer matrix composites, due to its high tensile strength, low density, and low cost [5]. Beyond the conventional application of sisal fiber, it is also used as a reinforcement for polymer composites [6]. Polymer matrix composites are used in several applications because they have important properties. Polymer matrixes are easily available in the market and its values are attractive to allow to work in the manufacture of composites [7].

10.3 Polyester Matrix Composites

Polymers are macromolecules made up of monomers. The monomers are the smallest repeating unit in a polymer. The monomers combine together to form polymers by the process called polymerization. Orthophthalic polyester resin, when used in composites, ensures

Biobased Composites: Processing, Characterization, Properties, and Applications, First Edition.
Anish Khan, Sanjay Mavinkere Rangappa, Suchart Siengchin and Abdullah M. Asiri.

lightness due to its low density, has durability, and shows weather resistance [7]. A composite was developed for the alternative use of the sisal powder residue contaminated with an oil softener, derived from the spinning process of string and ropes produced by the company supplying the material. This contamination prevents the use for animal feed, as is done with the sisal residue from the defibration process, which has no oil. So, the powder was then used as a reinforcement in a polymer matrix of orthophthalic polyester resin.

10.4 Manufacture of Composites

The sisal fiber residue was supplied by a ropes manufacturer in the region. This residue comes from the process of spinning ropes and strings. Initially, the residue was in the form of a bushing (Figure 10.1a). It was then cut and shredded in an industrial blender. Then, the material was subjected to the particle separation process through a sieve of 16 mesh (1190 μm), as shown in Figure 10.1b.

The composite was fabricated using the particle size corresponding to the residue of Figure 10.1b. In some specimens, the styrene monomer was used to dissolve the vinyl unsaturations contained in the unsaturated orthophthalic polyester resin. The curing of the composite was performed with the aid of the methyl ethyl ketone peroxide catalyst. The specimens were fabricated in several percentages (fiber, resin, and styrene monomer) presented in Table 10.1, for the study of the physical and mechanical properties. The molds used were gypsum with a layer of carnauba wax to facilitate the demolding of the samples [8].

10.5 Physical–Mechanical Tests

The tensile tests were performed using the Tensolab 3000 dynamometer from MESDAN, according to the current standard [9]. Samples (Figure 10.2) were tensioned at a rate of 10 mm/min. The MOE, the specific MOE, the stress–strain curve, and the tenacity were analyzed from this test.

The Young MOE was calculated through the ratio between the rupture of maximum stress (T) and strain (ε), as shown in Equation (10.1).

$$\mathrm{MOE} = \frac{T}{\varepsilon} \tag{10.1}$$

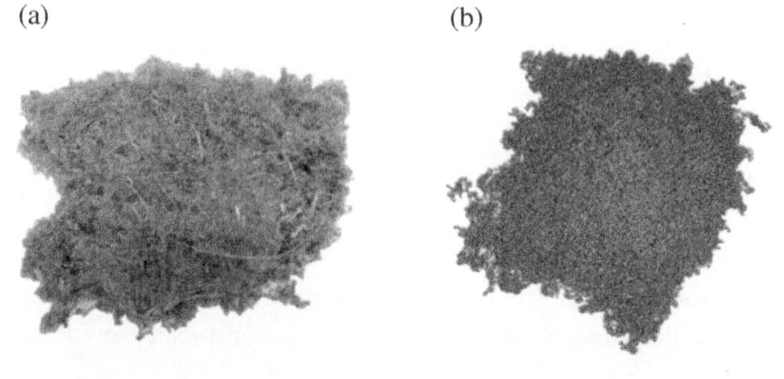

Figure 10.1 (a) Residue; (b) residue after crushing and sifting (particles).

Table 10.1 Classification of the samples.

Sample	Sisal powder (%)	Styrene monomer (%)
Standard CE[a]	0.00	2
2.5 FCE	2.50	2
3.75 FCE	3.75	2
5.0 FCE	5.00	2

[a] In which: CE means with styrene and the means standard CE no fiber.

The specific modulus at break was calculated through the ratio between the Young MOE and specific density of the material, as shown in Equation (10.2).

$$\text{MOE specific} = \frac{\text{MOE}}{\text{specific density}} \tag{10.2}$$

The Young modulus at break is the parameter that indicates how rigid the material is. But the rigidity can also be calculated, as shown in Equation (10.3).

$$k = \frac{A \times \text{MOE}}{L}, \tag{10.3}$$

where A is the area of the transversal section of the sample, and L is the length of the sample.

As can be seen, the MOE is directly proportional to the rigidity, that is, the greater the MOE of the material, the more rigid it will be and, therefore, the lower its strain.

The physical tests carried out contemplate the absorption of water (only for the composite) and the density of the sisal powder. The density of the sisal powder was $1.17 \pm 0.8\,\text{g/cm}^3$, performed according to the ASTM standard that uses the pycnometer method [10]. For the water absorption test, the samples were submitted to humidity and Equation (10.4) was used to determine the percentage of water absorbed [11].

Figure 10.2 Samples for tensile tests with different concentrations of fiber/resin and with a styrene monomer: (a) 2.5% FCE; (b) 3.75% FCE; (c) 5% FCE.

$$\text{Abs} = \frac{M_\text{f} - M_\text{i}}{M_\text{i}} \times 100, \tag{10.4}$$

where Abs is the water absorption in %, M_i is the initial mass, and M_f is the final mass.

The moisture absorption was performed for a period of 24 hours with a total immersion of the sample in water. The microscopic view of the composite sample is to verify the existence of voids (air bubbles). Such bubbles are unavoidable and originate during the production of polymeric composites and can compromise the mechanical properties [12–14]. Finally, the curing time is important for the consolidation of the composite and depends on the choice of parameters such as temperature and time. It is responsible for the final properties of the polymeric composite [15]. In this analysis, the curing time required for each composite was observed in a function of the percentage of fiber in the composition.

10.6 Analysis of Physical and Mechanical Properties

From the stress and strain test, a graph which shows the results for each sample of styrene composite was plotted, in addition to the stress and strain (Figure 10.3).

To investigate the effect of the sisal fiber powder content on the mechanical properties of styrene polymer matrix composites, the tensile strength test was performed. From the graphs of Figure 10.3a and Figure 10.3b, it is possible to perceive that the results of stress tend to increase with the addition of sisal powder. However, there is an optimum point for this result, because after the fiber content reaches 3.75%, there is a decrease in stress and strain values. The optimum point in the percentage of fibers in the composites was observed by Facca et al. [16] in their work with natural fiber composites [17]. From the graph of Figure 10.3b, it is also possible to perceive a similarity between the values of rupture stress and rupture strain of the samples with 2.5 and 5% of the fiber, evidencing that the lower and the greater value of fiber powder of sisal represent the minimum points of the mechanical properties [8, 18].

The MOE represents the rigidity of a material. The specific MOE is the relationship between the MOE and the specific density of the material. This value is mostly used when materials with high MOE value and low density are desired. Figure 10.4 shows the results of MOE and specific MOE at break for the composites with styrene [18].

From the results of Figure 10.4a, it is perceived that the MOE increases as the powder of the sisal fiber is increased. Thus, the rigidity of the composites also increases, because they are directly proportional quantities as shown in Equation (10.3). The increase in the MOE of the composite of the sisal fiber may be related to the high Young modulus of the sisal fiber (9000–20000 MPa) relative to the value found for the polyester resin (approximately 11 MPa) [3, 16].

It is also possible to perceive that as the specific MOE increases, that is, the ratio between the MOE and the density is increasing as the powder content of the sisal fiber is increased.

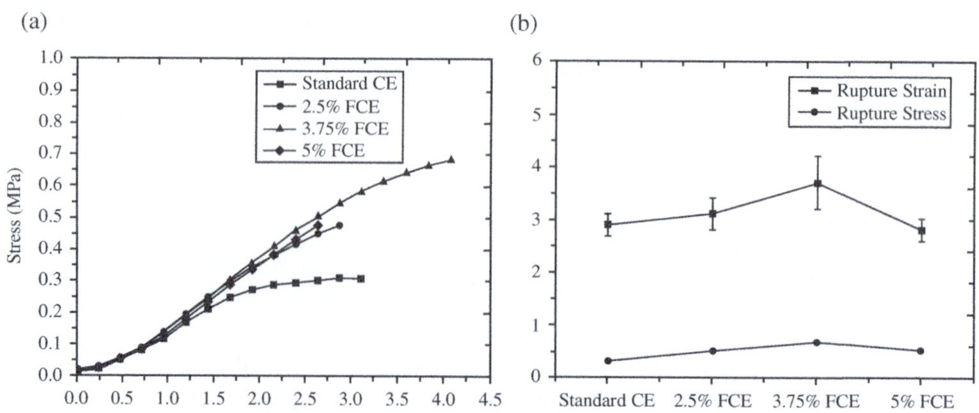

Figure 10.3 (a) Curved stress and strain of samples with styrene; (b) stress and strain of the samples with styrene.

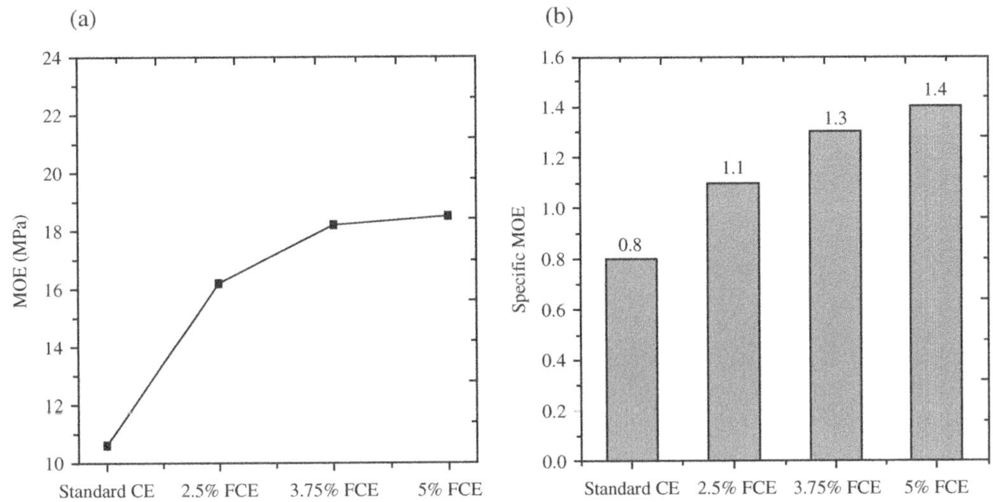

Figure 10.4 Results of (a) Young modulus at break and (b) specific modulus at break of the samples with styrene.

Thus, it is evident that the increase of sisal fiber powder improves the specific MOE results without compromising the density of the composite as shown in Figure 10.4a, b.

Tenacity is the measure of energy absorbed during a tensile test before the break. Figure 10.5 shows the graph of tenacity for the styrene composites.

It is noticed that tensile tenacity values are very similar for samples with 0% (standard), 2.5%, and 5% of sisal fiber powder. However, for the sample with 3.75% sisal fiber powder, the toughness increases by 150% in relation to the standard sample, and for the sample with 2.5% sisal fiber powder, it is 117.4% in relation to the standard sample, as shown in Figure 10.5. Since the stress and the strain of the composite were high (as shown in the graph of Figure 10.3), the tenacity (which is the area of the graph below the stress and strain curve) is also of the highest value. Therefore, the composite with 3.75% fiber powder has the highest energy absorbed until the sample ruptures. Thus, this composite is the most tenacious among the samples tested as show Figure 10.5 [18].

Comparing the MOE value of the graph of Figure 10.4b, it is observed that the highest value is for the 5% composite, due to the high Young MOE of the sisal fiber. However, it is important to remember that this value is also the highest due to the low strain of the composite. Therefore, for the tenacity test, the 3.75% composite has high stress and strain values. The samples with styrene thus presented the best results when compared with the literature.

A morphological analysis of the particulate composite with styrene was realized aiming to evaluate the presence of voids in the samples studied.

Figure 10.6 shows the composite with the presence of a few voids, or air bubbles. Thus, it can be observed that the production process practically did not intervene in the results of the mechanical properties of the composite, because there is no significant presence of the air bubbles that decrease the mechanical resistance of the material [19, 20]. Figure 10.7 shows the curing time results for each styrene composite as a function of the amount of the powder of the sisal fibers.

As can be observed, all the composite samples showed a significant increase in curing time, directly proportional to the amount of fiber powder (2.5, 3.75, and 5%) when compared with the standard sample CE [21]. This phenomenon is important for large-scale

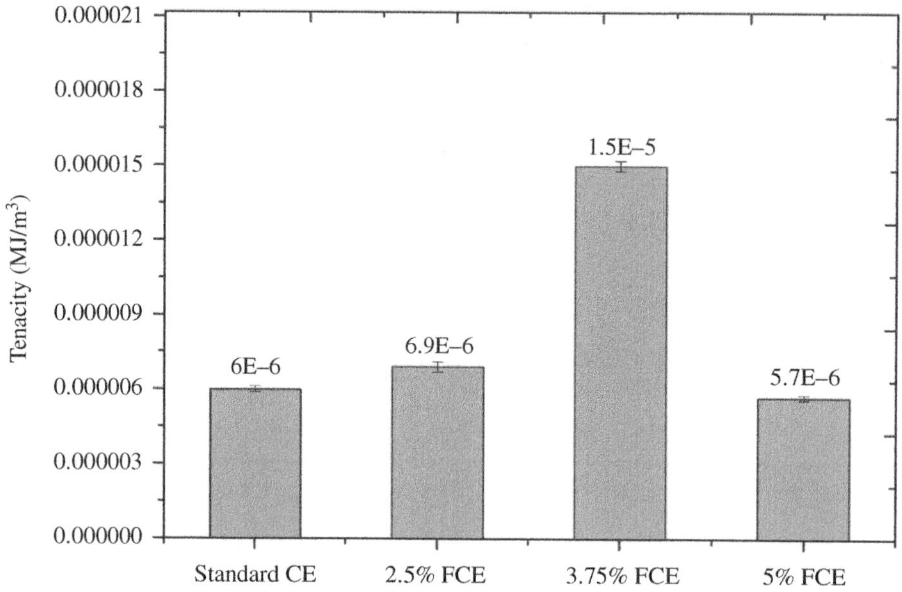

Figure 10.5 Results of tenacity of the samples studied.

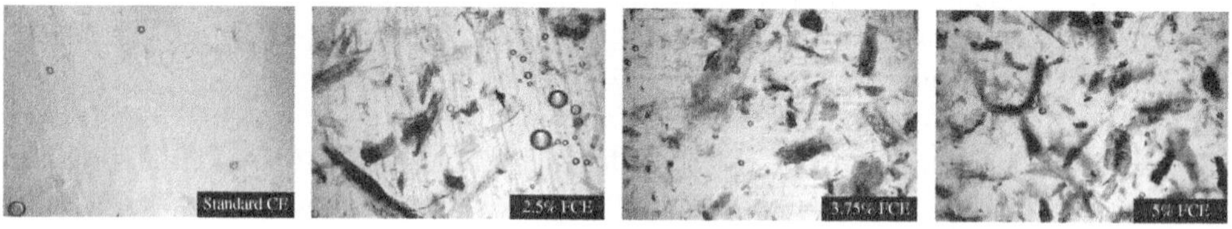

Figure 10.6 Morphology of the samples with styrene.

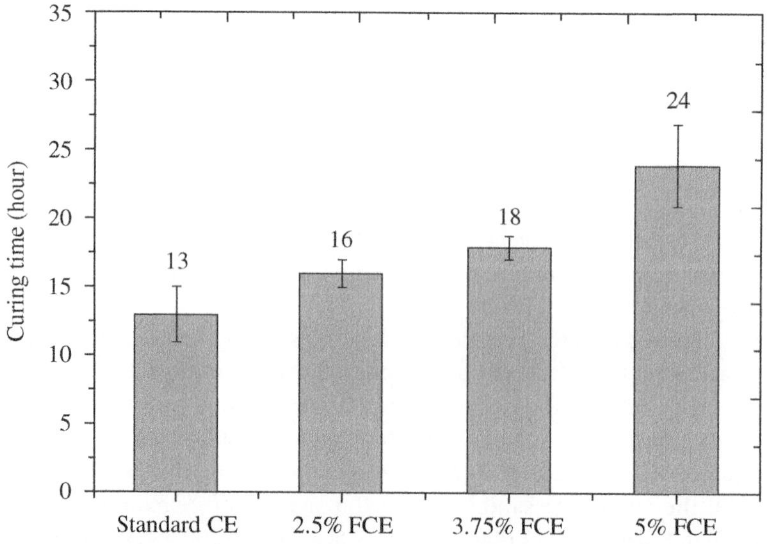

Figure 10.7 Curing time of the samples with styrene.

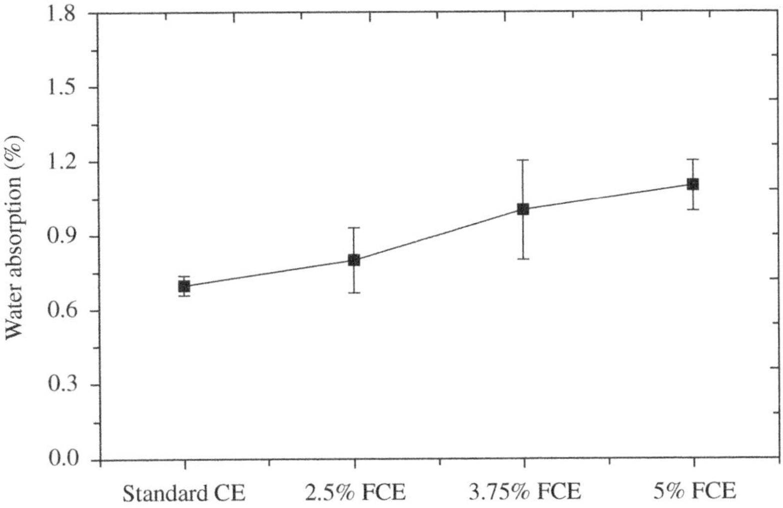

Figure 10.8 Results of water absorption of the samples studied.

production, since it directly influences the productivity of the company. Figure 10.8 shows the water absorption results for each styrene composite in a function of the amount of the powder of the sisal fibers.

The percentage of water absorption (weight gain) in various composites (CE standard, 2.5% FCE, 3.75% FCE, and 5% FCE) for 24 hours is shown in Figure 10.8. It is noted that the increase of the percentage of sisal fiber powder favors the water absorption when compared to the standard sample CE. This behavior was expected because the water absorption of these composites occurs mainly due to the hydrophilic nature of the sisal fibers and this increases with the increase in the content of the sisal fiber powder [22–25]. However, the percentage of water absorption was not very significant, since the unsaturated polyester resin has a hydrophobic character (Ref. [16]). Moreover, the fiber of the sisal powder used has an industrial oil layer that contributes to a certain hydrophobicity on its surface, facilitating the adhesion of the fiber to the resin, reducing the absorption of water by the composite [26–28].

10.7 Conclusions

The mechanical properties of the three concentrations of fiber powder (2.5, 3.75, and 5%) with a particle size of 16 mesh of polymeric composites were studied. Insertion of the powder from the sisal fibers into the unsaturated polyester resin significantly improved the properties studied. It was also observed that the addition of 2% of styrene monomer promotes the improvement of the mechanical strength of the composite as shown in the stress and strain graph. The physical and mechanical performance of the composite samples is substantially dominated by the fiber content. All the composites presented low water absorption and also low voids, which is an excellent result when using natural fiber composites in a polymer matrix. Microstructural analysis and durability assessment will be the scope of future investigations. However, the best results of the physical (water absorption) and mechanical properties (stress, strain, tenacity, Young modulus at break and specific modulus at break) were obtained for 3.75% of sisal fiber powder content in weight.

Acknowledgments

The authors thank the textile technology study group (GETTEX), Textile Quality Control Laboratory (LABCTEX), and (PPgET) Textile Engineering Post Graduate Program at the Federal University of Rio Grande do Norte for their support and assistance.

References

1 Väisänen, T., Haapala, A., Lappalainen, R. et al. (2016). Utilization of agricultural and forest industry waste and residues in natural fiber–polymer composites: a review. *Waste Management* 54: 62–73.

2 de Silva, F.A., RDT, F., de Filho, J.A.M. et al. (2010). Physical and mechanical properties of durable sisal fiber–cement composites. *Construction and Building Materials* 24: 777–785.

3 Joseph, P.V., Joseph, K., and Thomas, S. (1999). Effect of processing variables on the mechanical properties of sisal-fiber-reinforced polypropylene composites. *Composites Science and Technology*: 1625–1640.

4 Saxena, M., Morchhale, R.K., Asokan, P. et al. (2008). Plant fiber – industrial waste reinforced polymer composites as a potential wood substitute material. *Journal of Composite Materials* https://doi.org/10.1177/0021998307087014 (Epub ahead of print).

5 Jacob, M., Thomas, S., and Varughese, K.T. (2004). Mechanical properties of sisal/oil palm hybrid fiber reinforced natural rubber composites. *Composites Science and Technology* 64: 955–965.

6 Li, Y., Mai, Y.-W., and Ye, L. (2000). Sisal fibre and its composites: a review of recent developments. *Composites Science and Technology* 60: 2037–2055.

7 de Melo, K.M., dos Santos, T.F., da Santos, C.M.S. et al. (2019). Study of the reuse potential of the sisal fibers powder as a particulate material in polymer composites. *J Mater Res Technol.* https://doi.org/10.1016/j.jmrt.2019.07.010 (Epub ahead of print July).

8 Khan, A. et al. (eds.) (2020). *Biofibers and Biopolymers for Biocomposites*. https://doi.org/10.1007/978-3-030-40301-0_11. Springer Nature Switzerland AG.

9 Santos Júnior, Z.J. (2017). *Fabricação e caracterização de um compósito com matriz de resina poliéster ortoftálica e carga de fibras de agave angustifolia haw*, Natal, Brazil. https://repositorio.ufrn.br/jspui/handle/123456789/24658.

10 ASTM D3039/D3039M-17 (2017). *Standard Test Method for Tensile Properties of Polymer Matrix Composite Materials*. West Conshohocken, PA: ASTM International https://doi.org/10.1520/D3039_D3039M-17.

11 ASTM D2320-98 (2017). *Standard Test Method for Density (Relative Density) of Solid Pitch (Pycnometer Method)*. West Conshohocken, PA: ASTM International https://doi.org/10.1520/D2320-98R17.

12 ASTM D570-98 (2018). *Standard Test Method for Water Absorption of Plastics*. West Conshohocken, PA: ASTM International https://doi.org/10.1520/D0570-98R18.

13 Oliveira, A., Becker, C.M., and Amico, S.C. (2014). Efeito de aditivos desaerantes nas características de compósitos de epóxi/fibras de vidro. *Polímeros Ciência e Tecnologia* 24: 117–122.

14 Grunenfelder, L.K. and Nutt, S.R. (2010). Void formation in composite prepregs – effect of dissolved moisture. *Composites Science and Technology* 70: 2304–2309.

15 He, S., Wen, Y., Yu, W. et al. (2017). Study on voids of epoxy matrix composites sandwich structure parts. *IOP Conference Series: Materials Science and Engineering* 182: 012031.

16 Facca, A.G., Kortschot, M.T., and Yan, N. (2007). Predicting the tensile strength of natural fibre reinforced thermoplastics. *Composites Science and Technology* 67: 2454–2466.

17 Dmitriev, O. and Mischenko, S. (2011). Optimization of curing cycles for thick-wall products of the polymeric composite materials. In: *Advances in Composites Materials – Eco Design and Analysis*, 141–160. IntechOpen https://doi.org/10.5772/14926.

18 Khan, A. et al. (eds.) (2020). *Hybrid Fiber Composites Materials, Manufacturing, Process Engineering*. Wiley-VCH Verlag GmbH.

19 Sanchez, E.M.S., Zavaglia, C.A.C., and Felisberti, M.I. (2000). Unsaturated polyester resins: influence of the styrene concentration on the miscibility and mechanical properties. *Polymer* 41: 765–769.

20 Madsen, B. and Lilholt, H. (2003). Physical and mechanical properties of unidirectional plant fibre composites – an evaluation of the influence of porosity. *Composites Science and Technology* 63: 1265–1272.

21 Senthilkumar, K., Saba, N., Rajini, N. et al. (2018). Mechanical properties evaluation of sisal fibre reinforced polymer composites: a review. *Construction and Building Materials* 174: 713–729.

22 Becker, O., Cheng, Y.-B., Varley, R.J. et al. (2003). Layered silicate nanocomposites based on various high-functionality epoxy resins: the influence of cure temperature on morphology, mechanical properties, and free volume. *Macromolecules* https://doi.org/10.1021/ma0213448 (Epub ahead of print).

23 Naik, L.L., Gopalakrishna, K., and Yogesha, B. (2016). Physical and water absorption behaviour of sisal–glass fiber reinforced polymer based composites. *American Journal of Materials Science* 6: 91–94.

24 Gupta, M.K., Srivastava, R.K., Kumar, S. et al. (2015). Mechanical and water absorption properties of hybrid sisal/glass fibre reinforced epoxy composite. *Gupta MK American Journal of Polymer Science & Engineering* 3: 208–219.

25 Haghdan, S. and Smith, G.D. (2015). Natural fiber reinforced polyester composites: a literature review. *Journal of Reinforced Plastics and Composites* 34: 1179–1190.

26 Prasanna Venkatesh, R., Ramanathan, K., and Srinivasa Raman, V. (2016). Tensile, flexual, impact and water absorption properties of natural fibre reinforced polyester hybrid composites. *Fibres and Textiles in Eastern Europe* 24: 90–94.

27 Saheb, D.N. and Jog, J.P. (1999). Natural fiber polymer composites: a review. *Advances in Polymer Technology* 18: 351–363.

28 Ali, A., Shaker, K., Nawab, Y. et al. (2018). Hydrophobic treatment of natural fibers and their composites – a review. *Journal of Industrial Textiles* 47 (8): 2153–2183.

11

Physico-Mechanical Properties of Biobased Composites

A. V. Kiruthika

Seethalakshmi Achi College for Women, Pallathur, Tamil Nadu, India

11.1 Introduction

Composite materials have been an integral part of human civilization since its inception. It is a heterogeneous material, composed of at least two or more immiscible components, which combine to provide superior properties to those of the individual constituents. Composite materials can be classified according to their fabrication process (Figure 11.1), such as open mold, closed mold, and cast polymer molding. In all fabrication methods, the use of tools like the dye, mold, and mandrel is common, and they provide the structural shape to the composite material.

Biobased composites are a combination of novel type natural cellulose fibers reinforced with matrix (resins) materials manufactured from recycled raw materials in whole or in part. Biocomposites are an enticing combination of outstanding mechanical characteristics, are light weight, and have low impacts on the environment during processing, fabrication, and production time. In recent years, the efficiency and applicability of these composites have increased significantly with the growing availability of biobased (matrix and fiber) materials for composite applications. The demand for these composites has increased along the decades, leading to their considerable usage in the automobile, construction, packaging, biomedical, military, and agricultural fields. Currently, many scientists and technologists have shifted their research toward biobased composites, because (i) it is recyclable; (ii) it has the ability to absorb vibration and has sound proofing properties [1]; (iii) the cost of biocomposites reinforced with natural fiber is 2–3 times lesser than polymers reinforced with synthetic fiber; (iv) the incorporation of natural fibers with natural resins yields increased tensile strength and also results in 100% biodegradable material; (v) when burnt, the composites containing natural fibers produce less CO_2, CO, and toxic gases; (vi) it has low maintenance; and (vii) it has better design flexibility. The classification of biobased composites according to their origin is given in Figure 11.2.

Biobased composites are biocompatible, biodegradable, and are not accumulated in the environment. However, their disposal can cause problems when biobased matrices or fibers are chemically modified or embedded and mixed in synthetic matrices. At the end of the product's life cycle, disposal options require the materials to be completely degraded by environmental influences including temperature, UV light, water, and microbial attack.

Biobased Composites: Processing, Characterization, Properties, and Applications, First Edition.
Anish Khan, Sanjay Mavinkere Rangappa, Suchart Siengchin and Abdullah M. Asiri.

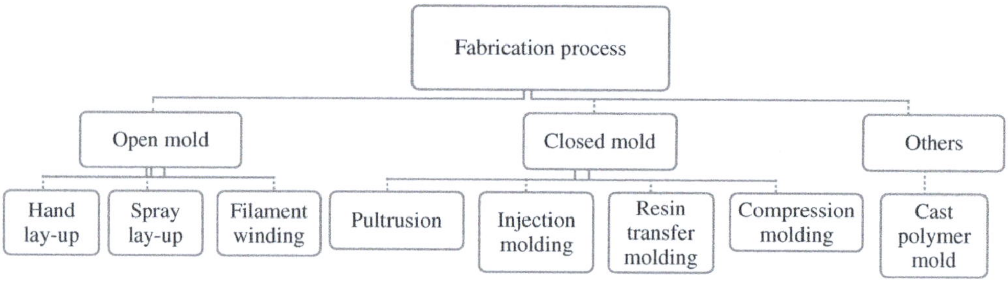

Figure 11.1 Various types of fabrication processes of composite materials.

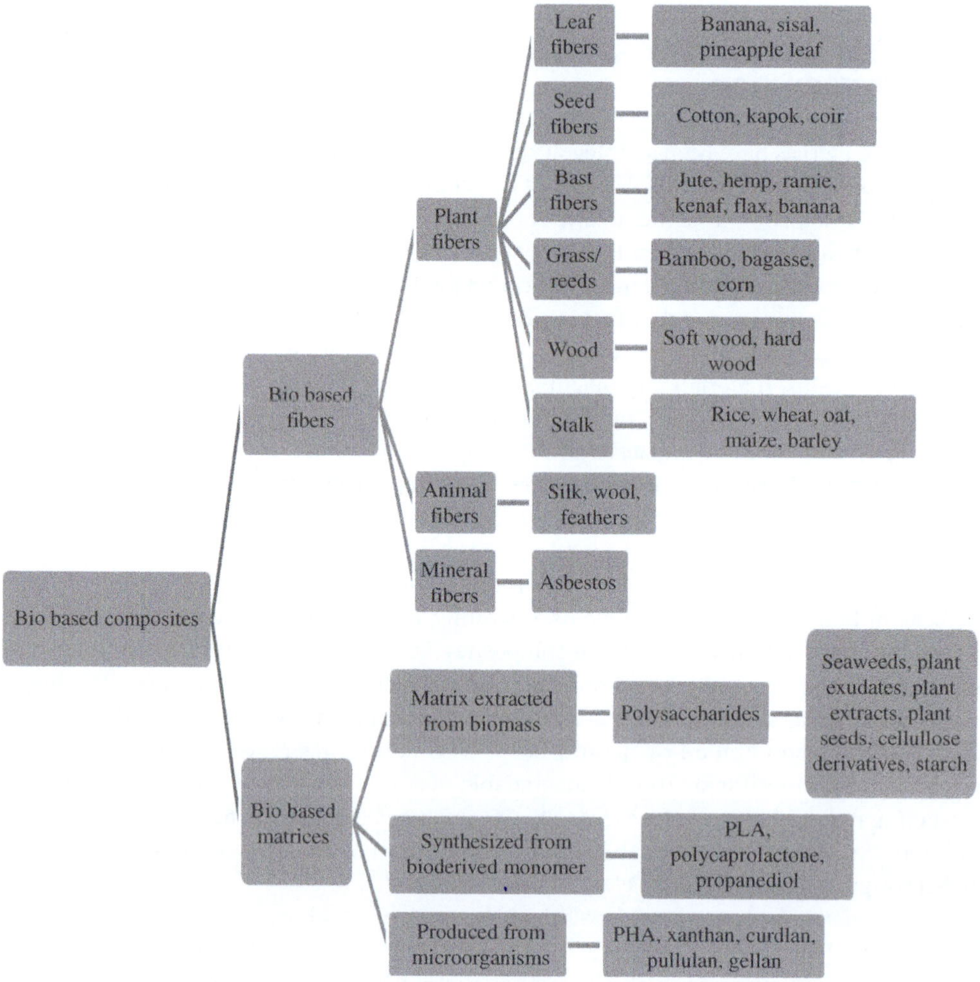

Figure 11.2 Classification of biobased composites according to their origin.

11.1.1 Biobased Fibers

Over the past few decades, human beings have used glass fibers as a reinforcement for polymer composites, but they have some drawbacks such as being nonrenewable, cause health and safety issues, and they cannot be recycled thermally by incineration. Instead of synthetic fibers, the use of natural fibers as reinforcement materials has received greater attention in the composite material manufacturing industries. Also, there is an increasing demand in environmental awareness among mankind, which makes them extensively use natural products. Currently, there are various kinds of natural fibers, often known as biobased fibers, which are available abundantly in nature, such as banana, sisal, coir, kenaf, flax, bagasse, jute, bamboo, ramie, hemp, pineapple leaf, straw, cotton, and wood for the preparation of composites. Variations in the physical properties of the various biobased fibers are highly dependent upon the chemical ingredients, geometry of the fiber, degree of polymerization, and the weather conditions at the time of growth.

11.1.2 Biobased Matrices

A biobased composite's mechanical characteristics also depend on the properties of thermoplastic or thermoset polymer matrices. The raw material for the manufacture of the above-said matrix may be either biobased natural (plant or animal), i.e. completely from renewable materials, or synthetic from oil resources [2]. Biobased matrix (resins) is defined as the combination of petroleum-based resins as the primary component and natural fibers as the secondary component. The products extracted from vegetable oils are sustainable and more biodegradable, hence they have less impact on the environment. Many studies have been reported on bioresins, such as epoxidized palm oil [3], soybean oil [4, 5], hempseed oil [6], CNSL or cardanol [7–9], starch [10, 11], natural rubber (NR) [12–14], etc. and on partly ecofriendly resins such as PLA [15] (Aliotta et al. [16]) and PHA [17] (Jiang et al. [18]).

11.2 Physico-Mechanical Property of the Biobased Composites

The simplest and easiest quality control experiment on composite materials is the determination of physical properties like density, porosity, etc.

11.2.1 Density of Biobased Composites

Density is an important feature that should be taken into consideration during the process of material selection and it is calculated with the mass of the specimens by its volume. This section discusses the density and void content of fiber, matrix, and composites.

Renewable foams formed from cassava starch and grape stalks with glycerol as a plasticizing agent were analyzed. In order to determine the true density for the prepared samples by helium gas pycnometry process, the following porosity formula was used

$$\text{Porosity}(\%) = 1 - \frac{\rho_b}{\rho_t} \times 100,$$

where ρ_b denotes bulk density and ρ_t denotes true density. The foam blend with resin and grape stalks gave a low density value of 0.18 g/cm^3, because of the hydrophilic nature of grape stalks. At the same time, the virgin cassava starch has 0.21 g/cm^3 and this is attributed to the hydrophobic nature of starch.

Biodegradable starch reinforced with date palm and flax fiber hybrid composites were prepared with varying fiber content and resin ratios [11]. Mettler Toledo densitometer was used to measure the density of the fiber with the immersing liquid as xylene, and for composites the liquid taken is distilled water. The density of raw fibers of date palm was 0.67 g/cm^3, whereas it was raised to 1.35 g/cm^3 after treatment with an alkali. The density value for the neat starch resin was 1.45 g/cm^3 and for hybrid (50 wt%) composites it was 1.403 g/cm^3.A considerable reduction in density was reached while incorporating the sugar palm fiber (SPF) at various loadings of 0, 2, 4, 6, and 8% into the cornstarch-based composites. Also, the highest density value of 1.55 ± 0.08 g/cm^3 was obtained for the neat cornstarch film [19].

Novel trays from cassava starch reinforced with sugarcane bagasse (SB), corn husk (CH), malt bagasse (MB), and orange bagasse (OB) were prepared [20]. The thickness of these biodegradable trays is higher for various components varying from 3.50 to 4.21 mm when compared to commercial trays (2.43 ± 0.47 mm). MB trays have the highest thickness value whereas the other trays have lower thickness values. According to density measurement, the values varied from 0.12 ± 0.02 to 0.21 ± 0.01 g/cm^3 for all the novel trays. Only the OB tray differed considerably from all the other trays, because more lipids (above 2 kDa) were highly concentrated in it.

Composites made from PLA/wood and man-made cellulosic fibers were fabricated by the hot pressing method [21]. The density value is increased from 1.47 g/cm^3 for the birch fiber to 1.59 g/cm^3 for softwood fibers. Similarly for the composites, the birch (bleached) has 1.02 g/cm^3 at 60 wt% of fiber whereas it has 1.04 g/cm^3 for softwood (bleached) fibers. The void content was very high and it increases dramatically when the fiber content has been changed from 40 to 60%. There is no evidence in void content made from different fibers of composites.

The influence of sodium hydroxide treatment on kenaf fiber in addition with NR/thermoplastic polyurethane (TPU) composites were prepared by using the hot pressing method [14]. Density was measured as 1.13 g/cm^3 for TPU, 1.4 g/cm^3 for kenaf fiber and 0.93 g/cm^3 for NR. Compared to resins used, the kenaf fiber has the highest density value, but 12.5% of NaOH treated fiber does not have a significant impact on the density of TPU composites or NR composites.

NR films could be enhanced through incorporation of Sago starch (SS) dispersion which was usually prepared using various kinds of dispersing agents including sodium alginate (SA), anchoid, and carboxymethyl cellulose (CMC). The influence of these agents on crosslink density of NR latex films were analyzed [22]. Using Flory–Rehner theory, the crosslink density (X) value was found. The experiment results indicated that crosslink density was higher ($X = 0.0483$ mol/cm^3) for NR latex with SS prepared with SA. This is because of a strong interaction produced between the SS and the dispersing agents, which leads the NR compounds to be a stronger contact with rubber-filler.

Biobased composites are prepared from bamboo fibers (BF) as a filler and styrene-free soybean oil as resin. The influence of reactive diluents (RD) such as 1,4-butanediol dimethacrylate (BDDMA) and trimethylolpropane trimethacrylate (TMPTMA) on acrylated epoxidized soybean oil (AESO) were studied. The crosslink density of composites followed the order of BFS/TMPTMA-AESO > BF/BDDMA-AESO > BF/pure AESO > BF/st-AESO and the values are 206.3, 167.7, 157.3, and 129.2, respectively [4].

Several modifications are employed to enhance, via the inclusion of isophorone diisocyanate (IDPI), the properties of the biobased composites from hemp fibers (HFs) and acrylated epoxidized soybean oil (AESO). According to rubber elastic theory, the crosslink density (ν_e) of the composite were calculated by

$$\nu_e = \frac{E'}{RT}.$$

The ν_e values of control is 3.28×10^2, IPDI-10 is 2.99×10^2, and IPDI-20 is $4.15\times10^2\,mol/m^3$. Compared to other crosslinking density values, the IDPI-20 showed the highest value which is due to the significant enhancement in crosslinking density of the composites when 20% of IDPI was used for modification [5].

An ecofriendly composite from natural fibers such as BF and HF reinforced with four biobased thermosets like AESO, AESO with methacrylated isosorbide (MI-AESO), methacrylated AESO with MI (MI-MAESO), and MI matrices were fabricated [23]. The influence of two fibers and four matrices on the crosslinking density measurement was analyzed. Among all resins, MI resin showed the highest density. The addition of MI with AESO produced the cured MI-AESO resin with an improved density than AESO, because of its increased crosslinking degree. However, compared to MI-AESO resin, the increase in unsaturated AESO level due to methacrylic anhydride (MAA) modification did not alter the density of MI-AESO resin substantially.

The density measurements on new biodegradable composites made from peach palm fibers and poly(butylene adipate-co-terephthalate) (PBAT) were evaluated [24]. The fibers were given with or without surface treatment in the presence of glycidoxypropyl-trimethoxysilane (GPTMS) for the matrix. Nonfoamed composites showed a slight rise in density ($1.31\,g/cm^3$) as the volume of fiber increased and this property was not influenced by the functionalization of the fiber. Obviously, due to the presence of pore derived from the foaming phase, the foamed samples showed a significant decrease in density ($1.07\,g/cm^3$). For the foamed materials, the influence of the fiber content on the density of the materials is more emphasized.

11.2.2 Mechanical Properties of Biobased Composites

It is one of the properties which plays a prominent role in the composite world. The mechanical properties of various fiber reinforced polymer matrix composites were discussed as follows:

A nontoxic hybrid composite materials from CS blended with fillers including corn husk (CHF) and SPF by solution casting techniques were evaluated [19]. Fructose plasticizer with 25% conc. along with CHF of 8% conc. was included to the matrix. SPF at various dry starch-based loadings (0, 2, 4, 6, 8%) was used as a hybridizing agent. The blending of SPF (80%) with CS-CHF increased the thickness value and similar results were observed for moisture content, water solubility, and water absorption (WA) characteristics. The influence of SPF with different conc. increased the mechanical properties of hybrid composites. For the films (6% conc. SPF), the tensile strength were raised from 6.8 to 19.05 MPa whereas elasticity modulus increased from 61.15 to 1133.47 MPa, respectively. Due to the hydrophilic nature of starches, the highest rate of water vapor permeability was observed in the neat CS film and the value is $2.28\times10^{-10}\,mmgs^{-1}\,m^{-2}\,Pa^{-1}$.

Renewable resource-based composites from Cassava starch reinforced with different amounts of coir fibers (5, 10, 15, 20, 25, and 30%) were prepared [25] by thermal molding technique using glycerol as a plasticizer (for starch). The addition of coir fiber content increased the mechanical properties of the cassava starch. The tensile strength, Young modulus, and maximum load were 1.56 MPa, 14.56 MPa, and 50.77 N, respectively, for the untreated composites, whereas it was higher for treated ones (3.24 MPa, 59.81 MPa, and 112.68 N). Similar to the Young modulus, with increasing fiber content, the maximum values of ~373.5 and ~176 MPa were noted for treated and untreated composites, respectively. Elongation at break (%) decreased with increasing fiber content for both types of composites due to the excellent adhesion between the matrix and coir fibers, water uptake, swelling, and moisture absorption. At two hours immersion period, the percentage of water uptake is 60 and 38% for TPS matrix and its composites, respectively, with 30% coir content.

Lomelí-Ramírez et al. [26] fabricated an ecofriendly green composite from cassava starch reinforced with green coconut fiber (up to 30 wt%) by thermal molding process. TGA curves predicted a gradual loss of mass for cassava starch and its composites at 25–200 °C, which occurs due to the loss of water molecules. The result also revealed that with the increasing content of the coir fiber from 0 to 30 wt%, there occurred an increase in relative crystallinity (from 39 to 62%), an increase in enhanced thermal stability (indicated by decreasing degradation temperature), and an increase in storage modulus (from 2027 to 3215 MPa).

DMTA showed that the storage modulus and glass transition temperature was increased, while damping decreased with increasing fiber content.

Rodney et al. [27] investigated the characteristics of tea tree (Melaleuca alternifolia) fiber reinforced with tapioca starch (TS) composites. Different parts of the tree trunk, such as tea tree leaf (TTL), a tea tree branch (TTB), and a tea tree trunk (TTT), were extracted and treated with 17.5 and 8.3% NaOH solution at 25 °C. According to the chemical composition of the tea tree fiber, the highest cellulose content (33.9%) for TTT and 0.42 g/cm^2 for TTL was noted. For composites, higher thickness (1.70 mm) and density (1.13 g/cm^3) is noted for TTL/TS, followed by TTB/TS and TTT/TS. The percentage of WA and thickness swelling rate was lower for composites with the increasing pattern of tea tree fiber contents. But it was higher for TS, i.e. 37.64 ± 0.66 for WA and 16.96 ± 0.74 for thickness swelling. This is attributed to the hydrophobic characters of tea tree fiber compared to the matrix taken. The percentage of cellulose content (33.9%) present in TTT was higher, resulting in highest tensile strength (203.18%), while for TTL/TS and TTB/TS it was 34.39 and 82.80%, respectively.

Thermoplastic cassava starch (TPCS)-based composites reinforced with cogon grass fiber (CGF) with different fiber loadings at 1, 3, and 5 wt% were obtained by the compression molding process [10]. The moisture content rate was decreased from 6.4% for TPCS solution to 5.9% for TPCS/CGF (at 1%). Also an increase in moisture was decreased from 5.9 to 7.9% for TPCS/CGF (at 5%). WA results for all the samples indicated the increasing with decreasing manner from 48.22 wt% (TPCS/CGF at 1%) to 45.23 wt% (TPCS/CGF at 5%) at 0.5 hours. A reduction in water solubility of the composites was observed with the introduction of CGF into TPCS. The neat TCPS has 36.3% solubility whereas it was 28.70% for TPCS/CGF at 1%, 28.90% for TPCS/CGF at 3, and 28.48% for TPCS/CGF at 5%, respectively. Higher water solubility of neat TPCS than the composites were due to the disintegration and dissolving capacity of matrix materials.

The effect of crystallinity, WA, thermal, and mechanical properties on natural fiber (sisal and hemp) reinforced thermoplastic starch (TPS) were characterized by the melt processing technique [28]. An increase in mechanical analyses such as Young modulus and tensile strength was noted for natural fiber reinforced TPS with 20% fiber content. A similar effect was observed for fiber reinforced TPS with the inclusion of natural latex (plasticizer) in the composites. The percentage of elongation at break was higher for neat TPS ($20 \pm 6\%$) and for TPS with latex ($27 \pm 7\%$). HF enhanced excellent mechanical characteristics in the composites than those of sisal fiber reinforced TPS composites. A higher crystallinity of 0.477 is noticed for neat TPS and 0.512 for TPS with the addition of latex. This behavior was followed to TPS with 10% of HF and TPS with 5% of sisal fiber. Diffusion coefficient of TPS with 5% of hemp has 3.37 cm^2/s whereas it was 2.40 cm^2/s for neat TPS with latex-treated composites. The influence of natural latex as a plasticizer decreased the WA characteristics without affecting the thermal and mechanical properties of the composites.

A new methodology was applied for preparing the thermoplastic starch (TPCS) reinforced kenaf fibers (at 2, 4, 6, 8, 10 wt%) using solution casting techniques [29]. To obtain cellulosic nanocrystals (CNCs), the fibers were first treated with NaOH, bleached with sodium chloride, then with an acetic buffer solution, and finally hydrolyzed. Plasticizers such as glycerol/sorbitol with various filler compositions (0–10 wt%) blended with starch was used for the composite preparations. TPCS-CNCs (6%) showed an increase in tensile strength (8.2 MPa) with increase of fiber loadings. Neat TPCS film had the lowest tensile strength of 3.5 MPa and acid hydrolysis had an enhancement of 91–134%. An increasing trend in the tensile modulus was noticed for the TPCS-CNCs composites with the addition of nanocellulose. The highest stiffness for all fiber composition with 326% MPa was noted for the same composites. A gradual increment in percentage of crystallinity for the kenaf treated fibers and CNCs (81.8%) were observed, followed by bleached kenaf (72.8%), acid hydrolyzed kenaf (68.2%), and raw kenaf fibers (60.8%), respectively.

Balaji et al. [7] studied the incorporation of treated and untreated bagasse fibers (with a length of 10 and 20 mm) into cardanol and characterized the mechanical, morphological, and thermal properties of the green composites. Compared to 0, 5, 10, 15, 20 wt% of composites, the highest tensile strength (28 MPa) and impact strength (2.5 kJ/m^2) were achieved for NaOH-treated fiber.

BF was used as a filler to make a new biodegradable composite using the cardanol–formaldehyde thermoset resin as a polymer matrix. The composites were fabricated by the hand layup process and the performance was achieved by modifying the fibers with alkali treatment (NaOH at 5 and 10%) and bleaching with sodium hypochlorite at 60–75 °C. The alkali-treated fibers have the highest tensile strength, breaking load, tensile modulus, and crystallinity index, i.e. 212 MPa, 358 N, 5.3 GPa, and 48.28%, respectively. Contrarily, elongation at break (4.82%) is higher for 5% alkali-treated fibers. The XRD pattern of the treated (NaOH) and untreated BF showed the major peaks at 2θ diffraction angles of 18.2° and 26.0° for all fiber samples, respectively [30].

Coir fiber reinforced cashew nut shell liquid (CNSL) composites have been processed at different fiber weight fractions (0 to 40%) by the hand layup technique [8]. Studies revealed that the tensile strength of biocomposites was increased, as the fiber content increases. Especially, the alkali-treated coir fiber reinforced CNSL composites had a higher tensile strength than untreated ones. At 30% wt of fiber content, the tensile strength was raised to 70.9% for alkali-treated fibers from 39.0% for untreated fibers. In the same way, the tensile modulus of the composites increased, as the fiber content increases and the optimum strength is attained at 30% wt fraction of coir fibers.

The effect of fiber content variation (0, 5, 10, 15, 20, and 25% wt) and particle sizes (400–600 μm) on the mechanical properties of oil palm mesocarp particle (OPMP) reinforced cashew nut shell novalac resin (CNSR) composites by compression molding process was studied [9]. In accordance to ASTM D638 and ASTM D790, tensile and flexural properties were performed. The good interfacial bonding between the reinforced OPMP and CNSR matrix results in an enhancement in tensile strength with the increasing content of OPMP. The maximum tensile strength (~60.48 MPa) was obtained for OPMP/CNSR composites with 20 wt%. reinforcing particles (of 400 μm). A similar trend was observed for Young modulus and flexural properties of OPMR/CNSR composites. At 20% OPMP reinforcement, a maximum of Young modulus of 940.57 MPa for 400 μm, 905.24 MPa for 500 μm, and 879.88 MPa for 600 μm of particle sizes, respectively, was obtained. Moisture absorption increases with increasing particle size from 400 to 600 μm. Hence, the absorption of water increases gradually with increasing content of particulate reinforcement and particle sizes until equilibrium is reached.

Udhayasankar and Karthikeyan [31] studied the effect of particle size at 25, 50, and 75 μm and particle loading of 0, 10, 20, 30, and 40% on the cardanol resin (CR) enhanced with coconut shell particle (CSP) by the compression molding method. The result suggested that compared to neat CR, tensile strength increased by 31, 25, and 19% for 25, 50, and 75 μm particle sizes, respectively, at 30% of CSP/CR composites. At the optimum weight of fiber, the same results were noted for flexural strength by 46, 37, and 27%, respectively, for the above said particle sizes.

The mechanical, dynamic mechanical, and morphological analysis of bleached jute fibers (BJF) reinforced with NR composites were studied [12]. A reduction in tensile strength (17–15 MPa) was observed on the addition of 10% bleached jute fibers to NR. Similarly, the tensile modulus values, $M_{100} = 2.10$ MPa and $M_{300} = 2.60$ MPa, indicated that the fibers increased the stiffness of NR. The treated fibers showed the highest crystallinity (66%) compared to the untreated (58%) fibers. The hardness value (5.13) was higher for treated fiber composites than the other.

Environmentally friendly materials from cogan grass fiber reinforced with epoxidized natural rubber (ENR) composites were prepared. The fiber was first treated with alkali (5% w/v) at various time intervals of 24, 48, and 72 hours and then it was treated with H_2SO_4 solution at 5% conc. at different time periods of 10, 15, and 20 hours. The result revealed that inclusion of grass fiber into ENR enhanced the mechanical properties of composites. Also, compared to untreated fiber composites, the alkali, followed by acid treatment, gave higher tensile strength with increasing fiber content. The addition of coupling agent (4-amino-6-hydroxy-2-mercaptopyrinidine monohydrate) with ENR showed higher tensile strength and percentage of elongation at break. This may be due to the high potential of the binding agent that can interfere with the ENR epoxy group leading to an increase in the crosslink density. It resulted in a higher tensile strength, a modulus with an elongation of 200%, and an elongation at break of ENRC relative to ENR.

An experimental study was conducted by Pappu et al. [32] on hybrid (sisal and HF) composites reinforced with polylactic acid (PLA) via the extrusion and injection molding process. The incorporation of sisal and HF with PLA improved the general properties of the composites than the neat PLA. It has been found that the average diameters of single sisal and HFs are 269.79 ± 78 and 105 ± 49 μm, respectively. HF density was found to be 1.48 g/cm^3, while sisal fiber density was found to be 1.42 g/cm^3. The average tensile strength and modulus of sisal fiber was 294 ± 113 MPa and 9.8 ± 0.9 GPa, whereas for HF it was 423 ± 186 MPa and 14.0 ± 3.8 GPa, respectively. Overall, HF's tensile strength/tensile modulus was found to be higher than the sisal fiber. The results also revealed that, compared to neat PLA, the tensile strength (46.25 ± 6.75 MPa), flexural strength (94.83 ± 11.21 MPa), Young modulus (6.1 ± 0.58 GPa), flexural modulus (6.04 ± 0.55 GPa), specific tensile strength (38.86 ± 5.0), specific flexural strength (79.76 ± 8.80), density (1.14 ± 0.047 g/cm^3), elongation at break (0.93 ± 0.35%), and WA (1.06 ± 0.18%) were higher for hybrid fiber reinforced PLA matrix composites. An alternate to petroleum-based products, this hybrid composites showed greater potential for use in automotive, packaging, electronics, interiors, and agricultural applications as an environmentally friendly material.

Xu et al. [33] focused on the mechanical properties of HF reinforced PLA composites. HF as a filler and PLA as a matrix were mixed with a specific weight ratio to produce needled hybrid fiber mats of 20 mm × 175 mm. According to ASTM D638 and D790, the tensile and flexural properties of composites were measured with the influence of hybrid ratio, forming temperature and pressure. The experimental analysis described that the tensile strength increased up to 52% and flexural up to 100% for the green composites. An increase in the interfacial bonding between the fiber and PLA matrix improved the mechanical properties of the composites. Especially, the needle punched compounding technology has improved the mechanical properties of PLA fiber/HF hybrid composites. The inclusion of 10% of PLA fiber, at a temperature of 195 °C and a forming pressure of 9 MPa showed the largest mechanical properties.

Yusefi et al. [34] examined the strengthening effects of cow dung (CD) on the properties of PLA. Using an internal Brabender mixer and compression molding process, the PLA/CD biocomposite mixture with various CD ratios (0–50 wt%) was prepared. In accordance to ASTM D638, the tensile test were performed. It is well known that in natural fiber composites, the incompatibility of the two constituents, which weakens the interface between the molecules, resulting in the tensile strength, was reduced as the CD was applied to the PLA. Contrarily, the neat PLA showed the tensile strength of 53.4 MPa whereas 10 wt% CD composite showed a 17% decrease in the tensile strength and at 50 wt% CD loading, the decrease in PLA/CD biocomposite's tensile strength was reduced by only 23%. The flexural modulus (using ASTM D790) of neat PLA was recorded to 3934 MPa, while the 50 wt% CD load revealed a 4080 MPa. Compared to neat PLA, the impact strength was lower for these biocomposites. Nevertheless, filler agglomeration coupled with microvoids served as high stress concentration areas at 50 wt% of CD content. This resulted in lower energy starting the propagation of crack, thereby reducing impact strength (25%). At 50 wt%, CD content of biocomposites suggested significantly greater WA. This was because the amount of voids and air bubbles present in the biocomposites were increased. Compared to pure PLA, the 50 wt%, CD shows the highest WA of 13.62%.

The mechanical properties of biobased composites comprising poly(ε-caprolactone) PCL reinforced with HF with the influence of different fiber aspect ratio 19, 26, 30, and 38 was fabricated by twin-extrusion techniques [35]. In order to generate the above said fiber aspect ratios, the element length l (μm) as 432, 568, 708, and 845 μm and element diameter d (μm) as 22.4, 21.7, 23.6, and 22.5 μm, respectively, are used for the HFs. At dry condition, the highest tensile strength, tensile modulus, flexural strength, and modulus was noticed for PCL blend with 26% fiber of 12.25, 0.40, 25.68, and 0.605 MPa, respectively, whereas the elongation at break (mm) and deformation at peak load (mm) was lower for the same. But at saturated conditions, the flexural properties showed higher value than the tensile characteristics of PCL with 26% of hem fibers. It is well known that with the increase of fiber ratio, the WA of composites was also increased. An outset value for 26% HF aspect ratio was noted for the samples in dry and wet conditions.

The effect of different wt% of wood fibers, talc reinforced with polyhydroxybutyrate-co-valerate (PHBV) on the mechanical properties of biocomposites were analyzed [36]. From the result it was indicated that the density of the talc/PHBV with 40/60% contained a higher ($1.54 \pm 0.02\,g/cm^3$) value compared to virgin PHBV, wood/PHBV, and hybrid composites. PHBV's Young modulus was improved by the inclusion of wood, talc, and also for both fibers to some degree. An inadequate interfacial bonding between the matrix (PHBV) and the wood fibers observed a reduction in tensile strength of the composites. There is an improvement in tensile strength (due to the talc reinforced with PHBV offered an advantage over the PHBV wood composites) that has a lower resistance than the hybrid composites. At 40 wt% talc in PHBV a leap of 270 and 22% in flexural modulus and strength was noticed. A reduction in impact strength of the PHBV is noted with the incorporation of both talc and wood fibers.

An environmentally friendly composite derived from polyhydroxybutyrate (PHB) and carnauba fibers were prepared by hot pressing techniques. In order to enhance the good adhesion between the fiber and the matrix, various treatments such as alkali, peroxide, $KMnO_4$, and acetylation were given to the fibers from carnauba. The effect of 10 wt% of fiber and various treatments were used to develop the mechanical properties of composites. The density of untreated (FNT) and peroxide-treated (FTPH) carnauba fiber was 1.344 and 1.44 g/cm^3, respectively. The elongation at break is similar to 1.7–2.6% for the FNT and FTPH fibers. Alkali-, acetic acid-, and $KMnO_4$-treated fibers showed the lowest tensile strength whereas the FTH and FNT have the highest strength [37].

Ju et al. [38] described the evaluation of PHBV-based PCL composites through melt compounding techniques. The static mechanical properties such as strength at yielding and tensile modulus for virgin PCL were ranged from 13.6 and 205 MPa whereas for 40 wt% PHBV it was increased from 15.1 and 469 MPa, respectively. Moreover, the elongation at break (%) and strength at break (MPa) was lower for increasing the fiber wt% of 5, 10, 20, 30, and 40. The same trend was observed for the LA and PHBV matrices blended with man-made cellulose, jute, and abaca fiber composites prepared by injection molding process. A considerable increment in tensile strength (>50%) and a notch impact strength (>250%) was observed for the prepared biocomposites. Due to an advantageous structure as well as its overall mechanical performance, man-made cellulose, in particular, has proved to be a better reinforcement when compared to jute and abaca fibers [39].

Natural fibers such as hemp, jute, and lyocell, and polyhydroxybutyrate (PHB) are used to improve the mechanical performance of biocomposites through twin- screw extrusion – injection molding process [40]. Compared with neat PHB, there is a significant reduction in the tensile strength of all the prepared composites. The order of the individual fiber's tensile strength is as follows: Lyocell (1 GPa) > hemp (600 MPa) > jute (370 MPa). At 30 wt% fiber loading, 37.37 and 4.72% of reduction in tensile strength was observed for lyocell and HF PHB composites. At the same time, as fiber loading increases, the strength of jute fiber composites also increases. The average fiber length (AFL) being below the critical fiber length (CFL), lack of interaction and dispersion of the fibers used and the PHB were the reasons for decrement in tensile strength of lyocell fiber composites. Similarly, the flexural modulus of the fiber follows the order as jute (10–40 GPa) > hemp (10 GPa) > lyocell (9 GPa). At 30 wt% fiber loading, there occurs a large increase in modulus of 591 and 246% for the jute fiber PHB and HF PHB composites, respectively. According to impact strength, the lyocell fiber and HF composites showed poor performance compared to the flexural modulus results.

Biocomposites obtained from PLA and modified (by butyl acrylate BA) rice straw fiber (MRSF) were prepared and the mechanical properties studied [41]. At first, the tensile strength of PLA/RSF composites increases and then decreases with an increase of wt%, while elongation at break increases gradually. The tensile strength of the composite increases significantly around 6 MPa, when the BA content increases, which could be attributed to the small PBA coatings and adsorption on rice straw fiber. Therefore, a little PBA played an eminent role as a compatibilizer. However, the tensile strength of PLA/MRSF composites diminishes rapidly by further increment of PBA composites. This may be due to the fact that the excess PBA content could not be well distributed in the matrix and caused stress deficiencies and induced poor interfacial bonding between the PLA and MRSF which reduced strength.

11.3 Applications of Biobased Composites

In recent years, the polymer composites have attracted substantial importance as a potential structural material for residential housing construction and are largely found in nonstructural housing components. Research and development efforts are being made to use cellulosic fibers as reinforcement for the automotive and building industries. Also biobased composites remain an important part of our everyday lives and they have a wide variety of applications, ranging from sports equipment to transportation vehicles. Most automotive companies in Germany (such as Volkswagen, Audi, and BMW) use natural fiber composites for interiors, door lining, and paneling; wood fibers are used to enclose the back of the seat. Fibers from cotton are used as a substitute for sound proofing and for vehicles, coconut fibers are used for interior trimming and seat trimming [42]. Composites have enhanced water barrier properties that are beneficial for future use as a packaging material [10].

Automotive applications include door panels, car roofs, package trays, load floors, and spare tire covers. In the building sector, typical applications are decking, fencing, railing, windows, partition boards, and doors. Other potential areas of application of wood fiber composites are the packaging and furniture industries. For applications such as building materials, sisal-fiber reinforced gypsum and cement composites have long been studied before 1994 [43, 44]. The majority of these works are focused on interface, mechanical, fire, and environmental properties and their applications.

Sisal-Cashew Nut Shell Liquid (CNSL) composites show potential for roofing applications in tropical countries such as Tanzania where the sisal fiber and cashew nut shell liquid are freely available [45]. Wambua et al. [46] have compared mechanical properties of several natural fibers/polypropylene composites. Their study shows that hemp or sisal/polypropylene composites have the potential to replace glass in many applications that do not require very high load-bearing capabilities. Plant fiber reinforced polymer composites have been continuously used for low-tech applications due to their superior mechanical properties, relative low cost, and higher environmental friendliness [47]. Automotive and aerospace industries have both demonstrated an interest in using more natural fiber reinforced composites in order to reduce vehicle weight. Automotive industries have already shifted from steel to aluminum and are now shifting from aluminum to fiber reinforced composites for some applications. Thus, in the near future, plastics and polymer composites will comprise approximately 15% of the total automobile weight [48].

11.4 Conclusions

In this chapter, recorded researches in the field of biobased polymers have been brought to light. Fibers obtained from agricultural waste residues and matrices from renewable resources can be turned into new composite products through proper technology. The physical and mechanical properties of natural rubber, cardanol, starch, soybean oil, etc., based kenaf, jute, sisal, banana, and coir fiber composites were reviewed. From this work, it was well understood that the mechanical performance of biocomposites showed better value when the resin content increased. Also, positive environmental benefits with respect to ultimate disposability and raw material utilization lead to the biodegradable composites. Hence, the utilization of this more environmental friendly composite material can make a revolution instead of nonbiodegradable materials in the composite world.

References

1 Kozlowski, R. and Mieleniak, B. (2000). New trends in the utilization of by-products of fibre crops residue in pulp and paper industry, building engineering, automotive industry and interior furnishing. *Proceedings from the Third International Symposium on Natural Polymers and Composites*, 504–510. Sao Paulo.

2 Siti Noor, H.M., Abdul, R.R., and Agus, A. (2014). Bio-based thermoset nanocomposite derived from vegetable oil: a short review. *Reviews in Chemical Engineering* 30 (2): 167–182.

3 Dong, J.L. and Sung, H.S. (2019). Investigation of epoxidized palm oils as green processing aids and activators in rubber composites. *International Journal of Polymer Science* 2019: 1–7. (Article ID 2152408).

4 Fei, M.-e., Wendi, L., Anming, J. et al. (2018). Bamboo fibers composites based on styrene-free soybean-oil thermosets using methacrylates as reactive diluents. *Composites Part A* 114: 40–48.

5 Liu, W., Xie, T., and Qiu, R. (2016). Improvement of properties for biobased composites from modified soybean oil and hemp fibers: dual role of diisocyanate. *Composites: Part A* 90: 278–285.

6 Manthey, N.W., Francisco, C., Gaston, F., and Aravinthan, T. (2013). Thermo-mechanical properties of epoxidized hemp oil-based bioresins and biocomposites. *Journal of Reinforced Plastics and Composites* 32 (19): 1444–1456.

7 Balaji, A., Karthikeyan, B., and Swaminathan, J. (2019). Comparative mechanical, thermal, and morphological study of untreated and NaOH-treated bagasse fiber-reinforced cardanol green composites. *Advanced Composites and Hybrid Materials* 2: 125–132.

8 Njoku, R.E., Ofilib, I., Agbiogwuc, D.O., and Agu, C.V. (2012). Effect of alkali treatment and fiber content variation on the tensile properties of coir fiber reinforced cashew nut shell liquid (CNSL) composite. *Nigerian Journal of Technology* 31 (2): 107–110.

9 Obasi, H.C., Igwe, I.O., Onyeagoro, G.N. et al. (2014). Oil palm mesocarp particles reinforced cashew nut shell novolac resin composites. *Applied Polymer Composites* 2 (1): 27–36.

10 Jumaidin, R., Zulhelmi, A.S.S., Rushdan, A.I. et al. (2019). Characteristics of cogon grass fibre reinforced thermoplastic cassava starch biocomposite: water absorption and physical properties. *Journal of Advanced Research in Fluid Mechanics and Thermal Sciences* 62 (1): 43–52.

11 Ibrahim, H., Farag, M., Megahed, H., and Mehanny, S. (2014). Characteristics of starch-based biodegradable composites reinforced with date palm and flax fibers. *Carbohydrate Polymers* 101: 11–19.

12 Correia, C.A., de Letícia, M.O., and Ticiane, S.V. (2017). The influence of bleached jute fiber filler on the properties of vulcanized natural rubber. *Materials Research* 20 (Suppl. 2): 466–471.

13 Ruksakulpiwat, C., Wasaphon, W., Apikiat, S., and Yupaporn, R. (2013). Cogon grass fiber-epoxidized natural rubber composites. *Advanced Materials Research* 747: 375–378.

14 Noor Azammi, A.M., Sapuan, S.M., Ishak, M.R., and Sultan, M.T.H. (2019). Physical and damping properties of kenaf fibre filled natural rubber/thermoplastic polyurethane composites. *Defence Technology* https://doi.org/10.1016/j.dt.2019.06.004.

15 Kamarudin, S.H., Abdullah, L.C., Aung, M.M., and Ratnam, C.T. (2017). A study of mechanical and morphological properties of PLA based biocomposites prepared with EJO vegetable oil based plasticiser and kenaf fibres. *IOP Conference Series: Materials Science and Engineering* 368 (2018): 012011.

16 Aliotta, L., Vito, G., Maria, B.C. et al. (2019). Evaluation of mechanical and interfacial properties of bio-composites based on poly (lactic acid) with natural cellulose fibers. *International Journal of Molecular Sciences* 20 (4): 960.

17 Volova, T.G., Zhila, N.O., Shishatskaya, E.I. et al. (2013). The physicochemical properties of polyhydroxyalkanoates with different chemical structures. *Polymer Science Series A* 55: 427–437.

18 Jiang, G., David, J.H., Marek, K. et al. (2016). Carbon sources for polyhydroxyalkanoates and an integrated biorefinery. *International Journal of Molecular Sciences* 17 (7): 1157.

19 Ibrahim, M.I.J., Sapuan, S.M., Zainudin, E.S., and Zuhri, M.Y.M. (2019). Preparation and characterization of cornhusk/sugar palm fiber reinforced cornstarch-based hybrid composites. *Journal of Materials Research and Technology* https://doi.org/10.1016/j.jmrt.2019.10.045.

20 Danielle, C.M.F., Gustavo, M., and Franciele, M.P. (2020). Biodegradable trays based on cassava starch blended with agroindustrial residues. *Composites Part B Engineering* 183: 107682.

21 Pupure, L., Varna, J., Joffe, R. et al. (2018). Mechanical properties of natural fiber composites produced using dynamic sheet former. *Wood Material Science & Engineering* https://doi.org/10.1080/17480272.2018.1482368.

22 Ab Rahman, M.F., Norfaizal, N.S., and Azura, A.R. (2019). The influence of sago starch dispersion on mechanical properties of biodegradable natural rubber latex films. *Materials Today: Proceedings* 17: 1040–1046.

23 Liu, W., Chen, T., Fei, M.-e. et al. (2019). Properties of natural fiber-reinforced biobased thermoset biocomposites: effects of fiber type and resin composition. *Composites Part B* 171: 87–95.

24 Pereira da Silva, J.S., Juliana, M.F.S., Bluma, G.S., and Sebastien, L. (2017). Fully biodegradable composites based on poly(butylene adipate-co-terephthalate)/peach palm trees fiber. *Composites Part B: Engineering* 129: 117–123.

25 Ramírez, M.G.L., Satyanarayana, K.G., de SetsuoIwakiri, B.M. et al. (2011). Study of the properties of biocomposites. Part I. Cassava starch-green coir fibers from Brazil. *Carbohydrate Polymers* 86: 1712–1722.

26 Lomelí-Ramírez, M.G., Kestur, S.G., Ricardo, M. et al. (2014). Bio-composites of cassava starch-green coconut fiber: part II – structure and properties. *Carbohydrate Polymers* 102: 576–583.

27 Rodney, J., Japar, S., Mohd, S.M.K., and Salit, M.S. (2015). Thermochemical and mechanical properties of tea tree (*Melaleuca alternifolia*) fibre reinforced tapioca starch composites. *e-Polymers* 15 (6): 401–409.

28 Gironès, J., López, J.P., Mutjé, P. et al. (2012). Natural fiber-reinforced thermoplastic starch composites obtained by melt processing. *Composites Science and Technology* 72: 858–863.

29 Zainuddin, S.Y.Z., Ahmad, I., and Kargarzadeh, H. (2012). Potential of using multiscale kenaf fibers as reinforcing filler in cassava starch-kenaf biocomposites. *Carbohydrate Polymers* 92: 2299–2305.

30 Costa Junior, A.E., Barreto, A.C.H., Rosa, D.S. et al. (2014). Thermal and mechanical properties of biocomposites based on a cashewnut shell liquid matrix reinforced with bamboo fibers. *Journal of Composite Materials* 49 (18): 2203–2215.

31 Udhayasankar, R. and Kathikeyan, B. (2019). Preparation and properties of cashew nut shell liquid-based composite reinforced by coconut shell particles. *Surface Review and Letters* 26 (04): 1850174.

32 Pappu, A., Kim, L.P., and Vijay, K.T. (2019). Manufacturing and characterization of sustainable hybrid composites using sisal and hemp fibres as reinforcement of poly(lactic acid) via injection moulding. *Industrial Crops and Products* 137: 260–269.

33 Xu, Z., Yang, L., Ni, Q. et al. (2019). Fabrication of high-performance green hemp/polylactic acid fibre composites. *Journal of Engineered Fibers and Fabrics* 14: 1–9.

34 Yusefi, M., Mohammad, K., Faizah Md, Y. et al. (2017). Physico-mechanical properties of poly(lactic acid) biocomposites reinforced with cow dung. *Materials Research Express* 4: 025302.

35 Dhakal, H.N., Sikiru, O.I., Zhongyi, Z. et al. (2018). Development of sustainable biodegradable lignocellulosic hemp fiber/polycaprolactone biocomposites for light weight applications. *Composites Part A* 113: 350–358.

36 Singh, S., Mohanty, A.K., and Misra, M. (2010). Hybrid bio-composite from talc, wood fiber and bioplastic: fabrication and characterization. *Composites: Part A* 41: 304–312.

37 Melo, J.D.D., Carvalho, L.F.M., Medeiros, A.M. et al. (2012). A biodegradable composite material based on polyhydroxybutyrate (PHB) and carnauba fibers. *Composites: Part B* 43: 2827–2835.

38 Ju, D., Han, L., Li, F. et al. (2014). Poly(ε-caprolactone) composites reinforced by biodegradable poly(3-hydroxybutyrate-co-3-hydroxyvalerate) fiber. *International Journal of Biological Macromolecules* 67: 343–350.

39 Bledzki, A.K. and Jaszkiewicz, A. (2010). Mechanical performance of biocomposites based on PLA and PHBV reinforced with natural fibres – a comparative study to PP. *Composites Science and Technology* 70: 1687–1696.

40 Michael, A.G., Luke, M.G., John, A.K. et al. (2013). Mechanical and biodegradation performance of short natural fibre polyhydroxybutyrate composites. *Polymer Testing* 32: 1603–1611.

41 Lijun, Q., Jianhui, Q., Mingzhu, L. ct al. (2011). Mechanical and thermal properties of poly(lactic acid) composites with rice straw fiber modified by poly(butyl acrylate). *Chemical Engineering Journal* 166: 772–778.

42 Suddell, B.C. and Evans, W.J. (2005). *Natural Fibre Composites in Automotive Applications in Natural Fibres in Biopolymers & Their BioComposites*, 231–259. CRC Press.

43 Coutts, R.S.P. and Warden, P.G. (1992). Sisal pulp reinforced cement mortar. *Cement and Concrete Composites* 14: 17–21.

44 Singh, M. and Garg, M. (1994). Gypsum-based fibre-reinforced composites: an alternative to timber. *Construction and Building Materials* 8: 155–160.

45 Bisanda, E.T.N. and Ansell, M.P. (1992). Properties of sisal-CNSL composites. *Journal of Material Science* 27: 1690–1700.

46 Wambua, P., Ivens, J., and Verpoest, I. (2003). Natural fibres: can they replace glass in fibre reinforced plastics? *Composites Science and Technology* 63: 1259–1264.

47 Santulli, C. (2007). Impact properties of glass/plant fibre hybrid laminates. *Journal of Material Science* 42: 3699–3707.

48 Mohanty, A.K., Drzal, L.T., and Misra, M. (2002). Novel hybrid coupling agent as an adhesion promoter in natural fiber reinforced powder polypropylene composites. *Journal of Material Science Letters* 21: 1885–1888.

12

Synthesis and Utilization of Biodegradable Polymers

Lalit Ranakoti[1], Brijesh Gangil[2], Pawan Kumar Rakesh[1], and Nikita Agrawal[3]

[1] *Department of Mechanical Engineering, NIT, Srinagar, Uttarakhand, India*
[2] *Department of Mechanical Engineering, H.N.B. Garhwal University, Srinagar, Uttarakhand, India*
[3] *Department of Pedodontics, Peoples College of Dental Sciences, Bhopal, Madhya Pradesh, India*

12.1 Introduction

Today's world is concerned over the waste disposal issues due to the increase in the utilization of non-biodegradable entities. A greater percentage of polymers amplify this waste disposal problem due to its non-biodegradability. To tackle this issue, a gradual shift toward biodegradable polymers has been observed. Moreover, due to the constant hike in the prices of petroleum products, the usage of non-biodegradable polymers derived from petroleum products are now being avoided. Thus, a stimulating interest toward biodegradable polymers has been developed among the masses [1]. Biodegradable polymers are nature friendly and easy to dispose of, causing minimum harm to the environment. Biodegradable polymers can be derived either from synthetic or from natural sources as categorized in Figure 12.1. The mechanism of biodegradation depends on the action of enzymes or the chemical action of living organisms. In the beginning of biodegradation, a fragment of polymer is broken into smaller chains due to the abiotic or biotic reaction which is followed by the assimilation and mineralization of the fragments by the microorganism [2]. The abiotic degradation of a biodegradable polymer is shown in Figure 12.2. The origin of the polymer, its chemical structure, and the environmental conditions are the major factors which decide the biodegradation rate of polymers. Among the responsible factors for biodegradation, chemical structure of the polymers plays a vital role [3]. The prime objective of this study is to present an insight into the necessary operation required for the amalgamation of a polymer into biodegradable polymers. Hence, in the present audit, various mechanisms applied in the synthesis of biodegradable polymers have been discussed.

12.2 Synthesis Techniques of Biodegradable Polymers

12.2.1 By Modifying Natural Polymers

Naturally available polysaccharide polymers like cellulose, starch, chitin, etc. can be converted into fully biodegradable polymers with the help of coblending. One such example of coblending is the blending of starch (extracted from rice) and chitosan on a leveled

Biobased Composites: Processing, Characterization, Properties, and Applications, First Edition.
Anish Khan, Sanjay Mavinkere Rangappa, Suchart Siengchin and Abdullah M. Asiri.

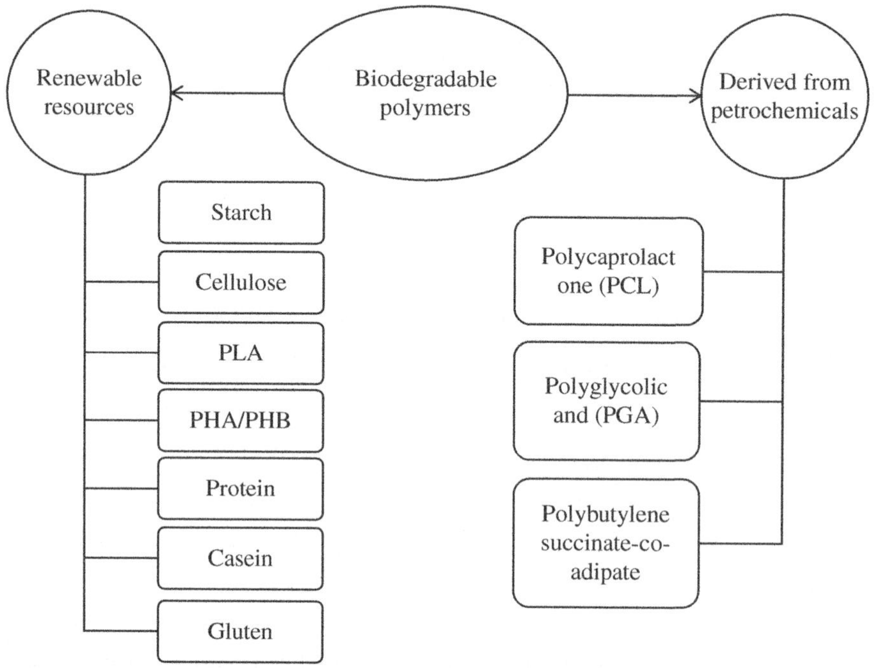

Figure 12.1 Biodegradable polymers derived from renewable resources and petrochemicals.

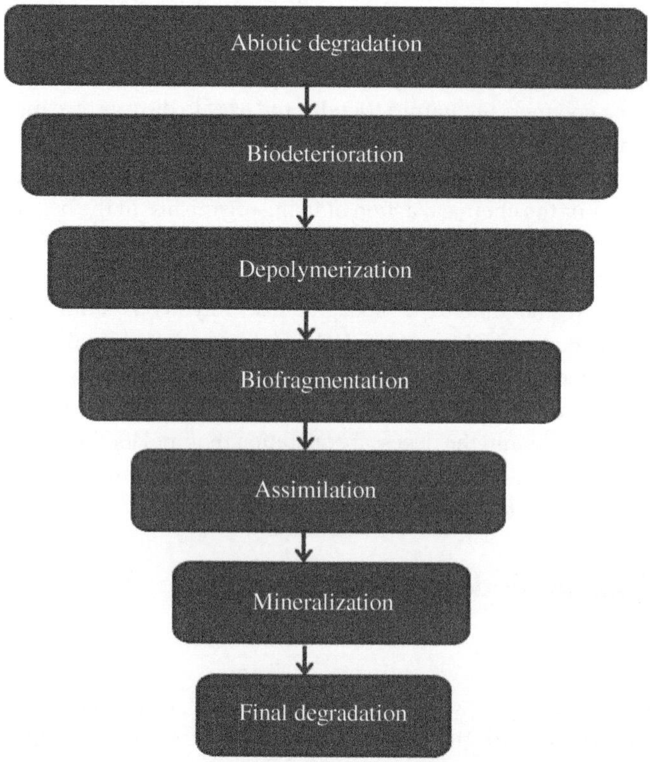

Figure 12.2 Abiotic degradation of biodegradable polymers.

tray. The blend is then casted into films of required dimensions. Modification leads to improved strength, reduces permeability, and decreases the tendency of solubility [4]. However, thermal properties of these blends are not satisfying, but efforts have been put in to develop a better material through different modifications [5].

12.2.2 Polymers Synthesized by Chemicals

Polymers having a chemical structure similar to natural polymers can be chemically synthesized. Polymeric chains like amides, peptides, and esters are easy to break and can be synthesized to form a biodegradable polymer. Upon synthesizing glycerol and lactic acid with optimal conditions, a biodegradable polymer called polylactic acid (PLA) is obtained [6]. During the process, formation of several harmful gases and, later on, various byproducts make this method unsuitable for the synthesis of biodegradable polymers.

12.2.3 Polymers Synthesized by Microorganisms

Microorganisms present in the environment feed themselves by consuming polymers and form a series of complex chains, for example, polyester, silk, and polysaccharides. The common source of food is starch and glucose which are abundant in nature. Due to the similarity in chemical structure, it is arduous to separate these polymers from one another. One of the most commonly utilized biodegradable polymers synthesized by microorganisms is polyhydroxyalkanoates (PHA) [7]. It is highly biodegradable and can be synthesized by several renewable resources.

12.2.4 Synthesis by Enzymes

Polymers are now being synthesized by enzymes. It is one of the latest methods in which a special type of enzyme can synthesize a special type of polymer. One of the important processes among enzymatic syntheses is the process of lipase-catalyzed polymerization, which produces the polymer aliphatic polyester. The reaction that takes place in this process is called polymeric reaction [8]. Due to the high specificity of enzymes, there is a less chance of obtaining byproducts, which makes this process easy to carry out. This process is also cost effective because of the recyclability of the enzymes and the conditions required for the process are normal ambient temperature and pressure [9]. Moreover, several different types of fully biodegradable polymers, such as polysaccharides, polyamides, polyester etc., can be obtained by this process.

12.2.5 Synthesis by Chemo-Enzymes

This is a process in which the specificity and stereoselectivity of the polymer have been taken care of. This process also enables the adjustment of the polymer's molecular weight. It is a combination of polymerization and region selective enzymatic strategy [10]. Thus, polymers with high molecular weight can be easily synthesized with high biodegradability rate, for example, nonsteroidal drugs with high molecular weight (which are anti-inflammatory) can be easily fabricated by optically active polymeric drugs through effective chemo-enzymatic process [11].

12.3 Biodegradable Polymers and Their Synthesis

12.3.1 Starch

Starch is a renewable polysaccharide mostly found in plants. Starch contains mainly two types of glucose, i.e. amylose and amylopectin. Amylose consists of D-glucose linked by α-1, -4 bonds while amylopectin comprises short D-glucose coupled by α-1, -6 bonds. Some encouraging characteristics of starch are that it is cheap, easy to handle, has a low manufacturing cost, and low oxygen permeability; but its brittle nature and hydrophilic property limit its applications [12]. Starch can be converted to thermoplastic starch with enhanced flexibility and low hydrophilicity through the process of extrusion. In the extrusion process, native starch is combined with plasticizers like glycol, sorbitol, and glycerol in the presence of heat and shear force [13].

12.3.2 Polylactic Acid

It is referred as PLA, produced from crops such as corn and sugar beets through the process of fermentation. Bacterial fermentation and synthesis of petrochemicals are the two major processes to synthesize lactic acid. Low-molecular-weight PLA is synthesized by the polycondensation of D/L lactic acid, while high-molecular-weight PLA is obtained by ring opening polymerization [14]. Normally, high-molecular-weight PLA possess higher mechanical properties. PLA can be blended with materials like clay and thermoplastic starch to enhance the properties of plastic deformation at a higher stress level. Physical and mechanical properties can also be enhanced by the blending of PLA with polyhydroxybutyrate (PHB) [15].

12.3.3 Polycaprolactone

Polycaprolactone, known as PCL, is a substitute of polyester, possessing low viscosity and low melting point. PCL is obtained by the synthesis of ε-caprolactone through the process of polymerization. With poor barrier resistance and low strength, PCL generally lacks importance in industrial applications. Blending of PCL with polymers like PLA, cellulose acetate butyrate, and cellulose propionate leads to the enhancement of properties like adhesiveness, crack resistance, and dye ability [16].

12.3.4 Polyhydroxyalkanoates/Polyhydroxybutyrate

Polyhydroxyalkanoates (abbreviated as PHA) has its applications in medical and agriculture fields. PHA comes in the category of polyesters composed of various hydroxyalkanoates. Microbial fermentation, a very common technique, can be used to synthesize PHA. PHA is a highly crystalline thermoplastic having low melting point. It is nontoxic and possesses high resistance against ultraviolet radiations [17]. Low strength, poor compatibility with processing methods, and high rate of thermal degradation are some of the drawbacks of PHA that limit its applications in industries. With a relatively high degree of crystallinity, PHB is a customary representation of the PHA [18]. Mechanical properties of PHB are as good as polyethylene provided with a crystallinity of 70%. Blending of PHB with PLA can be easily done due to a similar melting point [19]. The brittle nature of PHB, instability at higher temperatures, and poor permeability put a lot of constraint from the application point of view. The properties of PHB can be enhanced by melt mixing of the blend of PHB–PLA with the inclusion of catechin [20]. In fact, promising results have been reported in the literature in this regard [21].

12.3.5 Starch–Polyolefin Blends

A complete biodegradable polymer can be obtained by blending starch with the polyolefin (PE) polymers. To carry out the blending, a twin-screw extruder is normally employed. Compatibility issues between starch and PE can be handled with the addition of compatibilizers. Compatibilizers like maleic anhydride (MA) help in the easy dispersion of starch in the PE matrix. It has been observed that crystallization of PE gets enhanced with the compounding of starch. It should be noted that percentage of starch should be more than 20% in the starch–PE blend [22]. Resulting properties of the blend is the compromise between biodegradability and strength. Higher percentage of starch results in higher biodegradability but lower strength, and vice versa. Other PEs like high-density polyethylene (HDPE) and polypropylene (PP) can also be blended with native starch through the same twin-extruder process for various applications with higher strength and durability [23]. Improved thermal and mechanical properties of the blend can be achieved in the presence of modifiers, namely deep eutectic solvents. Crosslinkage between starch and PP can be obtained by reactive extrusion which can be used as a packaging material in the industry. Polystyrene, a well-known PE, has also shown positive results on compounding with starch. The mixing is carried out in reactive extrusion in the presence of zinc octanoate ($Zn(Oct)_2$) [24]. Observations reveal that ($Zn(Oct)_2$) strictly promotes catalytic reaction for the formation of crosslinking. Elevated thermal and impact properties have been reported in the literature for the blend of starch and high-impact polystyrene (HIPS) in the presence of glycerol as the plasticizer [25].

12.3.6 Starch–Polyester Blends

Starch possesses a lot of drawbacks like poor resilience, greater shrinkage, and high sensitivity toward moisture. These drawbacks can be mended by the blending of starch with PCL (a type of polyester) [26]. PCL makes the starch dimensionally stable and improves its impact strength. The final strength of the blend will depend on the percentage of starch, PCL, and the plasticizer. Sometimes, starch is blended to favor the other blending polymer, simultaneously exploiting its inheriting properties [27]. Blending of PHB with starch and ethylene vinyl acetate (EVA) in the presence of MA and benzoyl peroxide (BPO) can be prepared by in situ compatibilization [28, 29]. This led to the improvement in strength and toughness with the percentage of MA below 0.1% due to the enhancement in the molecular affinity between PHB–EVA and starch.

12.3.7 Starch–PLA Blends

High performance film for packaging can be easily prepared using the blend of starch and PLA. The blend of PLA and starch in proper proportion can lead to improvement in the tensile strength and resistance barrier against water vapor [30]. The biodegradability rate of starch and PLA reduced to some extent due to the plasticizing, but, on the other hand, the final blend comes out to be tougher than virgin starch. Cinnamaldehyde is normally used as the plasticizer in the blending of starch and PLA. It has been reported in the past that high transparency can be achieved in the blending of starch and PLA, but the resistance barrier was as similar as in case of neat starch [31].

12.4 Applications of Biopolymers in Industries

Through research and development, one can convert non-biodegradable polymers to biodegradable polymers. Conceptual design is needed to make the materials suitable for the required applications. It is expected that the material to be synthesized may act as an alternate to the already existing material or may replace it totally. Biodegradable polymers are not limited to medical, agriculture, and textile industries; in fact, their utilization has been expanding in many sectors. Biopolymers continue to secure more attention in the packaging industry, than any other sector. For example, Germany and China, two of the major consumers of packed commodities, are warranting higher utilization of biodegradable polymers all over the country. According to reports, around 50% of the commodities around the world are packed in polymer plastics [32]. Thus, the issues of renewability and biodegradability need to be resolved through novel and inventive ideas. It has been seen that strength of the biodegradable polymers are satisfactory and hence can be used in agricultural sectors. Ecoflex is the one such example where biodegradable polymers are used as a thin film to prevent the defrosting of plants, and after use, these thin films are buried in the soil and broken down into fragments by microorganisms. Bags used to store chemicals and fertilizers are also being made biodegradable as examined by scientists. From the view point of agriculture industries, biodegradable polymers have become a necessity to regulate the nutrient cycle in the environment [33]. Due to the constant change in the dynamics of the medical world, the immediate development in biopolymers is required. Compatibility of biodegradable polymers with medical drugs which are delivered, is the utmost priority for medical industries, ensuring the breakdown of the polymer not before the expected time period. In the field of tissue engineering, polymeric organs are now being developed which can be fit and can sustain in the human body [34]. The desirable characteristics of the polymer would be a high injection rate, promote growth of the cells, and should be compatible with blood vessels.

12.5 Conclusion

Biodegradable polymers have now become the topic of interest in research for medical, agriculture, and packaging industries. Conventional biopolymers are being synthesized and converted to biodegradable polymers through various methods. Blending of biodegradable polymers has been carried out to achieve the desired strength and functionality in the polymers. An increasing trend in the demand of biodegradable polymers has attracted various researchers to disinter several aspects to increase the applications of biodegradable polymers.

References

1 Lalit, R., Mayank, P., and Ankur, K. (2018). Natural fibers and biopolymers characterization: a future potential composite material. *Strojnícky casopis – Journal of Mechanical Engineering* 68 (1): 33–50.
2 Kyrikou, I. and Briassoulis, D. (2007). Biodegradation of agricultural plastic films: a critical review. *Journal of Polymers and the Environment* 15 (2): 125–150.
3 Azwa, Z.N., Yousif, B.F., Manalo, A.C., and Karunasena, W. (2013). A review on the degradability of polymeric composites based on natural fibres. *Materials & Design* 47: 424–442.
4 Andrade-Mahecha, M.M., Tapia-Blácido, D.R., and Menegalli, F.C. (2012). Development and optimization of biodegradable films based on achira flour. *Carbohydrate Polymers* 88 (2): 449–458.
5 Simi, C.K. and Abraham, T.E. (2010). Biodegradable biocompatible xyloglucan films for various applications. *Colloid and Polymer Science* 288 (3): 297–306.
6 Zeng, S. H., Duan, P. P., Shen, M. X., Xue, Y. J., & Wang, Z. Y. (2016). Preparation and degradation mechanisms of biodegradable polymer: a review. In IOP Conference Series: Materials Science and Engineering 137, 1, p. 012003). IOP Publishing.

7 Arrieta, M.P., López, J., Hernández, A., and Rayón, E. (2014). Ternary PLA–PHB–limonene blends intended for biodegradable food packaging applications. *European Polymer Journal* 50: 255–270.

8 Gross, R.A., Ganesh, M., and Lu, W. (2010). Enzyme-catalysis breathes new life into polyester condensation polymerizations. *Trends in Biotechnology* 28 (8): 435–443.

9 Zhang, J., Shi, H., Wu, D. et al. (2014). Recent developments in lipase-catalyzed synthesis of polymeric materials. *Process Biochemistry* 49 (5): 797–806.

10 Kobayashi, S. (2013). Green polymer chemistry: recent developments. In: *Hierarchical Macromolecular Structures: 60 Years After the Staudinger Nobel Prize II*, 141–166. Cham: Springer.

11 Yang, Y., Zhang, J., Wu, D. et al. (2014). Chemoenzymatic synthesis of polymeric materials using lipases as catalysts: a review. *Biotechnology Advances* 32 (3): 642–651.

12 Weber, C.J., Haugaard, V., Festersen, R., and Bertelsen, G. (2002). Production and applications of biobased packaging materials for the food industry. *Food Additives & Contaminants* 19 (S1): 172–177.

13 Isotton, F.S., Bernardo, G.L., Baldasso, C. et al. (2015). The plasticizer effect on preparation and properties of etherified corn starchs films. *Industrial Crops and Products* 76: 717–724.

14 Jamshidian, M., Tehrany, E.A., Imran, M. et al. (2010). Poly-lactic acid: production, applications, nanocomposites, and release studies. *Comprehensive Reviews in Food Science and Food Safety* 9 (5): 552–571.

15 Ma, Y., Li, L., and Wang, Y. (2018). Development of PLA–PHB-based biodegradable active packaging and its application to salmon. *Packaging Technology and Science* 31 (11): 739–746.

16 Navarro-Baena, I., Sessini, V., Dominici, F. et al. (2016). Design of biodegradable blends based on PLA and PCL: from morphological, thermal and mechanical studies to shape memory behavior. *Polymer Degradation and Stability* 132: 97–108.

17 Li, Z., Yang, J., and Loh, X.J. (2016). Polyhydroxyalkanoates: opening doors for a sustainable future. *NPG Asia Materials* 8 (4): e265.

18 Lin, X., Fan, X., Li, R. et al. (2018). Preparation and characterization of PHB/PBAT-based biodegradable antibacterial hydrophobic nanofibrous membranes. *Polymers for Advanced Technologies* 29 (1): 481–489.

19 Panaitescu, D.M., Frone, A.N., Chiulan, I. et al. (2018). Role of bacterial cellulose and poly(3-hydroxyhexanoate-co-3-hydroxyoctanoate) in poly(3-hydroxybutyrate) blends and composites. *Cellulose* 25 (10): 5569–5591.

20 Panaitescu, D.M., Nicolae, C.A., Frone, A.N. et al. (2017). Plasticized poly(3-hydroxybutyrate) with improved melt processing and balanced properties. *Journal of Applied Polymer Science* 134 (19).

21 Arrieta, M.P., Castro-Lopez, M.D.M., Rayón, E. et al. (2014). Plasticized poly(lactic acid)–poly(hydroxybutyrate)(PLA–PHB) blends incorporated with catechin intended for active food-packaging applications. *Journal of Agricultural and Food Chemistry* 62 (41): 10170–10180.

22 Peres, A.M., Pires, R.R., and Oréfice, R.L. (2016). Evaluation of the effect of reprocessing on the structure and properties of low density polyethylene/thermoplastic starch blends. *Carbohydrate Polymers* 136: 210–215.

23 Sabetzadeh, M., Bagheri, R., and Masoomi, M. (2015). Study on ternary low density polyethylene/linear low density polyethylene/thermoplastic starch blend films. *Carbohydrate Polymers* 119: 126–133.

24 Chen, X., Zhou, L., Pan, X. et al. (2016). Effect of different compatibilizers on the mechanical and thermal properties of starch/polypropylene blends. *Journal of Applied Polymer Science* 133 (17).

25 Mahmoudi Yayshahri, A., Peighambardoust, S.J., and Shenavar, A. (2019). Impact, thermal and biodegradation properties of high impact polystyrene/corn starch blends processed via melt extrusion. *Polyolefins Journal* 6 (2): 151–158.

26 Ortega-Toro, R., Morey, I., Talens, P., and Chiralt, A. (2015). Active bilayer films of thermoplastic starch and polycaprolactone obtained by compression molding. *Carbohydrate Polymers* 127: 282–290.

27 Guarás, M.P., Alvarez, V.A., and Ludueña, L.N. (2015). Processing and characterization of thermoplastic starch/polycaprolactone/compatibilizer ternary blends for packaging applications. *Journal of Polymer Research* 22 (9): 165.

28 Yeo, J.C.C., Muiruri, J.K., Thitsartarn, W. et al. (2018). Recent advances in the development of biodegradable PHB-based toughening materials: approaches, advantages and applications. *Materials Science and Engineering: C* 92: 1092–1116.

29 Ma, P., Xu, P., Chen, M. et al. (2014). Structure–property relationships of reactively compatibilized PHB/EVA/starch blends. *Carbohydrate Polymers* 108: 299–306.

30 Mangaraj, S., Mohanty, S., Swain, S., and Yadav, A. (2019). Development and characterization of commercial biodegradable film from PLA and corn starch for fresh produce packaging. *Journal of Packaging Technology and Research* 3 (2): 127–140.

31 Muller, J., González-Martínez, C., and Chiralt, A. (2017). Poly(lactic) acid (PLA) and starch bilayer films, containing cinnamaldehyde, obtained by compression moulding. *European Polymer Journal* 95: 56–70.

32 Zhang, Z., Ortiz, O., Goyal, R., and Kohn, J. (2014). Biodegradable polymers. In: *Handbook of Polymer Applications in Medicine and Medical Devices*, 303–335. William Andrew Publishing.

33 Dalla Rosa, M. (2019). Packaging sustainability in the meat industry. In: *Sustainable Meat Production and Processing*, 161–179. Academic Press.

34 Garcia-Gonzalez, C.A., Concheiro, A., and Alvarez-Lorenzo, C. (2015). Processing of materials for regenerative medicine using supercritical fluid technology. *Bioconjugate Chemistry* 26 (7): 1159–1171.

13

Forecasts of Natural Fiber Reinforced Polymeric Composites and Its Degradability Concerns – A Review

D. Divya[1], S. Indran[2], M. R. Sanjay[3], and Suchart Siengchin[3]

[1] *Research and Development Department, Pinnacle Bio-Sciences, Kanyakumari, Tamil Nadu, India*
[2] *Department of Mechanical Engineering, Rohini College of Engineering and Technology, Kanyakumari, Tamil Nadu, India*
[3] *Natural Composites Research Group Lab, King Mongkut's University of Technology North Bangkok, Bangkok, Thailand*

13.1 Introduction

The exploitation of natural resources for fibers instead of various synthetic materials is an eco-friendly approach in the context of improving environmental quality. Based on the sources from which fibers are extracted, natural fibers are of three types. They are plant fibers, animal fibers, and mineral fibers, of which plant fibers possess promising opportunities in the future since they are cellulosic fibers with good strength and are relevant to various infrastructure applications. Usually, plant fibers consist of cellulose in the range 60–80%, lignin 5–20%, and up to moisture of 20% [1, 2]. A variety of plants have been explored earlier for the production of natural fibers, whereas the efficacy of natural fibers depends on the chemical constituents of the fiber. Natural fibers having less lignin and hemicellulose content but high cellulosic content could be fine for a high degree of application in various fields. Therefore, the identification of plants which produce natural fibers possessing the above characteristics and good tensile properties have become an appealing trend of today [3–5].

Natural fibers also possess exclusive characteristics such as low density, renewability, recyclability, biodegradability, ecofriendliness, CO_2 neutral nature, and cost-effective nature. Some criteria of natural fibers enforce the production of fiber-based composites to reduce the limitations that appear when they are used alone. Moreover, natural fiber reinforced polymer composites reduce the health issues associated with the usage of synthetic fiber-encompassed materials. Hence, natural fibers would successfully substitute as healthy, sustainable, and high-tech choices for various purposes. The superior physical (density, diameter) and chemical (cellulose, hemicelluloses, lignin, wax, ash) properties of natural fibers compete with many of the synthetic fibers like glass, Kevlar, etc. Fibers with good mechanical and thermal properties are required to use with various polymer matrix in diverse environmental conditions. The tensile characteristics such as attractive tensile strength and Young modulus are desirable for the composites to contend with conventional materials along with offer for a wide range of application. Fibers with extensive thermal stability can process at high temperatures, which possesses an advantage of being used in high-temperature applications [6–8].

Biobased Composites: Processing, Characterization, Properties, and Applications, First Edition.
Anish Khan, Sanjay Mavinkere Rangappa, Suchart Siengchin and Abdullah M. Asiri.

Conversely, surface roughness is another feature to be considered to assure accurate adhesion between fiber and polymer matrix during reinforcement. Fibers with high surface roughness had shown good adhesion with the matrix through interfacial bonding, which resists easy degradation of composites. In some cases, incompatibility of natural fibers with matrices during composite preparation may lead to alteration in polymer composites; therefore, appropriate treatment is necessary to overcome these troubles. It is crucial to evaluate the degradability of natural fiber composites at different environmental conditions followed by manufacturing, before releasing it for specific purposes. Environmental factors such as temperature, moisture content, weathering, and biological agents cause harmful effects on natural polymer-based composites [9]. Likewise, its utilization in the marine environment causes degradation by microbes and moisture attack. From the above background, the present review aims to assess the current trends and future scenario increasing the utilization of natural fiber composites instead of synthetic fibers, by mounting its durability under saline environment through various prospects.

13.2 Recent Trends of Natural Fiber Production from Plants

Generally, fibers are categorized into two major groups – natural fibers and synthetic fibers. Plant fibers, animal fibers, and mineral fibers belong to natural fibers while man-made organic and inorganic fibers are synthetic fibers. Among different fibers, natural plant fibers were predominant over other fibers owing to the cellulosic nature and wide range of diversity. Based on the origin of the plant where they bear, plant fibers are classified into seven: seed fibers (cotton, kapok, etc.), leaf fibers (sisal, agave, banana, etc.), bast fibers (flax, hemp, jute, etc.), root fibers (pirandai and broom), fruit fibers (coir, tamarind, etc.), grass fibers (bamboo, bagasse, corn, etc.), and stalk fibers (rice, wheat, maize, etc.). The distributions of various fibers are demonstrated in Figure 13.1. There are few plants extensively studied previously for the production of fibers which include flax, cotton, hemp, kenaf, jute, sisal, coconut, bamboo, and banana, while bast fiber of plants showed higher tensile strength than others [2, 10, 11]. The origin of natural plant fibers are presented in Table 13.1.

A prosperous study by Sen and Jagannatha Reddy in 2011 [23] had shown that hybrid composites of prominent natural fibers from bark, for instance, hemp, kenaf, flax, and ramie could vastly recompense with synthetic fibers. Another aspect is that fibers from the stem of adult plants showed higher mechanical strength than that of fibers from the younger ones and this might be the reason that the stem of matured plants possess a secondary cell wall formed by close aggregation of ripened cellulosic fibers [12]. The fibers obtained from plants are lignocellulosic in nature which constitute cellulose, hemicelluloses, pectin, lignin, and waxy substances, where cellulose, hemicellulose, and lignin are the conspicuous portions (Figure 13.2) [24, 25].

The components of lignocelluloses are interlinked through covalent and noncovalent interactions [26]. Cellulose is the chief component, which interlinked with other components to form a fiber network, of which cellulose provides strength and stiffness to the fiber; hemicellulose acts as a cementing material for providing structural stability while lignin and pectin joined together to provide adhesive quality to all components mutually [25]. Few studies revealed that natural fibers have better strength when they possess more cellulosic content with a lower microfibrillar angle [27]. However, the exploration of new natural fibers with a high quality is the need of the hour to replace detrimental synthetic materials and facilitate its economical utilization along with retaining environmental quality.

Identification of plants producing fibers that compete with the physical, chemical, and mechanical properties of synthetic fibers has been paid attention by many researchers.

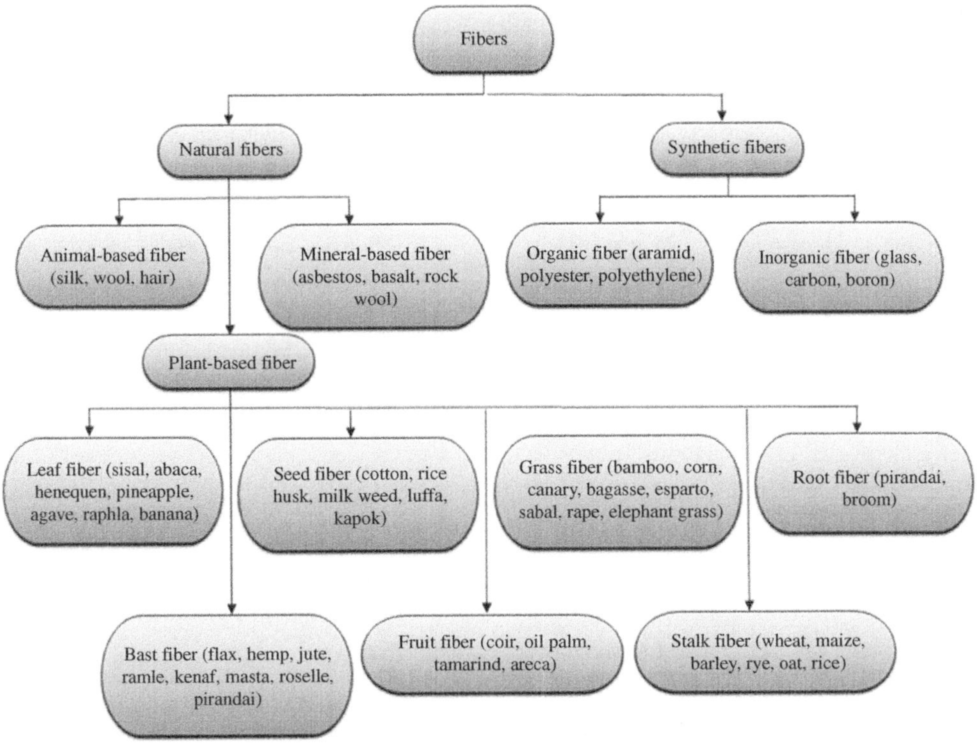

Figure 13.1 Distribution of diverse fibers.

Table 13.1 Origin of common natural plant fibers.

Origin/part of the plant	Fiber source	Species
Stem/bast	Abutilon	*Abutilon theophrasti*
	Cadillo/urena	*Urena lobata*
	China jute	*Abutilon theophrasti*
	Flax	*Linum usitatissimum*
	Hemp	*Cannabis sativa*
	Jute	*Corchorus capsularis*
	Kenaf	*Hibiscus cannabinus*
	Masta	*Polyalthia longifolia*
	Nettle	*Urtica dioica*
	Pirandai	*Cissus quadrangularis*
	Ramie	*Boehmeria nivea*
	Roselle	*Hibiscus sabdariffa*
	Sunn hemp	*Crotalaria juncea*
	Wood	(>10,000 species)
Leaf	Abaca	*Musa textilis*
	Banana	*Musa indica/Musa mannii*
	Cantala	*Agave cantala*

(*Continued*)

Table 13.1 (Continued)

Origin/part of the plant	Fiber source	Species
	Cocuiza	*Furcraea humboldtiana*
	Curaua	*Ananas erectifolius*
	Date palm	*Phoenix dactylifera*
	Fique	*Furcraea macrophylla*
	Gomuti palm	*Arenga pinnata*
	Istle	*Samuela carnerosana*
	Maguey	*Agave cantala*
	Mesta	*Hibiscus cannabinus*
	Phormium	*Phormium tenas*
	Piassava	*Attalea funifera*
	Pineapple	*Ananas comosus*
	Sisal	*Agave sisalana*
	Snake grass	*Sansevieria cylindrica*
Root	Pirandai	*Cissus quadrangularis*
	Broom root	*Muhlenbergia macroura*
Fruit	Areca (pan)	*Areca Catechu*
	Coconut husk/coir	*Cocos nucifera*
	Oil palm	*Elaeis guineensis*
	Palmyra palm	*Borassus flabellifer*
	Sponge gourd	*Luffa cylindrica*
	Tamarind	*Tamarindus indica*
Grass	Bagasse	—
	Bamboo	(>1250 species)
	Corn	*Zea mays*
Stalk	Straw (cereal)	—
	Palmetto	*Sabal palmetto*
Seed	Cotton	*Gossypium arboreum*
	Kapok	*Ceiba pentandra*
	Levant	*Gossypium hirsutum*

Source: Refs. [4, 7, 9, 12–22]

Studies on accessible plants for analyzing fiber production and further fiber characterization persuade much interest due to its wide range acceptance. India is the largest producer of jute and hemp where a total of about 6 million tons of natural fiber production per year was reported recently [28]. A review on plants yielding fibers of India by Pandey and Gupta [29] in 2003 exposed that 453 plant species belonging to a total of 82 families have been used all over India for the purpose of various fiber productions, of which the majority is plants belonging to *Malvaceae*. Most of the plant species are not studied well for fibers due to the lack of awareness of appropriate extraction and processing techniques. However, this is the time that various comprehensive approaches have been introduced to make the process effortless to study various plants for its fiber-bearing capability and aptness.

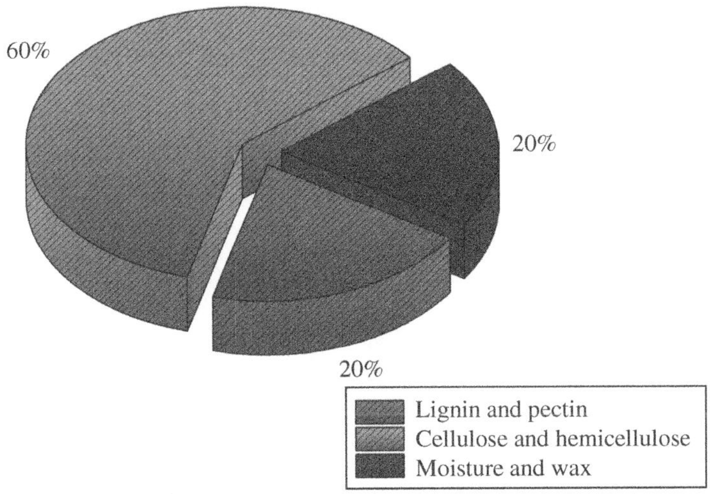

Figure 13.2 Range of major constituents in plant fiber.

Tensile properties of the fiber mainly depend upon the chemical nature of the fiber, age of the plant, and the site of the plant part where the fiber bears. Generally, plant fibers having high cellulose content increases the strength and durability of the fiber. A critical analysis revealed that few plants possess fibers with tensile properties equivalent to that of glass fibers. Few fibers experienced against synthetic fibers earlier were ramie, hemp, sisal, flax, jute, etc. Recently, a study on *Cissus quadrangularis* of the family *Vitaceae* reported that plant stem and root possess fibers having high cellulosic content of around 84 and 77%, respectively. Fiber obtained from this relevant plant showed properties approximately equal to that of glass fibers. It is also established that fibers from *Acacia planifrons, Cynara cardunculus L., Dichrostachys cinerea, Dracaena reflexa, Manicaria saccifera, Passiflora foetida*, and *Sida rhombifolia* possess high cellulose content in the range 72.4–77.9% and thermal stability in the range 200–320 (Table 13.2). Hence, such fibers with the relevant features could be successfully substituted for synthetic fibers and weak natural fibers [39, 40, 81].

13.3 Magnitude of Natural Fibers at this Juncture

Natural fibers are getting more attention today because of its admirable specific properties. The major advantage of a natural fiber is that it has comparable tensile properties of conventional fibers such as glass, aramid, and carbon fibers; hence, it can be substituted as a good alternative for them [1]. The values comparing physical and chemical properties of various natural fibers with that of glass fibers showed that most of the values are nearer to the average, except in the fact that the average of the extreme values was found in properties such as stiffness and density [12]. The physical, chemical, mechanical, and thermal properties of natural plant fibers are explored in Table 13.2.

The major disadvantage of synthetic fibers, mainly glass fibers, is that it has a high density of around 2500 kg/m^3 while plant fibers are enclosed with a low density of around 1200–1500 kg/m^3 [82]. Fibers from *Dracaena reflexa, Manicaria saccifera*, and red banana peduncle hold a low density of around 790–1070 kg/m^3 that could often be used for lightweight applications. The diameter of a fiber is another constraint to work with, where fibers with low diameter are recommended for fine mat formation. Studies regarding unwanted impurities found on the fiber surface suggest appropriate treatment processes to assure

Table 13.2 Properties of cellulose-based natural fibers.

	Physical properties		Chemical properties					
Fiber	**Diameter (μm)**	**Density (kg/m³)**	**Cellulose (wt%)**	**Hemi celluloses (wt%)**	**Lignin (wt%)**	**Wax (wt%)**	**Moisture content (wt%)**	**Ash (wt%)**
Acacia arabica	—	1028	68.10	9.36	16.86	0.49	—	—
Acacia leucophloea	—	1385	68.09	13.60	17.73	0.55	8.83	—
Acacia planifrons	—	660	73.1	9.41	12.04	0.57	8.21	4.06
Aerial roots of banyan tree	0.09–0.14	1234	67.32	13.46	15.62	0.81	10.21	3.96
Alfa	—	890	45.4	38.5	14.9	2	—	—
Althaea officinalis L.	156–194	1180	44.6	13.5	2.7	—	—	2.3
Areca palm leaf stalk	285–330	1090 ± 24	57.49 ± 0.66	18.34 ± 0.24	7.26 ± 0.12	0.71 ± 0.024	9.35 ± 0.15	1.43 ± 0.019
Artisdita hystrix	—	540	59.54	11.35	8.42	—	—	—
Arundo donax	—	1168	43.2	20.5	17.2	—	—	1.9
Bamboo	240–330	910	26–43	30	1–31	—	9.16	—
Borassus fruit	241.18	1256	68.94	14.03	5.37	0.64	6.83	—
Caster fiber	0.06–0.13	1181	65.5	—	5.56	0.87	10.17	4.86
Cissus quadrangularis root	610–725	1510	77.17	11.02	10.45	0.14	7.3	—
Cissus quadrangularis stem	770–870	1220	82.73	7.96	11.27	0.18	6.6	—
Coccinia grandis	27.33	1243	62.35	13.42	15.61	0.79	5.6	4.388
Coccinia grandis stem	543–621	1517.5	63.22	—	24.42	0.32	9.14	—
Coconut tree leaf sheath	140–990	1200	27	14	27.7	—	4.7	—
Cordia dichotoma	—	—	59.7	23.6	14.7	—	—	—
Cotton	—	1600	85–90	1–3	0.7–1.6	—	1.0	
Cynara cardunculus L.	—	1579 ± 8	75.3 ± 1.2	—	4.3 ± 0.5	—	—	2.2 ± 0.05
Cyperus pangorei	—	1102	68.5	—	17.88	0.17	9.19	3.56
Date palm rachis. Vascular bundles	520 ± 72	914	—	—	—	—	—	—
Dichrostachys cinerea	—	1240	72.4	13.08	16.89	0.57	9.82	3.97
Dracaena reflexa	176.20	790	70.32	11.02	11.35	0.23	5.19	6.23
Ferula communis	—	—	53.3	8.5	1.4	—	—	—
Flax	—	1500	85	09	4	1.7	10	—
Furcraea foetida	12.8	778	68.35	11.46	12.32	0.24	5.43	6.53
Grand Naine pseudostem	225	—	48.19	15.91	19.17	—	—	—

Crystalline properties		Mechanical properties				Thermal properties		
CI (%)	CS (nm)	Tensile strength (MPa)	Young modulus (GPa)	Elongation at break (%)	Microfibril angle (°)	Thermal stability (°C)	Max. degradation temperature (°C)	References
51.72	15	—	—	—	—	—	345.1	[30]
		317–1608	8.41–69.61	1.38–4.24	—	100	387	[31]
65.38	—	—	—	—	—	200	362.4	[30]
72.47	6.28	19.37 ± 7.72	1.8 ± 0.40	1.8 ± 0.40	10.88 ± 1.198	230	358	[32]
—	—	350	22	5.8	—	—	—	[33]
68	2.4	415.2	65.4	3.9	—	220	344	[34]
—	—	364.66 ± 21.46	9.39 ± 1.13	3.47 ± 1.15	—	110	365	[35]
44.85	—	440 ± 13.4	1.57 ± 0.04	—	12.64 ± 0.45	—	298.8	[36]
—	—	248	9.4	3.24	6.85 ± 1.23	275	320	[37]
—	—	503	35.91	1.4	—	—	—	[33]
—	—	175.52	—	31.34	—	—	—	[38]
48.88	4.8	356.3	34.931 ± 0.45	1.02 ± 0.053	—	—	—	[32]
56.6	7.04	1857–5330	68–203	3.57–8.37	5.89 ± 0.27	230	328.9	[39]
47.15	3.91	2300–5479	56–234	3.75–11.14	4.95 ± 0.32	270	342.1	[40]
52.17	13.38	273 ± 27.74	10.17 ± 1.261	2.703 ± 0.2736	13.25 ± 0.664	213.4	351.6	[41]
46.09	1.91	424.242	26.515	16	—	105	320	[42]
—	—	46.4	2.3	2.84	—	—	—	[33]
12.14	—	—	—	—	—	—	—	[43]
68	5–7	287–597	5.5–12.6	7–8	—	265	280	[44]
—	—	201	11.624	—	—	230	352	[45]
41	—	196 ± 56	11.6 ± 2.6	1.69	—	221	324	[46]
47.82	5.78	129.1 ± 47.3	5.88 ± 1.84	2.67 ± 0.57	—	230	360	[47]
57.82	—	873 ± 14	—	—	—	226	495.3	[48]
57.32	19.01	829.6	46.37	2.95	8.5–11.27	232.32	348.78	[38]
48	1.6	475.6 ± 15.7	52.7 ± 3.7	4.2 ± 0.2	—	200	313.5	[49]
70	5.4	400–938	61.4–128	3.6–3.8	—	—	—	[50]
52.6	28.36	623.52 ± 45	6.52 ± 1.9	10.32 ± 1.6	—	—	320.5	[51]
—	—	24.72	—	—	—	—	—	[52]

(*Continued*)

Table 13.2 (Continued)

	Physical properties		Chemical properties					
Fiber	Diameter (μm)	Density (kg/m³)	Cellulose (wt%)	Hemi celluloses (wt%)	Lignin (wt%)	Wax (wt%)	Moisture content (wt%)	Ash (wt%)
Grewia tiliaefolia	—	—	62.8	21.2	14.9		2.3	—
Hemp	—	1500	70.2	17.9	3.7	0.8	10.8	
Heteropogon contortus	—	602	64.87	19.34	13.56	0.22	7.4	—
Juncus	3300	385	40.99	27.84	18.54	—	—	7.3
Juncus effuses L.	1.139	280	—	—	—	—	—	—
Jute	26	1300	58–63	20–24	12–15	0.5	10.99	—
Kenaf	65–71	1400	53.14	14.33	8.18	0.8	6.2–12	2–5
Lygeum spartum L.	180–433	1499.7 ± 3.1	—	—	—	—	—	—
Manicaria saccifera palm	—	840–1070	74.1	12.0	31.1			
Mendong grass	33.4	892	72.14	20.2	3.44	—	4.2–5.2	4.2.
Mikania micrantha Kunth ex H.B.K	—	—	21.31 ± 1.56	14.61 ± 1.46	20.67 ± 0.80	—	34.30 ± 1.15	9.11 ± 0.23
Monthan pseudostem	170	—	48.55	15.75	21.56	—	—	—
Mulberry barks	—	—	37.38 ± 2.31	25.32 ± 2.45	9.99 ± 0.82	—	—	—
Napier grass strands	—	—	47.12	31.27	21.63	—	—	—
Nendran peduncle	129	—	60.41	10.20	17.56	—	—	—
Passiflora foetida	101 ± 40	1328	77.9	12.63	10.47	0.35	9.54	0.94
Pennisetum purpureum	210–270	—	—	—	—	—	—	—
Pergularia daemia	15–35	800–900	53 ± 2	26 ± 1	15 ± 0.8	4	10	1.2
Pergularia tomentosa L. seed fiber	35.6	1006	43.8	16	8.6	1.88	8.5	2.74
Perotis indica	—	785	68.4	15.7	8.35	0.32	9.54	4.32
Phoenix dactylifera L. (leaf sheath)	252.48	—	43.50	24	18	—	6.8	7.73
Piassava	—	1400	28.6	25.8	45	—	—	—
Pineapple leave fiber	20–80	1440	81.27	12.31	3.46	3.2–4.2	11.8	0.9–4.3
Poovan peduncle	146	—	56.24	14.89	19.17	—	—	—
Prosopis juliflora bark	20	580	61.65	16.14	17.11	0.61	9.48	5.2
Prosopis juliflora bark	20	580	61.65	16.14	17.11	0.61	9.48	5.2
Ramie	50	1500	76	15	1	—	8.0	—
Red banana peduncle	150–250	990	72.90	11.01	15.99	0.32	9.36	2.79
Rhectophyllum camerunense	—	947	68.2	16.0	15.6	—	—	—
Ricinus communis plant	60–130	1180	65.5		5.56	0.87	10.17	4.89

Crystalline properties		Mechanical properties				Thermal properties		
CI (%)	CS (nm)	Tensile strength (MPa)	Young modulus (GPa)	Elongation at break (%)	Microfibril angle (°)	Thermal stability (°C)	Max. degradation temperature (°C)	References
8.8	—	65.2	4.5	1.6			—	[43]
87.87	4.5	690	70	2.0–4	—	225	—	[53]
54.1	—	476 ± 11.6	48 ± 2.8	1.63 ± 0.06	—	220	337.7	[54]
43	—	31 ± 8	0.7 ± 0.1	—	—	—	—	[55]
33.4	3.6	113 ± 36	4.38 ± 1.37	2.75 ± 0.68	—	200	300	[56]
65.8	29.25	400–773	10–30	1.5–1.8	16.9	—	—	[38]
		470–785	25.1	1.75–1.9	—			[33]
46.19	—	64.63–280	4.47–13.27	1.49–3.74	12.65 ± 2.85	220	338.7	[57]
		72.09 ± 15.73	2.20 ± 0.44	—	—	220	370	[58]
58.6	14.3	452	17.4	—	22.9	—	—	[59]
11	—	—	—	—	—	220	420	[60]
—	—	49.67	—	—	—	—	—	[52]
58.8	—	—	—	—	—	210	250	[61]
62.4	2.83	13.15	88.40	0.99	—	220	383	[62]
—	—	252.69	—	—	—	—	—	[52]
67.36%	4.23	248–942	11–48	1.38–4.67	7.46–12.18	320	383	[63]
—	—	73 ± 6	5.68 ± 0.14	1.40 ± 0.23	—	230	364.7	[64]
56	—	—	—	1.8	—	240	360	[65]
52.02	2.41	—	—	—	—	—	315.48	[66]
48.3	15	—	—	—	—	—	339.1	[67]
69.2	—	—	—	—	—	245.1	—	[68]
—	—	134–143	1.07–4.59	7.8–21.9	—	—	—	[33]
58.6	—	413–1627	34.5–82.5	0.8–1	14	—	—	[69]
—	—	100.45	—	—	—	—	—	[52]
46	2.8	—	—	—	—	—	331.1	[70]
46	15	558 ± 13.4	—	1.77 ± 0.04	10.64 ± 0.45	217	331.1	[70]
58	16	8.0870	128	1.2	—	—	—	[33]
62.1	12.1	440 ± 13.4	12.41 ± 5	1.57 ± 0.04	12.64 ± 0.45	120	353	[51]
—	—	557.1	5.8	27.5	—	—	—	[38]
48.88	4.8	356 ± 23.87	34.931 ± 0.45	1.02 ± 0.053		225	326	[71]

(Continued)

Table 13.2 (Continued)

	Physical properties		Chemical properties					
Fiber	**Diameter (μm)**	**Density (kg/m^3)**	**Cellulose (wt%)**	**Hemi celluloses (wt%)**	**Lignin (wt%)**	**Wax (wt%)**	**Moisture content (wt%)**	**Ash (wt%)**
Root of *Ficus religiosa* tree	25.62	1246	55.58	13.86	10.13	0.72	9.33	4.86
Saharan aloe vera	80.61	1325.1	60.2	14.2	13.7	1.5	7.6	—
Salago fiber	0.00623	1023	79.05	13.6	10.13	—	50.21	60.93
Sansevieria ehrenbergii	20–250	887	80	11.25	7.8	0.45	10.55	0.6
Sansevieria cylindrical	6–30	915	79.7	10.13	3.8	0.09	6.08	—
Sida cordifolia	—	1330	69.52	17.63	18.02	0.42	8.51	2.62
Sida rhombifolia	—	1320.7	75.09	15.43	7.48	0.49	12.02	4.07
Sisal	50–300	1500	78	19	8	2	11	—
Tamarindus indica L.	564–789	1020–1270	72.84	11	15.38	0.2	6.35	0.51
Thespesia populnea	87–256	1412	70.2	12.64	16.34	0.76	10.83	1.80

proper adhesion. Thus, further research to improve adhesion property and interfacial bonding is a big challenge for researchers.

Natural fibers are obtained from a variety of sources since its production is cost effective and need total production energy of around 20–40% merely. Although the cost for glass fibers is around US$1200–US$1800 per ton, it is too high when compared to that of the natural fiber cost of around US$200–US$1000 per ton [83]. The overall production energy for natural fibers is approximately 60%, which is lesser than those used for manufacturing glass fibers [84]. Natural fibers are renewable resources that lessen energy consumption by reducing the usage of much energy for the production of synthetic fibers and therefore it can be used as a sustainable choice. The handling and processing of natural fibers are easy and require only a little investment. In addition, natural fibers have good insulating properties, high electrical resistance, depart little residue, and release less carbon dioxide during burning, which also possess high environmental impacts [85]. Natural fibers have some superior properties such as renewability, biodegradability, and nonabrasive nature [86]. Because of these properties, they replaced the conventional synthetic fibers nylon, polyester, and polypropylene to a great extent.

Previously, the use of natural fibers could be driven back mainly due to its low strength and instability in variable temperature. As a result, they could not be directly used in biomedical and engineering applications; at the same time synthetic fibers were of more interest for these purposes owing to their high strength, high heat resistance, and chemical properties [87]. At present, the striking plant fiber-based polymer composites are developed to improve the mechanical properties. For instance, studies demonstrated that fibers with a high degree of tensile strength of around 73–85% were significant over man-made fibers. Hence, natural fiber composites could successfully apply for both biomedical and engineering fields.

Crystalline properties		Mechanical properties				Thermal properties		
CI (%)	CS (nm)	Tensile strength (MPa)	Young modulus (GPa)	Elongation at break (%)	Microfibril angle (°)	Thermal stability (°C)	Max. degradation temperature (°C)	References
42.92	5.18	433.32±44	5.42±2.6	8.74±1.8	—	325	400	[33]
56.5	5.72	805.5	42.29	2.39	11.1	225	350	[72]
		1187	—	2.48	—	145	325	[73]
52.27	—	50–585	1.5–7.67	2.8–21.7	—	223.85	333.02	[74]
60	86	673.12±51	6.72±1.9	10.04±1.5	—	—	—	[75]
56.92	18	703.95±23.73	42.84±2.1	2.89±0.24	—	—	338.2	[76]
56.6	2.75	673±14	—	—	7.3	250	433.3	[77]
55	16.9	511–635	9.4–22	2.0–2.5	—	250	275	[78]
55	5.73	1137–1360	11.23–20.72	6.56–10.12	—	238	351.4	[79]
48.17	3.57	557.82	20.57	2.80	13.94	245.4	323.76	[80]

Composites of natural fibers are appropriate for various sectors including automotive, construction, sports, aerospace, and marine fields [88]. Plant fiber-based composites with high fiber content could be largely used for manufacturing interior parts of passenger cars, multiutility vehicles, and truck cabins. It is also applicable in railways for roof or floor panels, berths, lightweight coaches, and interior furnishing and seating [89]. Another interesting fact is that hemp composites were used as construction material in China for the Olympics of 2008. In addition, natural fiber composites are pertinent in marine engineering for the construction of shipping materials. However, the demand for natural fibers increased 60% annually in developing countries, mainly in the United States, due to its widespread applications in diverse fields [89]. Furthermore, the driving forces described above would endorse extensive research in the application of natural fibers in various sectors.

13.4 Constraints and Competence of Natural Fibers

Poor strength, high moisture absorption rate, quality variation, inferior fire resistance, restricted processing temperatures, and short durability are the major confines associated with weak natural fibers [12]. Natural fibers are vulnerable to biodegradation, which make them unsuitable for outdoor applications. Synthetic fibers are enriched with less moisture absorption, high strength, stiffness, thermal stability, chemical, and corrosion resistance. However, the restraints of synthetic fibers such as biodegradability, expensive processing, less recyclability, high energy conception, machine abrasion, and health hazards mainly due to carcinogenic effects, prompted to intent more compatible materials for the time being. The natural fibers with promising features are on consideration to replace synthetic fibers since the properties of renewability, less weight, high specific modulus, being cost effective, environmental friendliness, high crystallinity, low microfibrillar angle, higher length to diameter, and nonhazardous effects make them prominent over synthetic fibers [90]. The comparability between natural fibers and synthetic fibers are listed in Table 13.3.

Natural fiber composites are fabricated by polymeric binding materials and reinforcement natural fibers. Recently, biodegradable polymers are used as a matrix for the production of natural fiber composites. The matrix serves as a binding material to hold the fibers in proper position. Two kinds of matrixes that are in use for binding are thermoset and thermoplastic matrixes. The materials used for thermosets are epoxy resins, unsaturated polyesters, and phenolic resin, while thermoplastic matrix is contrived by elastomers, polypropylene, and polyethylene. During composite preparation, the successful adhesion between natural fibers with matrices is essential. Generally, natural fibers are hydrophilic in nature; hence, its adhesion with hydrophobic matrices may cause incompatibility, which is a typical constraint that hinders the application of natural fibers. However, appropriate modification techniques are needed to improve the binding of natural fibers with matrices in order to strengthen the composite [20]. It has been proven that physical (stretching, calendaring, thermo treatments, cold plasma treatment, hybrid yarn, electric discharge) and chemical (silane, alkaline, acetylation, benzoylation, and peroxide treatment) modification of fibers nullifies the negative impact and provides better mechanical and thermal stability [5, 91, 92].

Several factors affecting the properties of plant fiber composites include plant fiber structure, thermal stability, length, loading, orientation, moisture absorption, and presence of voids. The presence of hydroxyl groups in the crystalline region and waxy substances of surface of the fiber leads to distension of fiber frame work. Chemical treatment is effective for surface modification of fibers in order to remove hydroxyl group and surface impurities [93, 94]. Similarly, the chemical constituents of plant fiber cellulose, hemicellulose, and lignin are sensitive to thermal degradation. Earlier reports specify that thermal stability can be accomplished by getting rid of a fraction of lignin and hemicellulose through chemical modification [1]. The assets of plant fiber composites strongly hinge on the fiber length, loading percentage, distribution, and orientation into the matrix [18]. However, it is established that appropriate surface treatments boost the mechanical properties of natural fibers or composites in order to compete with synthetic fibers, in outlook.

Table 13.3 Advantages and disadvantages of natural plant fibers and synthetic fibers.

Natural plant fibers		Synthetic fibers	
Advantages	**Disadvantages**	**Advantages**	**Disadvantages**
Renewable	Poor bonding with matrix	High stiffness	High initial processing cost
Light weight	Poor/moderate strength	High chemical resistance	Moderate recyclable
High specific modulus	Absorbs moisture	High strength	High energy conception
Nontoxic	Biodegradable	High corrosive resistance	Toxic, health hazards
Cost effective	Unique extraction	Low water absorption	Nonbiodegradable
Easy processing	Less thermal stability	Good thermal stability	Machine abrasion

13.5 Degradability of Polymeric Natural Fiber Composites

Mostly, composites of biodegradable polymers have reduced economic impacts because of its highly degradative nature, and as a consequence, its utilization is limited to the interior portion. An important aspect of developing natural composites is to check its degradability and thermal stability in order to ensure its broad application [3, 95]. Generally, the major factors that influence the degradation of natural fibers and its polymer composites are identified as moisture absorption, temperature, weathering, and microbial activities [9]. Degradable polymeric composites possess a nature of high moisture absorption rate [96–98]. The physical, chemical, and thermal properties of natural fibers are influenced by moisture absorption of natural fiber composites [93]. The microvoids obtained during composite preparation may adversely affect the mechanical strength of the composites [99]. Fiber volume fraction, the orientation of reinforcement, diffusivity, permeability nature, the area of exposed surface, and temperature are persuaded moisture absorption rates of composite materials [81]. Hence, it is essential to study the moisture absorption rate and thereby biodegradability of natural fiber composites in different environmental conditions by measuring its weight loss at respective intervals during treatment.

A study on the degradability of jute fiber composite with polybutylene succinate (PBS) as matrix showed that PBS reinforced with 10% jute fiber showed 62.5% weight loss after 180 days of treatment in compost soil. Sometimes pores were generated on the surface of natural fiber composites due to poor adhesion between the fiber particles and polymer matrix that increases its moisture absorption capacity. Another point obtained from the study is that reducing moisture absorption by surface modification of PBS–jute fiber composites with alkali and coupling agent enhanced its hydrophobic behavior which reduces the degradation of composites in mild environmental conditions [28]. Recently few studies showed that degradability of composites is allied with high content of lignin in the fiber. Increased lignin content augments the uptake of moisture that further leads to its degradation, although the development of plant fiber-based polymer composites by fixing cellulosic content with reducing lignin content would diminish its degradability to a large extent [88].

Thermal resistance is an imperative criterion, a natural fiber composite to be obsessed. The nature of fiber and matrix of a composite determines its thermal resistance. The flammability of composites is mainly based on the variability of chemical components of fiber such as cellulose, lignin, wax, and ash content. Composites with less cellulose content and elevated lignin content showed low thermal resistance [86]. Correspondingly, the polymer matrix of composites may be susceptible to higher temperatures and is restricted to 200 °C. Reports show that a temperature between 300 and 500 °C causes the decomposition of polymer matrix [92, 100]. A number of methods are available to enhance the thermal resistance of composites, of which fire retardant coating on composites could be fundamentally effective. Thermogravimetric analysis and calorimetry are principally used to evaluate the thermal degradation behavior of natural fiber composites [9]. Sgriccia and Hawley in 2007 [91] and various researchers reported that the thermal stability of man-made fibers are good but natural fibers are extensively lower (Table 13.4).

Another factor which severely causes the degradation of natural polymer composites is weathering at different environments. Such degradation occurs through various phenomena like photo-oxidation, photoradiation/UV irradiation, hydrolysis, and thermal degradation on polymer composites and consequently, progressive alteration occurs in their tensile properties [19]. Photo-oxidation and thermo-oxidation proceed by the feat of light, temperature,

Table 13.4 Degradation temperature of plant fibers and man-made fibers.

Weight loss %	*T* (°C) 5%	*T* (°C) 25%	*T* (°C) 50%	*T* (°C) 75%
Acacia leucophloea	100	346.8	387	420
Aerial root of Banyan tree	230	300	358	370
Areca palm leaf stalk	110	250	365	410
Arundo donax	275	290	320	440
Bambusoideae	246	255	320	420
Cotton	55	280	330	410
Flax	88	322	338	441
Hemp	104	329	341	457
Henequen	61	301	337	391
Jute	60	290	340	470
Kenaf	64.3	313	340	370
Pennisetum purpureum	230	290	364	455
Pirandai root	88	328	395	481
Pirandai stem	91	345	405	495
Prosopis juliflora	110	331	380	480
Saharan aloe vera cactus	225	272	350	470
Sansevieria ehrenbergii	223	280	333	422
Sisal	52	275	345	465
Tamarind	84	319	381	479
Thespesia populnea bark	245	296	323	410
Tridax procumbens	195	170	330	492
Wood	107	270	367	400
Glass[a]	586			

[a] Ultimate fiber weight loss is 1.6%. Source: Refs. [1, 3, 9, 24, 82, 92, 97]

and oxygen, respectively [27], despite the fact that photodegradation of composites also depends upon its ability to absorb UV light. Lignin of fiber easily absorbs UV radiation that results in surface oxidation and such photo-radiation would change crystallinity of the matrix. Hence, the development of natural fiber composites with a property to overcome photodegradation is necessary [9]. Wallstrom et al., in 2005 [101] designed a microenvironmental chamber for controlling degradation of polymer composites under different circumstances such as photoradiation, thermo-oxidation, photo-oxidation, and biodegradation in order to establish the degradative activity of polymeric composites.

Biodegradation or biodeterioration refers to decomposition of materials with the help of microorganisms and their enzymes, which offers mineralization of materials into its smaller fractions. It is one of the disadvantages of polymer composites where microorganisms adhere to the surface of the polymeric material and initiate its degradation, which consequently leads to changes in their structural, physical, and chemical properties [101]. During degradation, microorganisms penetrate to the surface of polymeric materials and then secrete enzymes for its degradation [96]. In context, microorganisms aggregate together and form a structure called biofilms, identified as a major phenomena for the degradation of polymer structures recently. The action of fungal and algal species on different fiber composites was reported earlier by Wallström et al. in 2009 [101]. As a consequence, biofragmentation of polymeric molecules into oligomers and monomers resulted. However, environmental protection is the major agenda of the government for the past 25 years, for releasing materials which have no hazardous effects to the environment or human health [11]. By summarizing these studies, it is essential to perform degradability studies of natural fiber composites during its manufacture, in order to pledge its structural firmness at different environmental conditions.

13.6 Marine Application of Natural Fiber Composites and Its Degradation

Currently, natural fiber reinforced composites are relevant for the manufacture of marine structures instead of glass fiber reinforced composites [102]. Natural fiber composites are widely applicable in the marine environment for the reason that probable fibers encompass good mechanical properties similar to that of conventional materials used in the same rationale [98]. Good strength to weight ratio, thermal resistance, stiffness, and hardness properties could be pertinent for the manufacture of marine structures like heat exchangers, condensers, pumps, and shipboard [102, 103]. A few investigations revealed that a major drawback to the use of natural fiber composites in the marine environment is earlier corrosion with various external factors. Moisture invasion, osmotic blistering, cathodic reaction of dissolved oxygen, and marine microbial action are the dimensions affecting the degradation of composites in the marine environment [102, 104].

Attack of microorganisms on composites in natural waters is recognized as the major aspect that cause degradation [101]. Microorganisms, mainly bacteria and fungi, are capable of producing enzymes, in turn to uptake nutrients when they are exposed to suitable substrates. These enzymes could involve adversely in composite degradation once composites acquire an appropriate feedstock [105]. Molds are the common organism that causes degradation of polymeric composites in wet soil [106]. Experimental studies demonstrated that microbes attacked composites by three mechanisms: the action on composites by interfering fiber resin interface, action on fiber, or action on resin matrix. The former one was exemplified by Wagner et al. [103], in which fiber and vinyl ester resin interface bonding was disrupted by the action of hydrogen producing bacteria. A combined action of bacteria such as *Pseudomonas aeruginosa* and *Acinetobacter calcoaceticus* on graphite fibers showed considerable alteration in fiber configuration [107]. While the latter was verified by Gulati and Sain in 2005 [108] that fungi are the common organism favored on a resin matrix of various composites and consequently causes its degradation. The polyester and polyurethane resins are susceptible whereas vinyl ester and epoxy resins are opposed to the action of microbes [109]. Therefore, utilization of resins belonging to resistant category or modification of the whole composite by fiber or resin may be profitable in outlook.

Generally, synthetic organic polymers are stable in the marine environment while treating them in sea water for a long-term duration leads to biodeterioration and biodegradation [86, 102]. The debonding of fiber/matrix interface is identified as the major reason that causes degradation of flax/PLA composites when they are immersed in sea water for a long time. Along with structural changes by matrix hydrolysis, degradation of fibers are recognized as the motive for degradation of composites [110]. Once organic composites are immersed in water, living creatures will oxidize them, and as a result carbon dioxide concentration in water increases, which may lead to acidification. The anaerobic sulfate reducing bacteria are favorable to the former condition that metabolize the organic matter and finally decrease its strength and stiffness [109]. Hence, it is critical to demonstrate the durability of natural fiber composites in the sea water environment prior to bringing into play in order to accomplish its broad application in marine structures.

A surface treatment of composites for mounting hydrophobicity in order to reduce direct contact with microorganisms could be effective [111]. Singh et al. in 2007 [112] reported that corrosion on steel can be controlled by providing a hydrophobic surface coating of polyolefins and thermoplastic polyethylene. Biostabilization by coating antimicrobial agents on polymeric composites reduces the contact of biological agents with composites, and thereby, biodegradation. Nevertheless, special attention has to be paid to detect the durability of natural fiber reinforced polymer composites in the marine environment in sequence to ensure its extensive employment.

13.7 Conclusion

This contemporary review conferred the forecasts of natural plant fiber reinforced polymer composites and its degradability concerns. The density, crystallinity, tensile strength, and thermal stability of the plant fiber determines the potential of polymeric composites. The natural fibers with desired features that compete with the properties of artificial fibers are highly enviable for high tech choices. However, the hydrophilicity of typical fibers renowned a major constraint where modification techniques are adequate to attain comprehensive adhesion between the fiber and matrix. Yet, biodegradation studies after manufacturing offers a great outlook to establish the fragmentation frequency of polymeric composites. Appropriate surface coating is recommended to improve the durability of composite materials by preventing degradation under marine system. In conclusion, the identification of plants that afford fibers with promising properties along with the utilization of accessible technologies for diminishing the degradability of natural fibers are suggested to trim down the drawback which impedes natural fibers and fiber reinforced composites.

Acknowledgments

Authors acknowledge Rohini College of Engineering and Technology and Pinnacle Bio-Sciences for providing support to carry out this work.

References

1 Madhu, P., Sanjay, M.R., Senthamaraikannan, P. et al. (2019). A review on synthesis and characterization of commercially available natural fibers: part II. *J. Nat. Fibers* 16 (1): 25–36.
2 Yogesha, B., Sanjay, M.R., Senthamaraikannan, P., and Pradeep, S. (2018). A review on synthesis and characterization of commercially available natural fibers: part-I. *J. Nat. Fibers* 16 (8): 1–13.
3 Pickering, K.L., Efendy, M.G.A., and Le, T.M. (2016). A review of recent developments in natural fibre composites and their mechanical performance. *Compos. Part A Appl. Sci. Manuf.* 83: 98–112.

4 La Mantia, F.P. and Morreale, M. (2011). Green composites: a brief review. *Compos. Part A Appl. Sci. Manuf.* 42 (6): 579–588.
5 Väisänen, T., Das, O., and Tomppo, L. (2017). A review on new bio-based constituents for natural fiber–polymer composites. *J. Clean. Prod.* 149: 582–596.
6 Wambua, P., Ivens, J., and Verpoest, I. (2003). Natural fibres: can they replace glass in fibre reinforced plastics? *Compos. Sci. Technol.* 63 (9): 1259–1264.
7 Islam, M.S., Kovalcik, A., Hasan, M., and Thakur, V.K. (2015). Natural fiber reinforced polymer composites. *Int. J. Polym. Sci.* 2015: 1–7.
8 Sanjay, M.R., Madhu, P., Jawaid, M. et al. (2018). *Characterization and Properties of Natural Fiber Polymer Composites: A Comprehensive Review*. Elsevier B.V.
9 Azwa, Z.N., Yousif, B.F., Manalo, A.C., and Karunasena, W. (2013). A review on the degradability of polymeric composites based on natural fibres. *Mater. Des.* 47: 424–442.
10 Sanjay, M.R., Madhu, P., Jawaid, M. et al. (2018). Characterization and properties of natural fi ber polymer composites: a comprehensive review. *J. Clean. Prod.* 172: 566–581.
11 Madhu, P., Sanjay, M.R., Senthamaraikannan, P. et al. (2017). A review on synthesis and characterization of commercially available natural fibers: part II. A review on synthesis and characterization of commercially. *J. Nat. Fibers* 16 (8): 1–12.
12 Dittenber, D.B. and Gangarao, H.V.S. (2012). Critical review of recent publications on use of natural composites in infrastructure. *Compos. Part A Appl. Sci. Manuf.* 43 (8): 1419–1429.
13 Jawaid, M. and Abdul Khalil, H.P.S. (2011). Cellulosic/synthetic fibre reinforced polymer hybrid composites: a review. *Carbohydr. Polym.* 86 (1): 1–18.
14 Summerscales, J., Virk, A., and Hall, W. (2013). A review of bast fibres and their composites: part 3 – modelling. *Compos. Part A Appl. Sci. Manuf.* 44 (1): 132–139.
15 Holbery, J. and Houston, D. (2006). Natural-fiber-reinforced polymer composites in automotive applications. *JOM* 58 (11): 80–86.
16 Nyström, B. (2007). Natural fiber composites: a review. *Engineering* 15: 281–285.
17 Summerscales, J., Dissanayake, N.P.J., Virk, A.S., and Hall, W. (2010). A review of bast fibres and their composites. Part 1 – fibres as reinforcements. *Compos. Part A Appl. Sci. Manuf.* 41 (10): 1329–1335.
18 Yan, L., Chouw, N., and Jayaraman, K. (2014). Flax fibre and its composites – a review. *Compos. Part B Eng.* 56: 296–317.
19 Ku, H., Wang, H., Pattarachaiyakoop, N., and Trada, M. (2011). A review on the tensile properties of natural fiber reinforced polymer composites. *Compos. Part B Eng.* 42 (4): 856–873.
20 Faruk, O., Bledzki, A.K., Fink, H.P., and Sain, M. (2012). Biocomposites reinforced with natural fibers: 2000–2010. *Prog. Polym. Sci.* 37 (11): 1552–1596.
21 Sathishkumar, T.P., Navaneethakrishnan, P., Shankar, S. et al. (2013). Characterization of natural fiber and composites – a review. *J. Reinf. Plast. Compos.* 32 (19): 1457–1476.
22 Pacheco-Torgal, F. and Jalali, S. (2011). Cementitious building materials reinforced with vegetable fibres: a review. *Constr. Build. Mater.* 25 (2): 575–581.
23 Sen, T. and Jagannatha Reddy, H.N. (2013). Strengthening of RC beams in flexure using natural jute fibre textile reinforced composite system and its comparative study with CFRP and GFRP strengthening systems. *Int. J. Sustain. Built Environ.* 2 (1): 41–55.
24 Gowda, T.G.Y., Sanjay, M.R., Bhat, K.S. et al. (2018). Polymer matrix–natural fiber composites: an overview. *Cogent Eng.* 47: 1–13.
25 Kabir, M.M., Wang, H., Lau, K.T., and Cardona, F. (2012). Chemical treatments on plant-based natural fibre reinforced polymer composites: an overview. *Compos. Part B Eng.* 43 (7): 2883–2892.

26 Divya, D., Gopinath, L.R., and Merlin Christy, P. (2015). A review on current aspects and diverse prospects for enhancing biogas production in sustainable means. *Renew. Sust. Energ. Rev.* 42: 690–699.

27 Eichhorn, S.J., Baillie, C.A., Zafeiropoulos, N. et al. (2001). Current international research into cellulosic fibres and composites. *J. Mater. Sci.* 36 (9): 2107–2131.

28 Mngomezulu, M.E., John, M.J., Jacobs, V., and Luyt, A.S. (2014). Review on flammability of biofibres and biocomposites. *Carbohydr. Polym.* 111: 149–182.

29 Pandey, A. and Gupta, R. (2003). Genetic resources, perspective for collection and utilisation. *Fibre Yield. Plants India* 2 (4): 194–204.

30 Senthamaraikannan, P., Saravanakumar, S.S., and Arthanarieswaran, V.P. (2016). Physico-chemical properties of new cellulosic fibers from the bark of *Acacia planifrons*. *Int. J. Polym. Anal. Charact.* 21: 207–213.

31 Arthanarieswaran, V.P., Kumaravel, A., and Saravanakumar, S.S. (2015). Characterization of new natural cellulosic fiber from *Acacia leucophloea* bark. *Int. J. Polym. Anal. Charact.* 20: 367–376.

32 Ganapathy, T., Sathiskumar, R., Senthamaraikannan, P. et al. (2019). Characterization of raw and alkali treated new natural cellulosic fibres extracted from the aerial roots of banyan tree. *Int. J. Biol. Macromol.* 138: 573–581.

33 Moshi, A.A.M., Ravindran, D., Bharathi, S.R.S. et al. (2020). Characterization of a new cellulosic natural fiber extracted from the root of *Ficus religiosa* tree. *Int. J. Biol. Macromol.* 142: 212–221.

34 Sarikanat, M., Seki, Y., Sever, K., and Durmuşkahya, C. (2014). Determination of properties of *Althaea officinalis L.* (Marshmallow) fibres as a potential plant fibre in polymeric composite materials. *Compos. Part B Eng.* 57: 180–186.

35 Shanmugasundaram, N., Rajendran, I., and Ramkumar, T. (2018). Characterization of untreated and alkali treated new cellulosic fi ber from an Areca palm leaf stalk as potential reinforcement in polymer composites. *Carbohydr. Polym.* 195 (1): 566–575.

36 Kathiresan, M., Pandiarajan, P., Senthamaraikannan, P., and Saravanakumar, S.S. (2016). Physicochemical properties of new cellulosic *Artisdita hystrix* leaf fiber. *Int. J. Polym. Anal. Charact.* 5341.

37 Fiore, V., Scalici, T., and Valenza, A. (2014). Characterization of a new natural fiber from *Arundo donax* L. as potential reinforcement of polymer composites. *Carbohydr. Polym.* 106: 77–83.

38 Manimaran, P., Saravanan, S.P., Sanjay, M.R. et al. (2019). Characterization of new cellulosic fiber: *Dracaena reflexa* as a reinforcement for polymer composite structures. *J. Mater. Res. Technol.* 8 (2): 1952–1963.

39 Indran, S., Edwin Raj, R., and Sreenivasan, V.S. (2014). Characterization of new natural cellulosic fiber from *Cissus quadrangularis* root. *Carbohydr. Polym.* 110.

40 Indran, S. and Raj, R.E. (2015). Characterization of new natural cellulosic fiber from *Cissus quadrangularis* stem. *Carbohydr. Polym.* 117: 392–399.

41 Senthamaraikannan, P. and Kathiresan, M. (2018). Characterization of raw and alkali treated new natural cellulosic fiber from *Coccinia grandis. L. Carbohydr. Polym.* 186: 332–343.

42 Jebadurai, S.G., Raj, R.E., Sreenivasan, V.S., and Binoj, J.S. (2019). *Comprehensive Characterization of Natural Cellulosic Fiber from Coccinia grandis Stem*. Elsevier Ltd.

43 Jayaramudu, J., Guduri, B.R., and Varada Rajulu, A. (2010). Characterization of new natural cellulosic fabric *Grewia tilifolia*. *Carbohydr. Polym.* 79 (4): 847–851.

44 Teixeira, E.D.M., Corre, A.C., Manzoli, A. et al. (2010). Cellulose nanofibers from white and naturally colored cotton fibers. *Cellulose* 17: 595–606.

45 Fiore, V., Valenza, A., and Di Bella, G. (2011). Artichoke (*Cynara cardunculus* L.) fibres as potential reinforcement of composite structures. *Compos. Sci. Technol.* 71 (8): 1138–1144.

46 Mayandi, K., Rajini, N., Pitchipoo, P. et al. (2016). Extraction and characterization of new natural lignocellulosic fiber *Cyperus pangorei*. *Int. J. Polym. Anal. Charact.* 21 (2): 175–183.

47 Boumediri, A.H. and Bezazi, A. (2019). Extraction and characterization of vascular bundle and fiber strand from date palm rachis as potential bio-reinforcement in composite. *Carbohydr. Polym.* 222: 114997.

48 Baskaran, P.G., Kathiresan, M., Senthamaraikannan, P. et al. (2017). Characterization of new natural cellulosic fiber from the bark of *Dichrostachys cinerea* characterization of new natural cellulosic fiber from the bark. *J. Nat. Fibers* 20 (4): 1–7.

49 Seki, Y., Sarikanat, M., Sever, K., and Durmus, C. (2013). Extraction and properties of *Ferula communis* (chakshir) fibers as novel reinforcement for composites materials. *Compos. Part B* 44: 517–523.

50 Vijay, R., Lenin Singaravelu, D., Vinod, A. et al. (2019). Characterization of raw and alkali treated new natural cellulosic fibers from *Tridax procumbens*. *Int. J. Biol. Macromol.* 125: 99–108.

51 Manimaran, P., Senthamaraikannan, P., Sanjay, M.R. et al. (2018). Study on characterization of *Furcraea foetida* new natural fiber as composite reinforcement for lightweight applications. *Carbohydr. Polym.* 181: 650–658.

52 Preethi, P. and Balakrishna, G. (2016). Physical and chemical properties of banana fibre extracted from commercial agrotechnology physical and chemical properties of banana fibre extracted from commercial banana cultivars grown in Tamilnadu State. *Agrotechnology* 11: 1–3.

53 Taylor, P., Hajiha, H., Sain, M., and Mei, L.H. (2014). Modification and characterization of hemp and sisal fibers modification and characterization of hemp and sisal fibers. *J. Nat. Fibers* 11: 37–41.

54 Hyness, N.R.J., Vignesh, N.J., Senthamaraikannan, P. et al. (2017). Characterization of new natural cellulosic fiber from *Heteropogon contortus* plant characterization of new natural cellulosic fiber from *Heteropogon*. *J. Nat. Fibers* 15 (1): 1–8.

55 Naili, H., Jelidi, A., Limam, O., and Khiari, R. (2017). Industrial crops & products extraction process optimization of *Juncus* plant fibers for its use in a green composite. *Ind. Crop. Prod.* 107: 172–183.

56 Maache, M., Bezazi, A., Amroune, S. et al. (2017). Characterization of a novel natural cellulosic fiber from *Juncus effusus* L. *Carbohydr. Polym.* 171: 163–172.

57 Belouadah, Z., Ati, A., and Rokbi, M. (2015). Characterization of new natural cellulosic fiber from *Lygeum spartum* L. *Carbohydr. Polym.* 134: 429–437.

58 Porras, A., Maranon, A., and Ashcroft, I.A. (2015). Characterization of a novel natural cellulose fabric from *Manicaria saccifera* palm as possible reinforcement of composite materials. *Compos. Part B* 74: 66–73.

59 Suryanto, H., Marsyahyo, E., and Irawan, Y.S. (2013). Morphology, structure, and mechanical properties of natural cellulose fiber from mendong grass (*Fimbristylis globulosa*). *J. Nat. Fibers* 11: 37–41.

60 Lim, C.J., Arumugam, M., and Lim, C.K. (2018). Mercerizing extraction and physicochemical characterizations of lignocellulosic fiber from the leaf waste of *Mikania micrantha* Kunth ex H. B. K. *J. Nat. Fibers* 17 (5): 1–12.

61 Li, R., Fei, J., Cai, Y. et al. (2009). Cellulose whiskers extracted from mulberry: a novel biomass production. *Carbohydr. Polym.* 76 (1): 94–99.

62 Kommula, V.P., Reddy, K.O., Shukla, M. et al. (2016). Extraction, modification, and characterization of natural ligno-cellulosic fiber strands from napier grass. *Int. J. Polym. Anal. Charact.* 21 (1): 18–28.

63 Natarajan, T., Kumaravel, A., and Palanivelu, R. (2016). Extraction and characterization of natural cellulosic fiber from *Passiflora foetida* stem. *Int. J. Polym. Anal. Charact.* 21 (6): 478–485.

64 Ridzuan, M.J.M., Abdul Majid, M.S., Afendi, M. et al. (2015). Characterisation of natural cellulosic fibre from *Pennisetum purpureum* stem as potential reinforcement of polymer composites. *JMADE* 89: 839–847.

65 Karthik, T. and Murugan, R. (2013). Characterization and analysis of ligno-cellulosic seed fiber from pergularia daemia plant for textile applications. *Fibers Polym.* 14 (3): 465–472.

66 Sakji, N. (2016). Physico-chemical characteristics of a seed fiber arised from *Pergularia tomentosa* L. *Fibers Polym.* 17 (12): 2095–2104.

67 Prithiviraj, M., Muralikannan, R., and Senthamaraikannan, P. (2016). Characterization of new natural cellulosic fiber from *Perotis indica* plant. *Int. J. Polym. Anal. Charact.* 21: 669–674.

68 Alotaibi, M.D., Alshammari, B.A., Saba, N. et al. (2019). Characterization of natural fi ber obtained from different parts of date palm tree (*Phoenix dactylifera* L.). *Int. J. Biol. Macromol.* 135: 69–76.

69 Asim, M., Abdan, K., Jawaid, M. et al. (2015). A review on pineapple leaves fibre and its composites. *Int. J. Polym. Sci.* 2015: 1–16.

70 Saravanakumar, S.S., Kumaravel, A., Nagarajan, T. et al. (2013). Characterization of a novel natural cellulosic fiber from *Prosopis juliflora* bark. *Carbohydr. Polym.* 92 (2): 1928–1933.

71 Nijandhan, K., Muralikannan, R., and Venkatachalam, S. (2018). *Ricinus communis* fiber as potential reinforcement for lightweight polymer composites. *Mater. Res. Exp.* 5 (2018): 095307.

72 A.N, B. and K.J, N. (2017). Characterization of alkali treated and untreated new cellulosic fiber from Saharan aloe vera cactus leaves. *Carbohydr. Polym.* 174: 200–208.

73 Pouriman, M., Caparanga, A.R., and Ebrahimi, M. (2017). Characterization of untreated and alkaline-treated salago fibers (genus *Wikstroemia* Spp.) characterization of untreated and alkaline-treated salago fibers. *J. Nat. Fibers* 15 (2): 1–12.

74 Sathishkumar, T.P., Navaneethakrishnan, P., and Shankar, S. (2013). Characterization of new cellulose sansevieria ehrenbergii fibers for polymer composites. *J. Nat. Fibers* 20 (8): 575–593.

75 Sreenivasan, V.S., Somasundaram, S., Ravindran, D. et al. (2011). Microstructural, physico-chemical and mechanical characterisation of *Sansevieria cylindrica* fibres – an exploratory investigation. *Mater. Des.* 32 (1): 453–461.

76 Manimaran, P., Prithiviraj, M., Saravanakumar, S.S., and Arthanarieswaran, V.P. (2018). Physicochemical, tensile, and thermal characterization of new natural cellulosic fibers from the stems of Sida cordifolia. *J. Nat. Fibers* 15 (6): 860–869.

77 Gopinath, R., Ganesan, K., Saravanakumar, S.S., and Poopathi, R. (2016). Characterization of new cellulosic fiber from the stem of *Sida rhombifolia*. *Int. J. Polym. Anal. Charact.* 21 (2): 123–129.

78 Kaushik, V.K., Kumar, A., and Kalia, S. (2015). Effect of mercerization and benzoyl peroxide treatment on morphology. *Thermal Stab. Crystal. Sisal Fibers* 1 (6): 101–105.

79 Binoj, J.S., Raj, R.E., and Indran, S. (2018). Characterization of industrial discarded fruit wastes (*Tamarindus indica* L.) as potential alternate for man-made vitreous fiber in polymer composites. *Process. Saf. Environ. Prot.* 116: 527–534.
80 Kathirselvam, M., Kumaravel, A., Arthanarieswaran, V.P., and Saravanakumar, S.S. (2019). Characterization of cellulose fibers in *Thespesia populnea* barks: influence of alkali treatment. *Carbohydr. Polym.* 217: 178–189.
81 Indran, S., Edwin Raj, R.D., Daniel, B.S.S., and Binoj, J.S. (2018). Comprehensive characterization of natural *Cissus quadrangularis* stem fiber composites as an alternate for conventional FRP composites. *J. Bionic Eng.* 15 (5): 914–923.
82 Saba, N., Jawaid, M., Alothman, O.Y., and Paridah, M.T. (2016). A review on dynamic mechanical properties of natural fibre reinforced polymer composites. *Constr. Build. Mater.* 106: 149–159.
83 Satyanarayana, K.G., Guimara, J.L., and Wypych, F. (2007). Studies on lignocellulosic fibers of Brazil. Part I: source, production, morphology, properties and applications. *Compos. A Appl. Sci. Manuf.* 38: 1694–1709.
84 Mittal, V., Saini, R., and Sinha, S. (2016). Natural fiber-mediated epoxy composites – a review. *Compos. Part B Eng.* 99: 425–435.
85 Kim, N.K., Dutta, S., and Bhattacharyya, D. (2018). A review of flammability of natural fibre reinforced polymeric composites. *Compos. Sci. Technol.* 162: 64–78.
86 Fakhrul, T. and Islam, M.A. (2013). Degradation behavior of natural fiber reinforced polymer matrix composites. *Procedia Eng.* 56: 795–800.
87 Hallett, S.R., Green, B.G., Jiang, W.G. et al. (2009). The open hole tensile test: a challenge for virtual testing of composites. *Int. J. Fract.* 158 (2): 169–181.
88 Shahinur, S. and Hasan, M. (2019). *Natural Fiber and Synthetic Fiber Composites: Comparison of Properties, Performance, Cost and Environmental Benefits*. Elsevier Ltd.
89 Misnon, M.I., Islam, M.M., Epaarachchi, J.A., and Lau, K.T. (2014). Potentiality of utilising natural textile materials for engineering composites applications. *Mater. Des.* 59: 359–368.
90 Indran, S., Raj, R.E., Daniel, B.S.S., and Saravanakumar, S.S. (2016). Cellulose powder treatment on *Cissus quadrangularis* stem fiber-reinforcement in unsaturated polyester matrix composites. *J. Reinf. Plast. Compos.* 35 (3). 212–227.
91 Sgriccia, N. and Hawley, M.C. (2007). Science and Thermal, morphological, and electrical characterization of microwave processed natural fiber composites. *Compos. Sci. Technol.* 67: 1986–1991.
92 Monteiro, S.N., Calado, V., Rodriguez, R.J.S., and Margem, F.M. (2012). Thermogravimetric behavior of natural fibers reinforced polymer composites – an overview. *Mater. Sci. Eng. A* 557: 17–28.
93 Mohammed, L., Ansari, M.N.M., Pua, G. et al. (2015). A review on natural fiber reinforced polymer composite and its applications. *Int. J. Polym. Sci.* 2015: 1–15.
94 Sanjay, M.R., Siengchin, S., Parameswaranpillai, J. et al. (2019). A comprehensive review of techniques for natural fibers as reinforcement in composites: preparation, processing and characterization. *Carbohydr. Polym.* 207: 108–121.
95 Shekar, H.S.S. and Ramachandra, M. (2018). Green composites: a review. *Mater. Today Proc.* 5 (1): 2518–2526.
96 Adesina, O.T., Jamiru, T., Sadiku, E.R. et al. (2019). Mechanical evaluation of hybrid natural fibre-reinforced polymeric composites for automotive bumper beam: a review. *Int. J. Adv. Manuf. Technol.* 103 (5–8): 1781–1797.

97 Santhosh Kumar, S. and Hiremath, S.S. (2019). *Natural Fiber Reinforced Composites in the Context of Biodegradability: A Review*. Elsevier Ltd.

98 Mohanty, A.K., Misra, M., and Hinrichsen, G. (2000). Biofibres, biodegradable polymers and biocomposites: an overview summary of contents. 1. Introduction. 2. Reinforcing biofibres. 2.1 Chemical constituents and structural aspects. 2.2 Properties of biofibres. 2.3 Degradation properties of biofibres. 2.4 Cost aspects, availability and sustainable development of biofibres. *Macromol. Mater. Eng.* 276: 277.

99 Miao, C. and Hamad, W.Y. (2013). Cellulose reinforced polymer composites and nanocomposites: a critical review. *Cellulose* 20 (5): 2221–2262.

100 Das, M. and Chakrabarty, D. (2008). Thermogravimetric analysis and weathering study by water immersion of alkali-treated bamboo strips. *Bioresources* 3 (4): 1051–1062.

101 Wallström, S., Strömberg, E., and Karlsson, S. (2005). Microbiological growth testing of polymeric materials: an evaluation of new methods. *Polym. Test.* 24 (5): 557–563.

102 Summerscales, J. (2014). Durability of composites in the marine environment. *Solid Mech. Appl.* 208: 1–13.

103 Wagner, P.A., Little, B.J., Hart, K.R., and Ray, R.I. (1996). Biodegradation of composite materials. *Int. Biodeterior. Biodegrad.* 38 (2): 125–132.

104 He, W., Liu, J., and Xie, D. (2014). Numerical study on fatigue crack growth at a web-stiffener of ship structural details by an objected-oriented approach in conjunction with ABAQUS. *Mar. Struct.* 35: 45–69.

105 Karthika, M., Shaji, N., Johnson, A. et al. (2019). Biodegradation of green polymeric composites materials. *Bio Monomers Green Polym. Compos. Mater.* 7: 141–159.

106 Estabragh, A.R., Bordbar, A.T., and Javadi, A.A. (2013). A study on the mechanical behavior of a fiber–clay composite with natural fiber. *Geotech. Geol. Eng.* 31 (2): 501–510.

107 John, M.J. and Thomas, S. (2008). Biofibres and biocomposites. *Carbohydr. Polym.* 71 (3): 343–364.

108 Gulati, D. and Sain, M. (2006). Fungal-modification of natural fibers: a novel method of treating natural fibers for composite reinforcement. *J. Polym. Environ.* 14 (4): 347–352.

109 Ryan, C.A., Billington, S.L., and Criddle, C.S. (2017). Methodology to assess end-of-life anaerobic biodegradation kinetics and methane production potential for composite materials. *Compos. Part A Appl. Sci. Manuf.* 95: 388–399.

110 Witayakran, S., Smitthipong, W., Wangpradid, R. et al. (2017). *Natural Fiber Composites: Review of Recent Automotive Trends*. Elsevier Ltd.

111 Jasmin, M.Y., Syukri, F., Kamarudin, M.S., and Karim, M. (2020). Potential of bioremediation in treating aquaculture sludge: review article. *Aquaculture* 519: 734905.

112 Singh, S.K., Tambe, S.P., Raja, V.S., and Kumar, D. (2007). Thermally sprayable polyethylene coatings for marine environment. *Prog. Org. Coat.* 60 (3): 186–193.

14

Biofibers and Biopolymers for Biocomposites – in the Eyes of Spectroscopy

Madhu Yadav[1,2]*, Jamal Akhter Siddique*[3]*, Aftab Aslam Parwaz Khan*[4]*,*
Anish Khan[4]*, and Abdullah M. Asiri*[4]

[1] *Department of Chemistry, Bioorganic Research Laboratory, University of Allahabad, Allahabad, Utter Pradesh, India*
[2] *GGIC, Prayagraj, Utter Pradesh, India*
[3] *Department of Chemistry, School of Basic and Applied Sciences, Lingaya's Vidyapeeth (Formerly known as Lingaya's University), Faridabad, Haryana, India*
[4] *Center of Excellence for Advanced Materials Research and Chemistry Department, King Abdulaziz University, Jeddah, Saudi Arabia*

14.1 Introduction

Biocomposites are made from either natural fibers or biofibers along with petroleum-derived nonbiodegradable polymers like PP and PE or biodegradable polymers like polylactic acid (PLA) and polyhydroxyalkanoate (PHA) [1]. Biocomposites obtained from renewable resources received much attention worldwide due to their biodegradable nature. So the biocomposites are the most acceptable materials profound in life due to their ecofriendly advantages in various fields [2, 3]. Natural biofiber-derived biocomposites (BBC) are renewable, lightweight, energy proficient, biodegradable, environmentally friendly, and biocompatible as compared to other binder fabric composites [4]. The most abundant renewable natural polymer is cellulose, which has extensive applications because it is readily available and biodegradable [5]. The yearly biomass production of cellulose is estimated at roughly about 1.5 trillion tons, which makes it an unlimited source of raw material for environmentally friendly and biologically compatible products. Cellulose-containing biocomposites have been used in hemodialysis membranes [6, 7], for drug coatings, and drug-releasing compounds [8, 9]. Biobinders is commonly known as biopolymers. These compounds are usually obtained from natural resources and consist of monomeric units that are covalently bonded for making larger structures. Many applications of biobinders are also found in numerous fields such as in medicinal areas like wound healing and drug delivery system, agricultural films and food containers, soil retention sheeting, waste bags, filtration, hygiene, protective clothing, and automobile industries [10–12]. There are several types of biobinders, but the most common biobinders are categorized below (Figure 14.1).

Biodegradable polymers are classified into four families [13]. Most polymers are obtained from renewable resources (biomass), excluding the petrochemical family, which is of fossil origin; the polysaccharides from the second family, which are geopolymers, are obtained by fractionation of biomass. The third and fourth families are derived either from the

Biobased Composites: Processing, Characterization, Properties, and Applications, First Edition.
Anish Khan, Sanjay Mavinkere Rangappa, Suchart Siengchin and Abdullah M. Asiri.

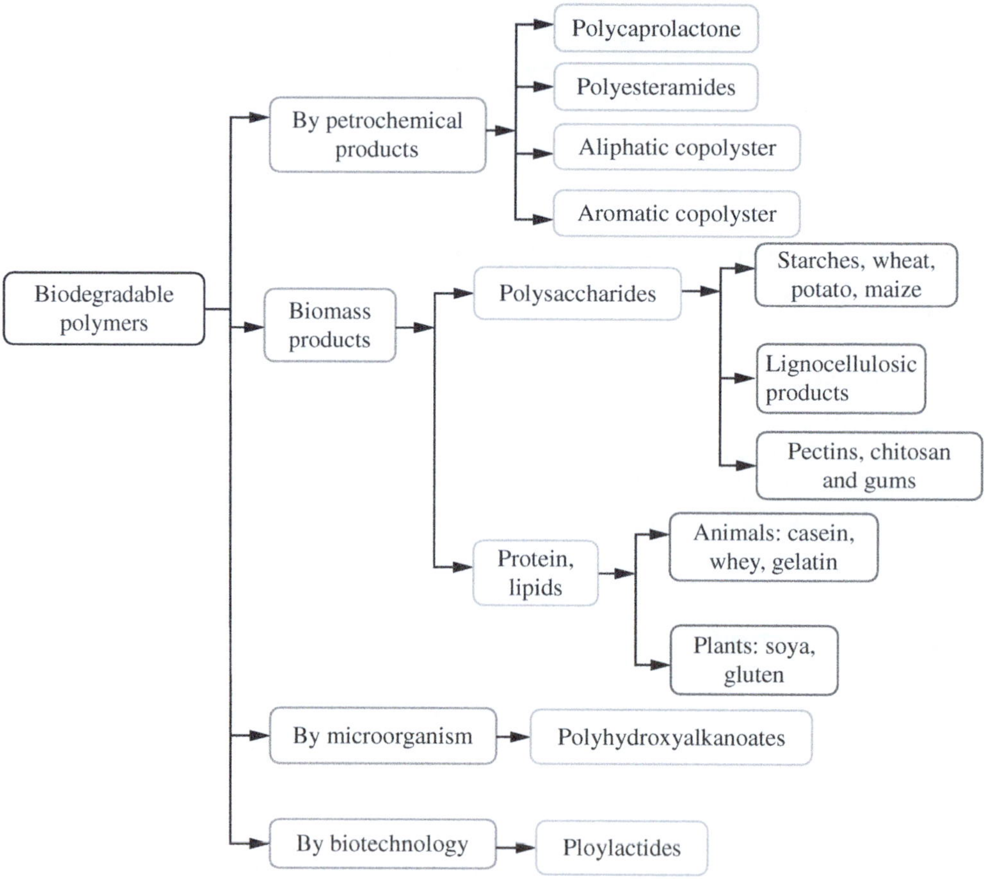

Figure 14.1 Classification of biodegradable polymers.

fermentation of biomass or from genetically modified plants (PHA or polyhydroxyalkanoate). Some family members (PLA or polylactic acid) of the fourth family are obtained by the synthesis of monomers, which are again extracted from biomass. The first family is obtained by petrochemical products, which are polyesters and can be synthesized by the petrochemical process (PCL or polycaprolactone, PEA or polyesteramide, aromatic and aliphatic copolyesters). The abundant availability of these biodegradable polymers (biopolymers) makes it a more suitable choice for commercial purposes. Also, they exhibit many properties by which they can compete with nonbiodegradable polymers in various industrial uses (packaging).

14.1.1 Polylactic Acid

A biodegradable polymer like PLA is obtained from natural resources that are renewable. PLA is obtained by polymerization (ring-opening) of lactides, while the lactic acid monomers are obtained via fermentation procedures [14]. Recently PLA has got more attention in the research area due to its availability from renewable resources like corn. PLA is hydrophobic in nature because of the presence of the CH_3 side groups in comparison to PGA.

14.1.2 Polyhydroxyalkanoates

In the synthesis of PHAs, biopolymers, energy is reserved by various microorganisms [15]. The polyhydroxybutyrate (PHB) of PHAs is composed of 3-hydroxybutyric acid monomeric units. This PHB is a frequently used polymer.

14.1.3 Polycaprolactone

PCL is obtained by ring-opening polymerization, where the monomeric unit is caprolactone. This caprolactone is prepared from petroleum resources [16]. PCL possess its biodegradable nature, and this is used for the controlled release of drugs as well as for packaging [17]. Biobinder molecules get to tie up itself to the fiber surface either by chemical reaction or adsorption, which determines the extent of interfacial adhesion. Some interfaces may constitute an additional component like an interlayer between the two ingredients of the composite.

In the manufacturing of biocomposites, the biopolymer is used as a binder and natural fiber as the reinforcement material. Natural fibers are categorized based on their origin, which is from plants (composed of cellulose), animals (consist of proteins), or mineral sources. So, biofibers are entirely derived from plants that are fully biodegradable in nature [18–20]. The main components of biofibers are cellulose, lignin, hemicelluloses, pectin, and wax (Figure 14.2). Recently, biofibers have attracted the attention of material scientists because of their advantages [21, 22]. These low-density biofibers are very economical, readily available, nonabrasive, biodegradable, noncorrosive, and health friendly in properties.

14.2 Characterization

The present chapter is intended to access a depth analysis of the characterization of biofibers. The characterization is essential for assessing the properties of the biocomposites. Characterization results provide knowledge about the chemical composition as well as the interaction present between the fiber and the matrix [23]. The fiber characterization includes different techniques like chemical composition analysis, DSC, Scanning electron microscopy (SEM), UV–vis diffuse reflectance spectroscopy crystallinity, tensile strength, TG, and ^{13}C CP-MAS NMR. Nearly all the techniques proved the modifications in the fibers also.

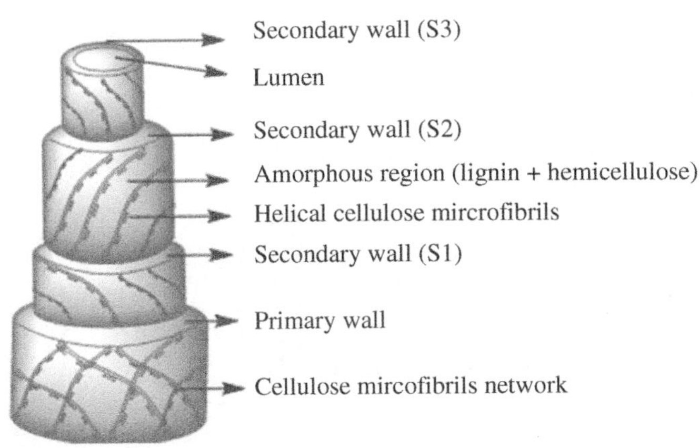

Figure 14.2 Structure of a biofiber.

14.2.1 Scanning Electron Microscopy

For more extensive morphological inspection and interfacial characterization, generally SEM is employed. This technique consists of monitoring the fractured surface films at liquid nitrogen temperature. SEM results usually reveal the homogeneity of the composite, dispersion level of the continuous matrix, presence of voids, the possible orientation of fibers, presence of aggregates, and sedimentation. The comparative study of the surface fractures can give detailed information about the interfacial adhesion prevailing in the composite.

14.2.1.1 Morphological Inspection by SEM

The structure of sisal fiber is assessed by SEM before and after the treatment (Figure 14.3) [24]. By the SEM micrographs, it is clear that the treatment (removing lignin and hemicelluloses) brought changes in the fiber surface but still did not damage the fiber structure, and the fibrils remained joined in a bundle.

So the analysis of the SEM micrographs characterizes the surface morphology and crystallinity of the treated and untreated wood fibers [25]. The SEM results also revealed that treatments caused the removal of noncrystalline constituents present in the threads. SEM of esterified and untreated wood fibers are shown in Figure 14.4. The micrographs reflect that in the untreated fibers, there is a layer, the composition of which is probably a waxy substance, as has been reported earlier [26]. The micrographs show that this layer is not

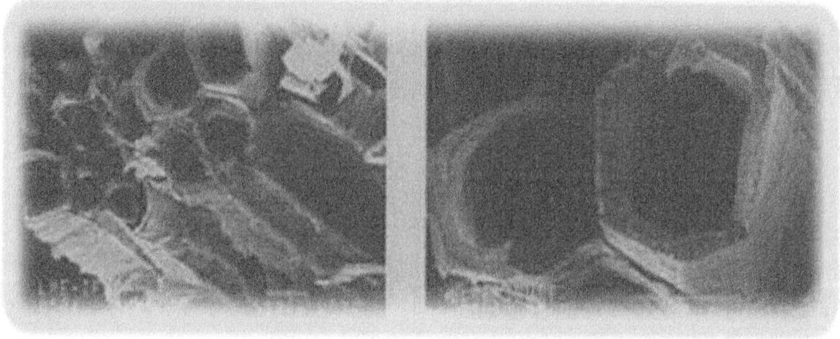

Figure 14.3 SEM micrographs of the sisal fibers. *Source:* Reproduced with permission from Ref. [24].

Figure 14.4 SEM micrographs of untreated and treated wood fibers. *Source:* Reproduced with permission from Ref. [25].

distributed evenly along the fiber surface, its thickness varying at all points. Analysis of micrograph results revealed that the surface of the esterified materials became smoother than that of the nonesterified materials. The reason behind the smoothening of the treated fiber surface is the removal of waxy substances from the lignocellulosic material's surface [27], and the surface hydroxyl group is also replaced by acetyl and propionyl groups [28].

14.2.1.2 Degree of Adhesion by SEM

SEM is used as a technique to characterize the degree of adhesion in banana fiber reinforced polyester composites (Figure 14.5) [29]. SEM is also used to characterize the nanofibrils extracted from curve fibers [30]. These fibers were used for the development of transparent polyvinyl alcohol (PVA) films that find applications in packing, biomedical, and optical devices. These extracted nanofibrils depict an aspect ratio of approximately 55 nm.

Alkali treatment provides better adhesion to composites in the fiber, which is also confirmed from SEM. The surface morphology of the fibers like sisal, hemp, kapok, and jute has been studied using SEM before and after treatments (Figure 14.6) [31]. For untreated fibers, SEM micrographs showed a relatively smooth surface, whereas alkalization treatment by sodium hydroxide, an uneven surface, appeared in all the threads. The SEM results reflected

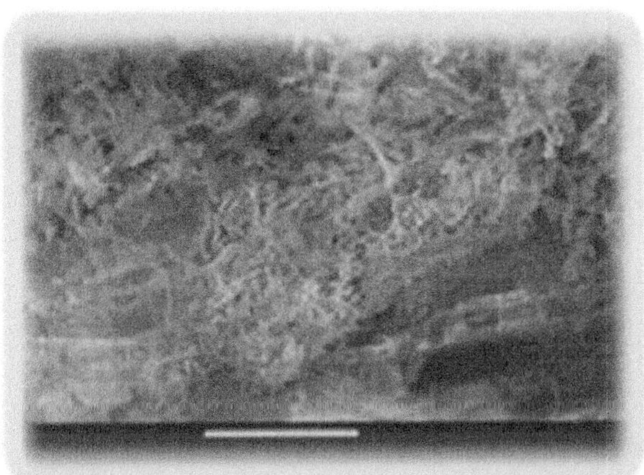

Figure 14.5 SEM of the composite showing excellent fiber–matrix adhesion. *Source:* Reproduced with permission from Ref. [29].

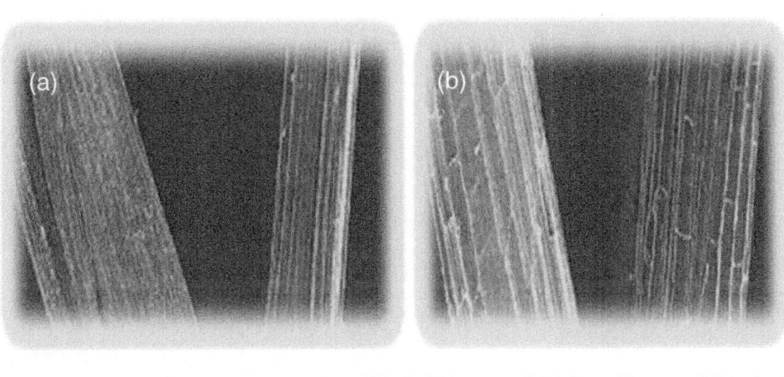

Figure 14.6 SEM of untreated and 8% NaOH treated (a) jute fiber and (b) sisal fiber. *Source:* Reproduced with permission from Ref. [31].

the changes in plant fibers like enhancing the development of adhesion in fibers after alkalization treatment, which in turn results in increased interfacial energy. Thus, there is an improvement in the thermal and mechanical stability of the fibers.

14.2.1.3 Water Absorption of Composites by SEM

By using the SEM technique, water absorption and the effect of chemical modification on water uptake can also be analyzed. Treatments like salination and mercerization are subjected to sisal and oil palm fibers and further analyzed by SEM [32]. Sometimes the water uptake is decreased in the composites by chemical modification.

14.2.2 Optical Microscopy

Optical transparency, high mechanical performance, and bond assessment in biofibers can be achieved by optical microscopy [33]. In Figure 14.7, optical films of the cellulose composites and filter paper show that the filter paper is white and opaque, but the cellulose composite became transparent in the solvent with increased immersion time. The reason behind the transparency was due to the nanodimensional property of the fibers [34]. Thus, optical clarity in the composite is thought to be due to the closure of the internal cell wall porosity by compression. Nanoscale cellulose fibrils of original sulfite cellulose macroscopic fibers of sizes in several tens of micrometers are also studied (Figure 14.8) [35, 36].

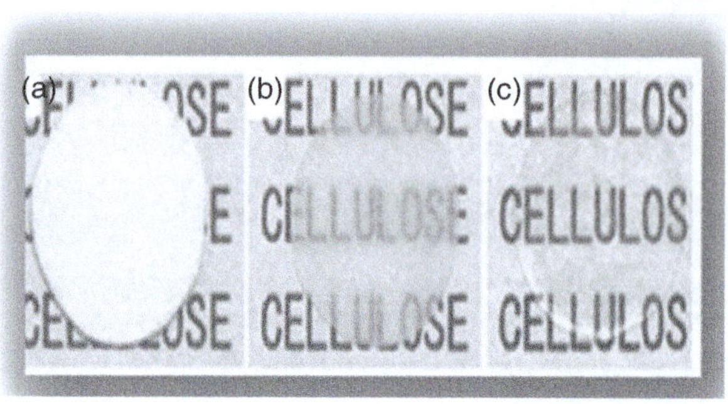

Figure 14.7 Optical photographs of (a) filter paper and all cellulose composites (immersion times: (b) 6 hours and (c) 12 hours). *Source:* Reproduced with permission from Ref. [33].

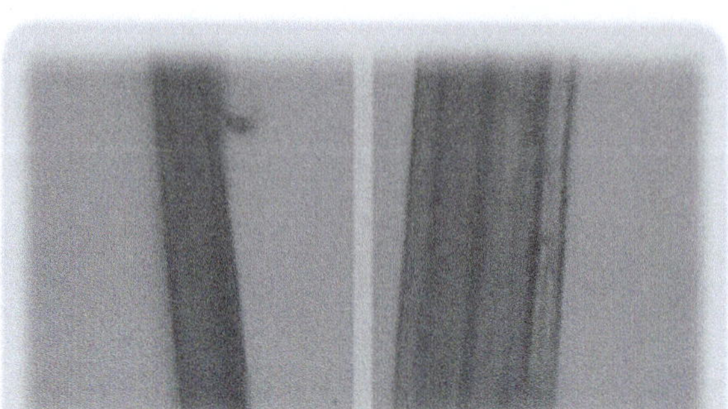

Figure 14.8 Optical micrographs of acrylate broom fibers. *Source:* Reproduced with permission from Ref. [36].

14.2.3 Atomic Force Microscopy

Atomic force microscopy (AFM) is a beneficial technique for measuring the roughness of the fiber surface. Its advantages like being nondestructive and high resolution offer a distinctive probability for repetitive examinations. The force modulation property of an oscillating cantilever tip, which indents into the sample surface, gives a piece of qualitative information for the local sample surface elasticity. The amplitude obtained by this deflection is studied as a function of tip position as the cantilever tip moves cyclically into the surface. AFM employs much smaller forces between the specimen and the tip. So the tip having a smaller radius of curvature gives better spatial resolution [37]. The microscopic image shows the surface where constant force is applied between the tip and the specimen. AFM has the ability to image nonconducting surfaces, which is its prominent feature. Scanning of the surface gives the information collected, and is quantitative in three dimensions. Figure 14.9 clearly shows the three-dimensional image of untreated and fibers treated with silane [38–42].

14.2.4 Transmission Electron Microscopy

Electron microscopy has proved an excellent tool for analyzing the morphology of the composite transmission [43]. Structure depiction of the fibers by transmission electron microscopy (TEM) showed that the fiber size is reduced to the nanometer range (less than 40 nm), and the resultant nanofibrils have a needle-like structure [27]. TEM images can also be used to confirm the nanostructure of the filler as well as the composite (Figure 14.10) [44, 45].

14.2.5 Spectroscopic Techniques

Electron spin resonance (ESR), nuclear magnetic resonance (NMR), and photoacoustic spectroscopy were shown to be successful in the characterization of biofibers [46].

14.2.5.1 NMR Analysis

The structure and the thermomechanical properties of biofiber can be investigated by NMR techniques [47]. The H^1 NMR analysis of cellulose triacetate (CTA) processed from the ramie

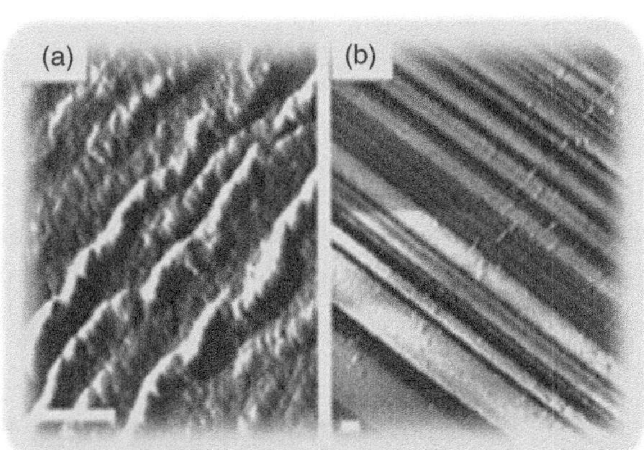

Figure 14.9 AFM image data showing the morphology of fiber surfaces before (a) and after (b) treatment. *Source:* Reproduced with permission from Ref. [39]

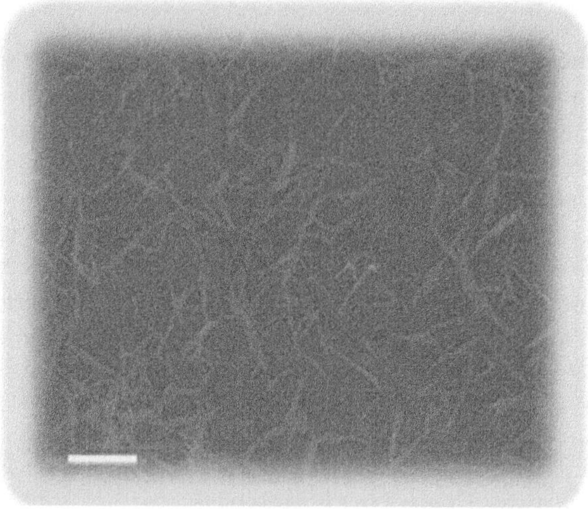

Figure 14.10 Transmission electron microscopy (TEM) images of cellulose nanofibrils. *Source:* Reproduced with permission from Ref. [45].

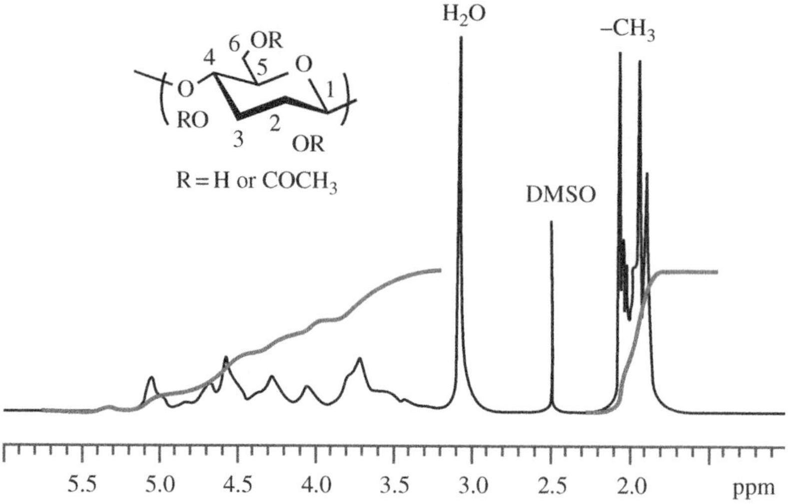

Figure 14.11 ^{1}H NMR spectrum of CTA obtained from ramie fiber. *Source:* Reproduced with permission from Ref. [47]. © 2009, American Chemical Society.

fiber biomass can be viewed, as shown in Figure 14.11. There are three types of hydrogen molecules in the linear molecular structure of CTA, whereas in spectra, only two groups of peak signals. Signal (δ) at 3.30–5.20 ppm shows the proton resonance of the glucose ring and the signal at 1.50–2.20 ppm correlating with the resonance for the methyl protons of the acetate group. The surface chemical modification of sisal fiber can also be observed by the characterization of treated and untreated sisal fibers by using ^{13}C NMR [48].

In Figure 14.12, the ^{1}HNMR spectra of isolated lignin shows the presence of methoxy and aromatic protons having signals at 3.8 and 6.8 ppm, respectively; the signal at 3.4 ppm showing the presence of residual water and signal at 2.6 ppm confirming the DMSO. Signals that present between 5.0 and 5.5 ppm are for an alpha, beta, and gamma photons of aryl ether

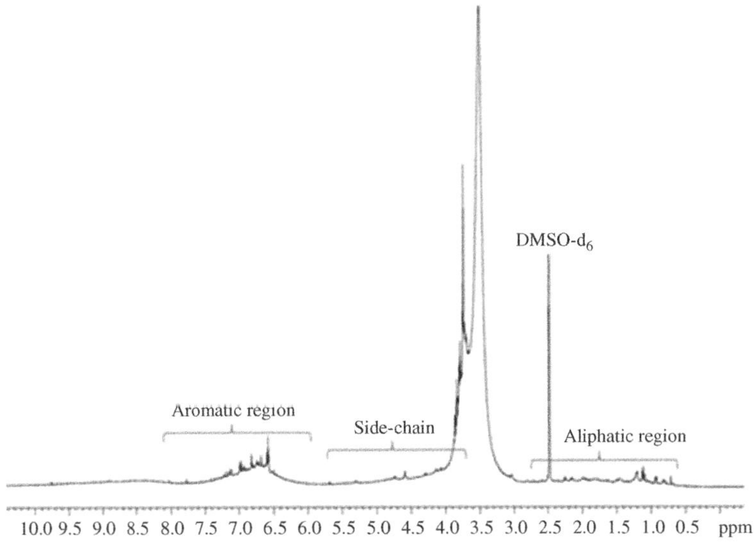

Figure 14.12 ^{1}H NMR spectrum of lignin processed from sisal fiber. *Source:* Reproduced with permission from Ref. [48]. © 2007, American Chemical Society.

units present in lignin. The signals lying between 0.8 and 1.4 ppm corresponds to methyl and methylene protons of saturated aliphatic side chains. Signals presenting between1.9 and 2.1 ppm are designated either for the doubly bonded carbon's proton or proton adjacent to the carbonyl group. These signals were less intense than the signals of aliphatic side chain protons.

Detailed information can be obtained from solid samples only. So for characterizing wood polymer composites (WPCs), solid-state ^{13}C NMR with cross-polarization and magic angle spinning (CP/MAS) is generally used. Numerous studies show that solid state NMR includes cross-polarization between protons and carbon to intensify the ^{13}C peaks, high power decoupling to get rid of dipolar line broadening because of protons, and there is sample spinning about a magic angle, i.e. 54.74° for reducing the anisotropic effects. A CP/MAS NMR spectrum of southern pinewood is shown in Figure 14.13 [49]. The aromatic ring carbons of lignin appear in the spectral region from 160 to 110 ppm, whereas the peaks between 160 and 143 ppm are allotted for oxygen-substituted aromatic carbons. Methoxy carbons of lignin show the peak at 56 ppm, whereas the two peaks at 122 and 135 ppm are assigned for alkylated and unsubstituted aromatic carbons, respectively. Signals at 21 ppm (methyls) and 172 ppm (carbonyls) are responsible for the acetyl groups of hemicellulose components. The C1 carbon of the cellulose anhydroglucose repeat unit appeared in a single peak at 105 ppm. So the spectrum data confirm the qualitative and quantitative identifications of the three major constituents of wood, i.e. cellulose, hemicelluloses, and lignin.

14.2.5.2 Infrared Spectroscopy (IR)

The primary chemical components of biofibers can be detected in vivo by using FT-IR spectroscopy [50]. This technique also provides information about the molecular changes in biofibers before and after chemical treatment, by ageing and mechanical processing [51]. The Perkin-Elmer system is used to obtain the infrared spectra of biofibers. This system is set at 2 cm^{-1} resolution to obtain the FT-IR spectra with eight scans per sample over the range

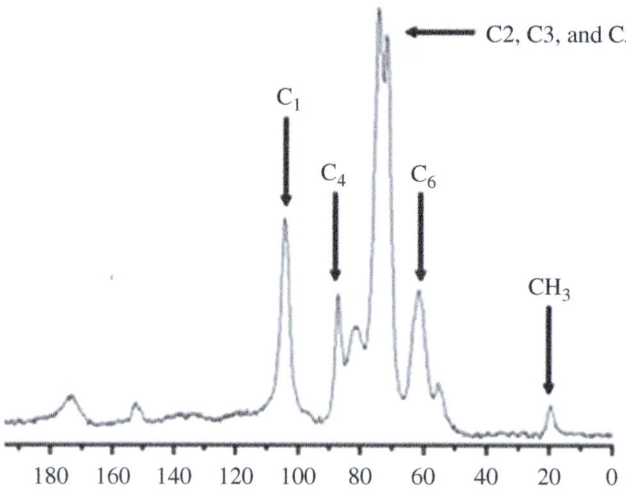

Figure 14.13 CP/MAS NMR spectrum of southern pinewood. *Source:* Reproduced with permission from Ref. [49]. © 2003, John Wiley and Sons.

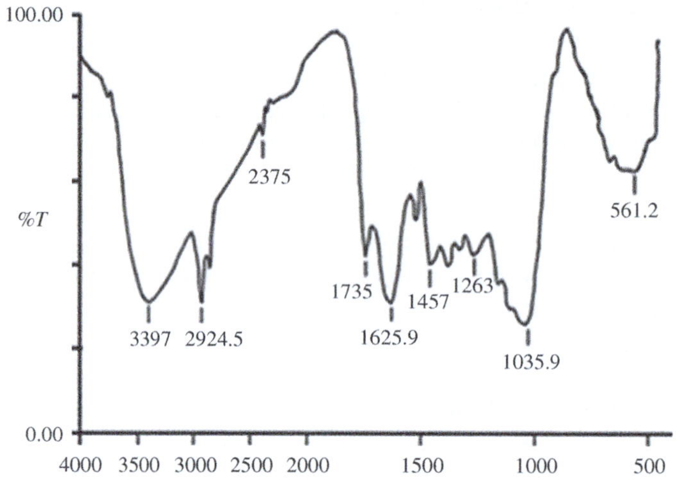

Figure 14.14 The FT-IR spectrum of raw cellulosic pine needles. *Source:* Reproduced with permission from Ref. [53]. © 2013, Elsevier.

of 4000–500 cm^{-1} [52]. In Figure 14.14, the FT-IR spectrum of raw cellulosic pine needles is shown. Here, a spectrophotometer is used to record this spectrum [53]. In this spectra, a broad peak at 3397 cm^{-1} represents bonded –OH groups. The bands present in the region 1240–1270 cm^{-1} and at 1737 cm^{-1} is due to the C–O ring of lignin and carboxylic groups of hemicelluloses and pectin present in the raw cellulosic pine needles. Mild peaks observed in the region 1372–1457 cm^{-1} may be due to the C–H, $-CH_2$, and $-CH_3$ bending vibrations.

So, the frequency obtained at 1735 cm^{-1} is due to the vibrations of the free carboxyl group, which is the characteristic band of pectin. The symmetrical stretching vibrations of $-COO^-$ ionic groups of pectins are observed at about 1610 cm^{-1} [54]. Spectra of the transgenic flax fiber (Figure 14.15) reflected broadband at 3400 cm^{-1} for the free –OH groups of cellulose and also for those –OH groups which are involved in the inter- and intramolecular hydrogen bondings [55].

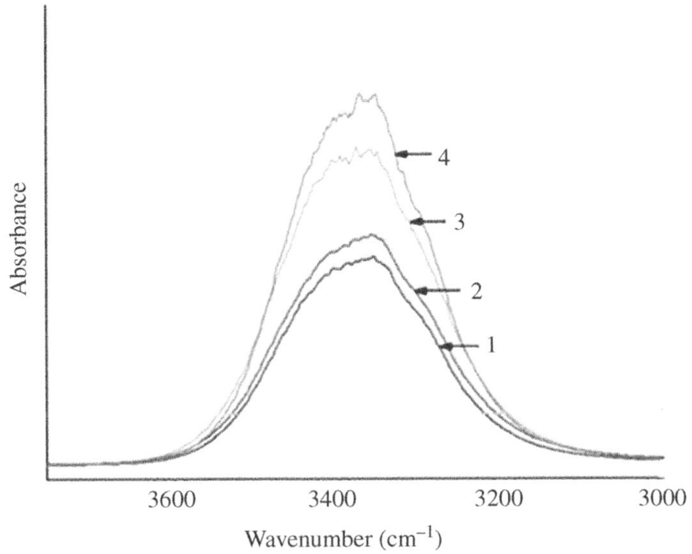

Figure 14.15 FT-IR spectra of transgenic flax fibers. *Source:* Reproduced with permission from Ref. [55]. © 2009, John Wiley and Sons.

14.2.5.3 Acoustic Emission Spectrometry

Traditional polymer composites like glass, reinforced carbon, and aramid fibers are mechanically tested using acoustic emission spectrometry (AES) [56–60]. This study involves a very distinct and comparatively better examination than the classical nondestructive testing (NDT), which had the main target of rejection or acceptance for the material used for observation and limited the gathered destruction. In some instances, AE signals are also responsible for absolute removal in composites, i.e. exfoliation and fiber failure. The mechanical properties of any fiber reinforced polymeric composites furnish a substantial impact on the final industrial application of fiber. Mechanical properties of composites are again influenced by several factors, i.e. the nature of the components (matrix and reinforcement), their shape, size, respective volume fraction, the geometry of the reinforcement phase, and the interfacial shear strength (IFSS) of support and matrix. So the IFSS plays a vital role in the transverse mechanical properties, interlaminar shear strength, and toughness of composites. A piece of detailed knowledge about the fiber–matrix interface is required for its different industrial applications [61]. The tensile and flexural strength of the composite can be increased by improving IFSS, whereas diminishing the impact of IFSS lowers the toughness and durability. Another property is the hydrophilic nature of natural fibers, which provides adverse implications on adhesion to the hydrophobic matrix. The weak adhesion results in low strength and compatibility [62].

Moreover, the durability of composites needs a strong interfacial bond. AE is widely used for the observation and minimization of fiber degradation during single fiber composite (SFC) tests used to analyze the interfacial behavior of natural fiber composites. It has been analyzed that mostly all fiber breakages were captured and related to a single AE event. So, there is a correlation between the AE results and the fiber damages observed [63]. AE technique is also employed to keep the damage and unsuccessful performance of biodegradable jute fabric reinforced thermoplastic polyester composites. AE amplitude is increased from lowest to highest in debonding, fiber fracture, and fiber-pull out during the tensile test [64].

AE technique also investigates nonperformance of oil palm and pineapple reinforced phenol formaldehyde composites. So a connection is observed between the AE modes and amplitude concerning the time during the loading implementation. Thus, AE is proved as a useful tool for observing the mechanical behavior of natural fiber composites and opening the future in fully biodegradable materials.

14.3 Conclusions

Fiber reinforced composites were a rapidly growing part of the revolution in advanced materials developed during the 1980s and 1990s. These fiber reinforced composites are proved as better potential choices than the traditional synthetic fibers. These new generation composites have various applications along with the specific motif which is moving from synthetic to natural. Natural fibers and their derived polymer composites provide numerous benefits like flexibility, considerable toughness, recyclability, secure processing, and ecofriendliness. The essential properties of the composite, such as interlaminar shear and flexural strength, can be monitored by different spectroscopic techniques. Adhesion between the fiber matrix and interface is a critical property of the composite that must be measured. So in this chapter, we have analyzed different natural fibers, their characterization, and in-depth analysis along with a comprehensive study.

References

1 Netravali, A.N. and Chabba, S. (2003). Composites get greener. *Materials Today* 6: 22–29.

2 Singha, A.S., Sharma, A., and Mishra, B.N. (2008). Pressure induced graft co-polymerization of binary vinyl monomer mixtures onto *Saccharum cilliare* fiber and evaluation of some physical, chemical and thermal properties. *Journal of Polymer Material* 25: 91–99.

3 Mohanty, A.K., Mishra, M., and Drazal, L.T. (2000). *SAMPE, Advanced Composite Technology for 21 Century Transportation, Midwest Advanced Materials and Processing Conference Proceedings*, 299. Dearborn (12–14 September 2000).

4 Ning, W., Xiang, X., Na, Z.H., and Jianming, F. (2010). Effects of water on the properties of thermoplastic starch poly(lactic acid) blend containing citric acid. *Journal of Thermoplastic Composite Materials*. 23: 19.

5 Zabetakis, D., Dinderman, M., and Schoen, P. (2005). Metal-coated cellulose fibers for use in composites application to microwave technology. *Advanced Materials* 17: 734.

6 Sevillano, G., Rodriguez, P.M., and Martos, R. (1990). Cellulose acetate membrane improves some aspects of red blood cell function in haemodialysis patients. *Nephrology Dialysis Transplantation* 5: 497.

7 Lonnemann, G., Bingel, M., and Koch, K.M. (1987). Plasma interleukin-1 activity in humans undergoing hemodialysis with regenerated cellulosic membranes. *Lymphokine Research* 6: 63.

8 Desai, S.D. and Blanchard, J. (1998). In vitro evaluation of pluronic F127-based controlled-release ocular delivery systems for pilocarpine. *Journal of Pharmaceutical Sciences* 87: 226.

9 Doheny, J.G., Jervis, E.J., and Guarna, M.M. (1999). Cellulose as an inert matrix for presenting cytokines to target cells: production and properties of a stem cell factor-cellulose-binding domain fusion protein. *Biochemical Journal* 339: 429.

10 Zitzmann, N.U., Rateitschak, P.E., and Marinello, C.P. (2003). Treatment of angular bone defects with a composite bone grafting material in combination with a collagen membrane. *Journal of Periodontology* 74: 687.

11 Ramshaw, J.A.M., Werkmeister, J.A., and Peters, D.E. (1990). Current Perspectives on Implantable Devices, vol. 2 (ed. D.F. Williams), 151. London: Jai Press Ltd.

12 Vert, M., Li, S.M., Spenlehauer, G., and Guerin, P. (1992). Bioresorbability and biocompatibility of aliphatic polyesters. *Journal of Materials Science: Materials in Medicine* 3: 432.

13 Averous, L. and Boquillon, N. (2004). Biocomposites based on plasticized starch: thermal and mechanical behaviours. *Carbohydrate Polymers* 56: 111.

14 Wu, T.M. and Wu, C.Y. (2006). Biodegradable poly(lactic acid)/chitosan-modified montmorillonite nanocomposites: preparation and characterization. *Polymer Degradation and Stability* 91: 2198.

15 Ojumu, T.V., Yu, J., and Solomon, B.O. (2004). Production of polyhydroxyalkanoates, a bacterial biodegradable polymers. *African Journal of Biotechnology* 3: 18.

16 Okada, M. (2002). Chemical syntheses of biodegradable polymers. *Progress in Polymer Science* 27: 87.

17 Albertsson, A.C. and Varma, I.K. (2002). Aliphatic polyesters: synthesis, properties and applications. *Advances in Polymer Science* 157: 1.

18 Singha, A.S. and Thakur, V.K. (2009). Synthesis, characterization and study of pine needles reinforced polymer matrix based composites. *Journal of Reinforced Plastics and Composites* 29: 700.

19 Singh, A.S. and Thakur, V.K. (2008). Mechanical properties of natural fiber reinforced polymer composites. *Bulletin of Materials Science* 31: 91.

20 Singh, A.S. and Thakur, V.K. (2009). Fabrication and characterization of *S. cilliare* fiber reinforced polymer composites. *Bulletin of Materials Science* 32: 49.

21 Sun, X.F., Sun, R.C., Fowler, P., and Baird, M.S. (2004). Physicochemical characterization of lignin isolated with high yield and purity from wheat straw. *International Journal of Polymer Analysis and Characterization* 9: 317.

22 Singh, A.S. and Thakur, V.K. (2008). Synthesis and characterization of *S. cilliare* fiber reinforced green composites. *International Journal of Plastic Technology* 11: 835.

23 Thomas, S., Pothan, L.A., and Cherian, B.M. (2009). Advances in natural fibre reinforced polymer composites: macro to nanoscales. *International Journal of Materials and Product Technology* 36: 317.

24 Martins, M.A., Kiyohara, P.K., and Joekes, I. (2004). Scanning electron microscopy study of raw and chemically modified sisal fibres. *Journal of Applied Polymer Science* 94: 2333.

25 Tserki, V., Zafeiropoulos, N.E., Simon, F., and Panayiotou, C. (2005). A study of the effect of acetylation and propionylation surface treatments on natural fibres. *Composites Part A Applied Science and Manufacturing* 36: 1110.

26 Zafeiropoulos, N.E., Williams, D.R., Baillie, C.A., and Matthews, F.L. (2002). Engineering and characterisation of the interface in flax fibre/polypropylene composite materials part I: development and investigation of surface treatments. *Composites Part A Applied Science and Manufacturing* 33: 1083.

27 Cherian, B.M., Pothan, L.A., Nguyen Chung, T. et al. (2008). A novel method for the synthesis of cellulose nanofibril whiskers from banana fibres and characterization. *Journal of Agricultural and Food Chemistry* 56: 5617.

28 Rana, A.K., Basak, R.K., Mitra, B.C. et al. (1997). Studies of acetylation of jute using simplified procedure and its characterization. *Journal of Applied Polymer Science* 64: 1517.

29 Pothan, L.A., Oommen, Z., and Thomas, S. (2003). Dynamic mechanical analysis of banana fiber reinforced polyester composites. *Composites Science and Technology* 63: 283–293.

30 Souza, S.F., Leao, A.L., Jenny, H.C. et al. (2010). Nanocellulose from curava fibres and their nanocomposites. *Molecular Crystals and Liquid Crystals* 522: 42.

31 Mwaikambo, L.Y. and Ansell, M.P. (2002). Chemical modification of hemp, sisal, jute, and kapok fibres by alkalization. *Journal of Applied Polymer Science* 84: 2222.

32 Jacob, M., Varghese, K.T., and Thomas, S. (2005). Water sorption studies of hybrid biofibre reinforced natural rubber biocomposites. *Biomacromolecules* 6: 2969.

33 Nishino, T. and Arimoto, N. (2007). All-cellulose composite prepared by selective dissolving fibre surface. *Biomacromolecules* 8: 2712.

34 Nogi, M., Handa, K., Nakagaito, A.N., and Yano, H. (2005). Optically transparent bionanofibre composites with low sensitivity to refractive index of the polymer matrix. *Applied Physics Letters* 87: 243110.

35 Pääkko, M., Ankerfors, M., Kosonen, H. et al. (2007). Enzymatic hydrolysis combined with mechanical shearing and high pressure homogenization for nanoscale cellulose fibrils and strong gels. *Biomacromolecules* 8: 1934.

36 Cassano, R., Trombino, S., Bloise, E. et al. (2007). New broom fibre (*Spartium junceum* L.) derivatives: preparation and characterization. *Journal of Agricultral and Food Chemistry* 55: 9489.

37 Mai, K., Mader, E., and Muhle, M. (1998). Interphase characterization in composites with new non destructive methods. *Composites Part A Applied Science and Manufacturing* 29: 1111.

38 George, J., Ivens, J., and Verpoest, I. (1999). Mechanical properties of flax fibre reinforced epoxy composites. *Angewandte Makromolekulare Chemie* 272: 41.

39 Pickering, K.L., Beckermann, G.W., Alam, S.N., and Foreman, N.J. (2007). Optimizing industrial hemp fibre for composites. *Composites Part A Applied Science and Manufacturing* 38: 461.

40 Fleming, K., Gray, D.G., and Matthews, S. (2001). Cellulose crystallites. *Chemistry European Journal* 7: 1831.

41 Lima, M.M.D. and Borsali, R. (2004). Rodlike cellulose microcrystals: structure, properties, and applications. *Macromolecular Rapid Communications* 25: 771.

42 Lowys, M.P., Desbrieres, J., and Rinaudo, M. (2001). Rheological characterization of cellulosic microfibril suspensions. Role of polymeric additives. *Food Hydrocolloids* 15: 25.

43 Wu, Q., Henriksson, M., Liu, X., and Berglund, L.A. (2007). High strength nanocomposite based on microcrystalline cellulose and polyurethane. *Biomacromolecules* 8: 3687.

44 Wågberg, L., Decher, G., Norgren, M. et al. (2008). The build up of polyelectrolyte multilayers of microfibrillated cellulose and cationic polyelectrolytes. *Langmuir* 24: 784.

45 Capadona, J.R., Shanmuganathan, K., Trittschuh, S. et al. (2009). Polymer nanocomposites with nanowhiskers isolated from microcrystalline cellulose. *Biomacromolecules* 10: 712.

46 Rajisha, K.R., Deepa, B., and Thomas, S. (2011). Thermomechanical and spectroscopic characterization of natural fibre composites. In: Interface Engineering of Natural Fibre Composites for Maximum Performance, 241–274. Elsevier Ltd.

47 Fan, X. (2009). Cellulose triacetate optical film preparation from ramie fiber. *Industrial & Engineering Chemistry Research* 48: 6212.

48 Megiatto, J.D., Hoareau, J.R., Gardrat, C. et al. (2007). Sisal fibres: surface chemical modification using reagent obtained from a renewable source; characterization of hemicellulose and lignin as model study. *Journal Agricultral and Food Chemistry* 55: 8576.

49 Wright, J.R. and Mathias, L.J. (1993). Physical characterization of wood and wood–polymer composites: an update. *Journal of Applied Polymer Science* 48: 2225.

50 Edwards, H.G.M., Farwell, D.W., and Webster, D. (1997). FT Raman microscopy of untreated natural plant fibres. *Spectrochimica Acta Part A* 53: 2383.

51 Schrader, B., Klump, H.H., Schenzel, K., and Schulz, H. (1999). Non-destructive NIR FT Raman analysis of plants. *Journal of Molecular Structure* 501: 201.

52 Jahn, A., Schroder, M.W., Futing, M. et al. (2002). Characterization of alkali treated flax fibres by means of FT Raman spectroscopy and environmental scanning electron microscopy. *Spectrochimica Acta, Part A* 58: 2271.

53 Thakur, V.K., Thakur, M.K., Gupta, R.K., and Kumar, V. (2013). Graft copolymers from cellulose: synthesis, characterization and evaluation. *Carbohydrate Polymers* 97: 18.

54 Yu, P., Block, H., Niu, Z., and Doiron, K. (2007). Rapid characterization of molecular chemistry, nutrient make up and microlocation of internal seed tissue. *Journal of Synchrotron Radiation* 14: 382.

55 Magdalena, W.K., Katarzyna, S.T., Lucyna, D. et al. (2009). Biochemical, mechanical, and spectroscopic analyses of genetically engineered flax fibers producing bioplastic (poly-b-hydroxybutyrate). *Biotechnology Progress* 25: 5.

56 Mizutani, Y., Nagashima, K., Takemoto, M., and Ono, K. (2000). Fracture mechanism characterization of cross-ply carbon–fibre composites using acoustic emission analysis. *NDT&E International.* 33: 101.

57 Barre, S. and Benzeggagh, M.L. (1994). On the use of acoustic emission to investigate damage mechanisms in glass fibre-reinforced polypropylene. *Composites Science and Technology* 52: 369.

58 Benevolenski, O. and Karger-Kocsis, J. (2001). Fracture and failure behavior of partially consolidated discontinuous glass fibre mat-reinforced polypropylene composites. *Macromolecular Symposia* 170: 165.

59 Margueres, P., Meraghn, F., and Benzeggagh, M.L. (2000). Comparison of stiffness measurements and damage investigation techniques for a fatigued and postimpact fatigued GFRp composite obtained by RTM process. *Composites Part A Applied Science and Manufacturing* 31: 151.

60 Davidovitz, M., Mittelman, A., Roman, I., and Marom, G. (1984). Failure modes and fracture mechanisms in flexure of Kevlar–epoxy composites. *Journal of Materials Science* 19: 377.

61 Luo, S. and Netravali, A.N. (1999). Interfacial and mechanical properties of environment friendly 'green' composites made from pineapple fibres and poly(hydroxybutyrateco-valerate) resin. *Journal of Materials Science* 34: 3709.

62 Mohanty, A.K., Misra, M., and Hinrichsen, G. (2000). Biofibres, biodegradable polymers and biocomposites: an overview. *Macromolecular Materials and Engineering* 1: 276.

63 Acha, B.A., Marcovich, N.E., and Karger-Kocsis, J. (2006). Biodegradable jute cloth reinforced thermoplastic copolyester composites: fracture and failure behaviour. *Plastics Rubber and Composites* 35: 73.

64 Sreekala, M.S., Czigany, T., Romhany, G., and Thomas, S. (2003). Investigation of oil palm and pineapple fibre reinforced phenol formaldehyde composites by acoustic emission technique. *Polymer Composites* 11: 9.

15

Environmental Impact Study on Biobased Composites Using Lifecycle Methodology

P. Ramesh[1], *H. Mohit*[2], *and V. Arul Mozhi Selvan*[2]

[1] *Department of Production Engineering, National Institute of Technology, Tiruchirappalli, India*
[2] *Department of Mechanical Engineering, National Institute of Technology, Tiruchirappalli, India*

15.1 Introduction

Sustainable materials and its initiatives are raised globally due to fossil fuel depletion and global warming. The highly investigated material is natural composites as a substitution for synthetic composites and other traditional engineering materials [1]. Natural composites are fabricated with degradable bioresources (wood, natural fibers) and biopolymers. Biobased composites are compounded with natural fibers and biopolymers. Biobased composites degrade after a certain period without producing any environmental impact to the atmosphere. Natural fiber composite products are environmentally friendly properties that contain less hazardous materials [1] and traditional polymer composites are replaced by biobased composites because of good environmental performance in end-of-life (EoL) scenario. Application of biobased composites had a massive demand in many fields such as automobile, construction, and aerospace industries. Due to environmental regulation of government policy it is necessary to find the environmental impact for a product or a technology. Global warming and environmental issues led researchers and industrials to develop alternative materials instead of fully petro-sourced materials [2]. Environmental impact analysis is performed using life cycle assessment (LCA) with guidelines of ISO 14040 and 14044 [3]. Le Duigou et al. [4] compared the environmental impact of four different biocomposites: (i) Flax/PLLA (Poly-L-(lactic)-acid) and biocomposites, glass/polyester composites; (ii) Flax/PLLA biosandwich and Glass/Polyester/Balsa sandwich; (iii) Flax/PLLA biocomposites had a lesser impact than Glass/Polyester composites; (iv) Biosandwich Flax/PLLA/Balsa had a lesser impact than sandwich Glass/Polyester/balsa. In the production of both composites, PLLA production had a higher impact than others, because of the synthesis and fertilizers used during corn cultivation. Batouli et al. [5] studied the mechanical properties and LCA of polyurethane reinforced with kenaf fiber (KF) with varying percentages of fiber content (5, 10, and 15%). From the results, less percentage of kenaf shows good mechanical and lesser environmental impact compared to other compositions. La Rosa et al. [6] compared cradle to cradle manufacturing for the eco sandwich (cork, hemp, and bioresin) and petroleum-based (glass and polyurethane) composite panels. SimaPro and CML baseline was used. From the

Biobased Composites: Processing, Characterization, Properties, and Applications, First Edition.
Anish Khan, Sanjay Mavinkere Rangappa, Suchart Siengchin and Abdullah M. Asiri.

results, the ecosandwich composite had a lower impact compared to the glass-based petroleum composite, but going for the EoL option for ecosandwiches had lesser impact in landfill compared to incineration. Korol et al. [3] did comparative LCA for biocomposites and traditional composites. Polypropylene (PP), glass fibers (GF), cotton fibers (CF), jute fibers (JF), and KF were used as materials with different percentage levels. From the results, PP/KF had a lesser impact than other combination composites. In total score, PP/CF had a higher impact than other composites. Günkaya and Banar [7] compared the environmental impact of biocomposite film and LDPE film. The results express that biocomposites had a higher impact than the LDPE, because of extrusion energy consumption, film forming, and starch preparation that increased the environmental impacts for biocomposites. Molins et al. [8] analyzed the environmental impact of the biocomposite process, made up of PLA and chicken feathers. From the results, PLA production had a significant impact than others (feather production and composite manufacturing). Maize production contributes a major impact because of fertilizers, water, energy, and land occupation. Khoshnava et al. [9] compared the environmental impact of biocomposites (kenaf bast fibers and oil palm EFB fibers) and petroleum-based composites (GFPP and GFPE). From the results, it is found that biocomposites had lesser impact than petroleum-based composites. Mansor et al. [1] performed LCA for an automotive steel component. The steel component is compared with a hybrid natural fiber (sugar palm/glass/polyurethane) composite. In all categories, steel had a higher impact than the hybrid composite. Beigbeder et al. [2] performed LCA for wood flour (WF) reinforced polypropylene (PP/WF) and flax fiber reinforced polylactic acid (PLA/Fl). The study was done only for EoL of biocomposites with four scenarios: incineration, landfill, composting, and recycling. From the results, PP/WF composite should be incinerated, and PLA/FI should be composted in the EoL scenario. In this chapter, biobased (fiber-based), and traditional composites (glass-based) are analyzed for environmental impact using LCA methodology. Cradle to gate and cradle to grave system was analyzed for biobased composites.

15.2 Lifecycle Assessment

LCA is a technique to find the environmental burdens of products, materials, and technology. LCA will allow manufacturers and researchers to find an environmental impact from raw material extraction to EoL (cradle to grave analysis). Many organizations like Society of Toxicology and Chemistry (SETAC) and Environmental Protection Agency (EPA) established LCA guidelines.

15.2.1 Goal and Scope

The objective of LCA is to be mentioned with appropriate system boundaries and functional units. System boundaries define the flow and extent of LCA, i.e. cradle to gate (raw production to factory gate), and cradle to grave (raw material production to EoL). The functional unit describes the quantity of the product to be performed for environmental impact.

15.2.2 Inventory Data

Input data like material, energy, and transport and output data of emissions for LCA is added in the section, e.g. for cradle to gate; it requires raw part production, processing, preparation, packing, and dispatch to factory gate.

15.2.3 Impact Assessment

Summarize the environmental impact based on its character (water, air, and soil). Impacts are evaluated based on input and output data information of inventory. Using impact assessment methods, different environmental indicators are identified (e.g. carcinogens, climate change, ozone layer, ecotoxicity, land use).

15.2.4 Interpretation

A brief explanation, discussion, and alternatives should be mentioned if there is a real possibility to reduce environmental burdens for products or technology from LCA results.

15.3 Simplified Case Study

Traditional polymer composites are replaced by a biocomposite due to poor environmental performance in EoL scenario. Waste management of traditional polymer composites leads to rising costs for manufacturers in terms of recycling process cost, energy, human resources, etc. In the case of biocomposites, it will degrade after use when undergoing EoL option. In this case study, jute and kenaf natural fibers were used as reinforcement material along with biopolymer PLA used for LCA.

15.4 Goal and Scope

The objective of this study is a cradle to gate and cradle to grave analysis. The impacts of biobased composites are compared with glass-reinforced composites. Both natural fibers and biopolymer materials are degraded after over a period of use without any environmental burdens. Production of 1 kg biobased polymer composite is assumed as a functional unit.

Materials for biobased composite – Jute, kenaf, and PLA.

Materials for glass-based composite – Glass fiber/Polyester (GF/Polyester).

15.5 System Boundary

A cradle to gate and cradle to grave analysis is evaluated for biocomposites and glass-based composites. Raw material extraction, i.e. crop cultivation of fibers, processing, fiber making, and PLA production to pellet is considered. Incineration, landfill, and recycling are considered as EoL scenarios. Figure 15.1 depicts the system boundary considered for this simplified case study.

15.6 Inventory Analysis

In this step, according to ISO standards, input and output element data necessary for raw material production is collected. Production of biobased composites using an injection molding process is assumed. JF and KF are used as reinforcing materials with a total quantity of 0.5 kg and production data of fibers obtained from EcoInvent database and NatureWorks PLA are chosen as a biopolymer for environmental assessment (Table 15.1).

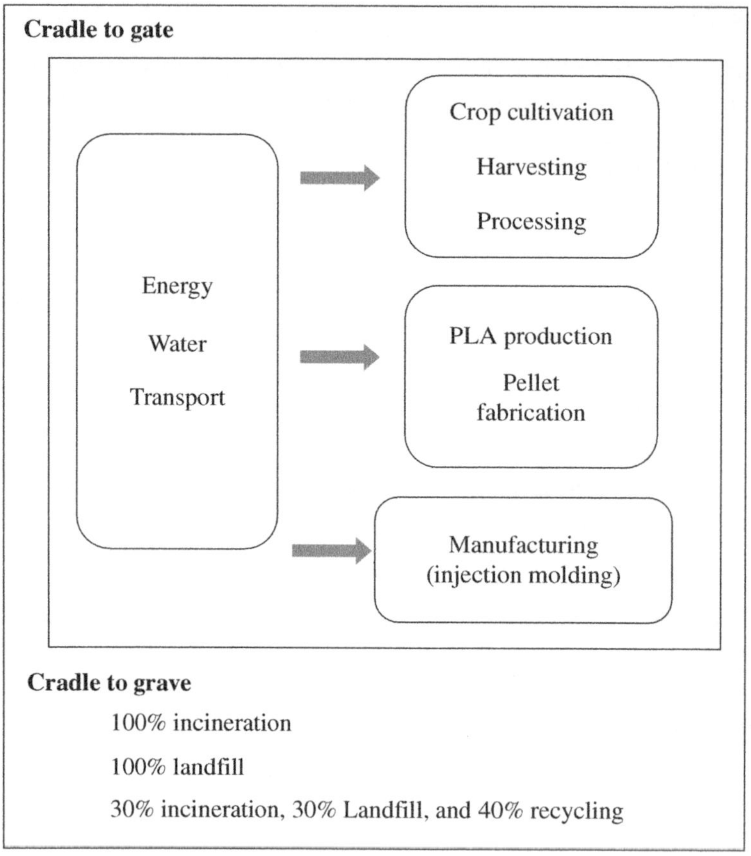

Figure 15.1 System boundary for LCA.

Table 15.1 Inventory data for Lifecycle Assessment.

Category	Input	Remarks
Biobased composite		
Jute fiber production	0.250 g	Ecoinvent data
Kenaf fiber production	0.250 g	Ecoinvent data
Nature works PLA production	0.5 kg	Ecoinvent data
Electrical energy	1.5 kWh	Assumed
Glass-based composite		
Glass fiber	0.5 kg	Ecoinvent data
Polyester	0.5 kg	Ecoinvent data
Electrical energy	1.5 kWh	Assumed
End of life		
100% incineration		
100% landfill		
50% of recycling and 50% landfill		

Table 15.2 Environmental damage of Eco-indicator 99.

Categories	Definition	Reference
Human health	The number and duration of diseases and loss of life-years owing to deaths caused by environmental degradation	Singh et al. [10]
Ecosystem quality	The impact on species diversity, acidification, ecotoxicity, eutrophication, and land use	
Resources	The depletion of raw materials and energy resources	

Table 15.3 Environmental indicators.

Label	Unit
Carcinogens	DALY
Resp. organics	DALY
Resp. inorganics	DALY
Climate change	DALY
Radiation	DALY
Ozone layer	DALY
Ecotoxicity	PAF*m2yr
Acidification/eutrophication	PAF*m2yr
Land use	PAF*m2yr
Minerals	MJ surplus

15.7 Impact Assessment

SimaPro module is used to find the environmental impact with Eco-indicator 99 (I) impact assessment method. Eco-indicator 99 enables access to environmental damage in the three categories mentioned in Table 15.2. Using Eco-indicator 99 (I), a total of ten environmental impacts can found for the process or product. Table 15.3 depicts various environmental indicators that can compute using Eco-indicator 99.

15.8 Results

A cradle to gate and cradle to grave analysis was done using LCA methodology for a biocomposite and traditional composite. JF and KF, along with the biopolymer PLA, were used to perform environmental impact assessment. On the other hand, traditional composites (GF/polyester) were used to compare the environmental impact with biobased composites. From the results, it is found that biobased composites had a lesser impact than the traditional composites. The first phase of this case study is the cradle to gate computation.

Figure 15.2 depicts the cradle to gate assessment of biobased composites. Totally ten environmental indicators are found through Eco-indicator. From Figure 15.2, PLA had a higher

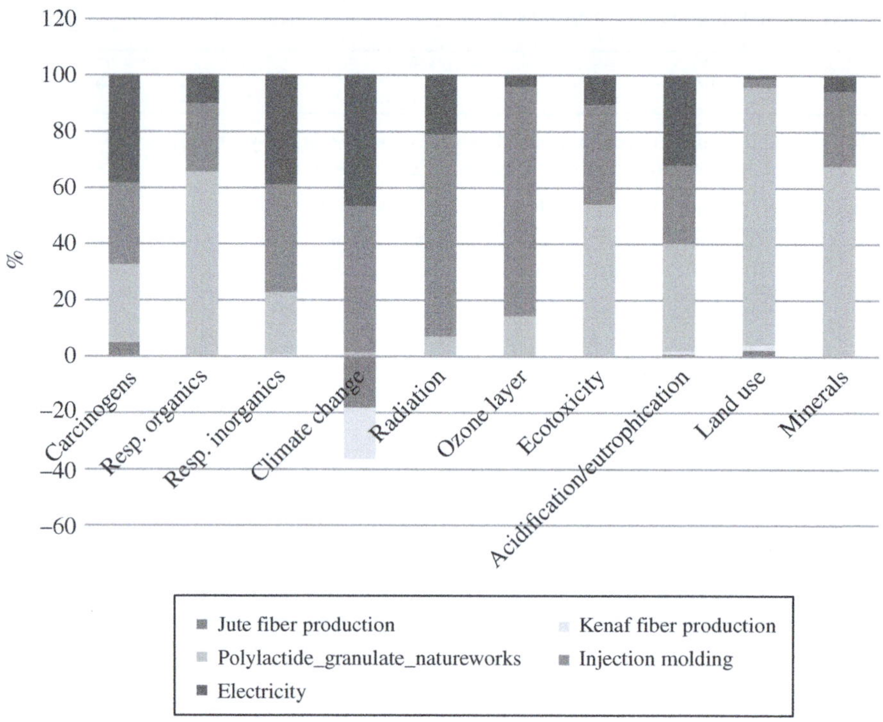

Figure 15.2 Environmental impact of biocomposite fabrication.

impact in four categories (land use, minerals, resp. organics, and ecotoxicity) because of fertilizers, irrigation, and cultivation of plants, and 92 % of land use is observed in the impact category. After the fiber production, injection molding had a higher impact. Electricity grid mix was added for both biobased and traditional composites. Production of electricity causes more impact than fibers.

15.8.1 Normalization

Indicators from the impact category results were compared with the reference value. Computation done by impact category is divided by the reference value [11]. Figure 15.3 expresses the normalized value of biobased composites. From Figure 15.3, in ecosystem quality and resource, PLA had a higher impact than others because of cultivation, irrigation, corn separation, and synthesis of lactic acid.

The damage category indicator scores are multiplied by the weighting factors and are added to form a total score [11]. Like normalization in weighting, PLA production produces more environmental burdens. From Figure 15.4, it is clearly identified that PLA consumes more resources and damages the ecosystem.

From Figures 15.3 and 15.4, it is understood that PLA production damages the environment, after injection molding and electricity. From the study of Le Duigou et al. [4] and Molins et al. [8], it is observed that PLA production occupies more land use for cultivation of plants, and composite manufacturing had lesser impact than PLA production. In cradle to gate study, biobased composite is capered with glass-based composite. Table 15.4 and Figure 15.5 depicts

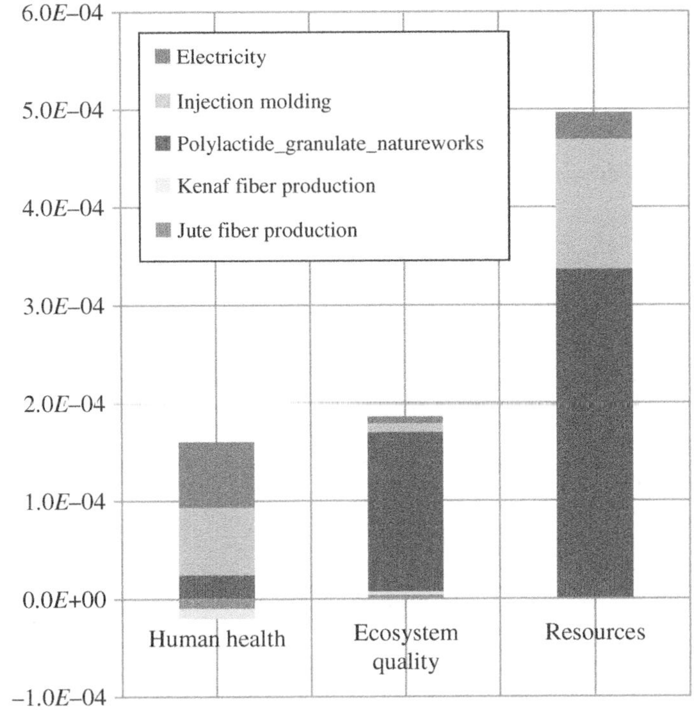

Figure 15.3 Normalized value of biocomposite fabrication.

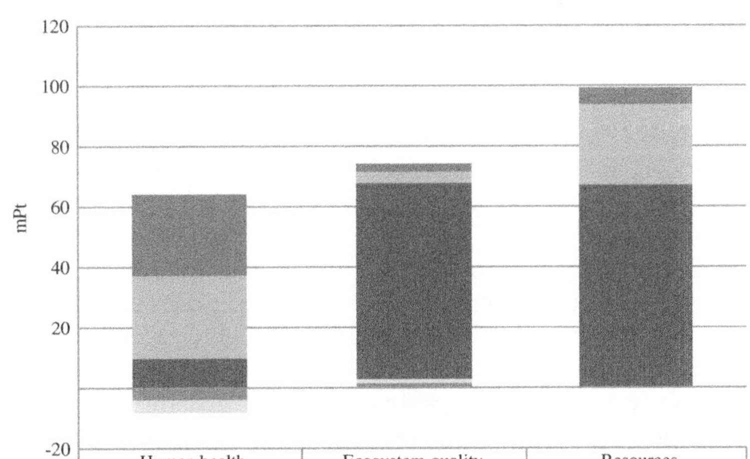

	Human health	Ecosystem quality	Resources
■ Electricity	26.7308	2.5521	5.4065
■ Injection mouding	27.5838	3.6817	26.5496
■ Polylactide_granulate_natureworks	9.7913	64.9495	67.0692
■ Kenaf fiber production	–4	1.388	0.1218
■ Jute fiber production	–4	1.6133	0.101

Figure 15.4 Weighting of biocomposite fabrication.

Table 15.4 Weighting comparison of a biocomposite and a GF/polyester.

Damage category	Unit	Biocomposite	GF/polyester
Human health	mPt	56	108
Ecosystem quality	mPt	74.2	10.1
Resources	mPt	99.2	227
Total	mPt	229.4	345.1

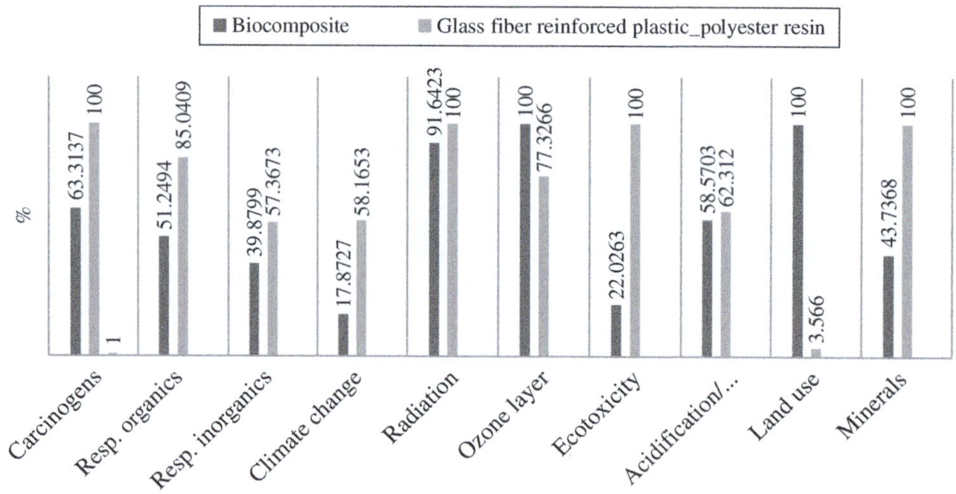

Figure 15.5 Comparison chart of a biocomposite and a GF/polyester.

the weighting and environmental indicators of biobased and glass-based composites, respectively, for the cradle to gate scenario.

From Figure 15.5, it is understood that biobased composites have a lesser impact than glass-based composites. Of the ten environmental indicators, except in two indicators (land use and ozone layer), PLA had a lesser impact than glass-based composites. Environmental impact of PLA production increased due to cultivation of plants, irrigation, and synthesis process. Many authors [4, 8] accepted and concluded that PLA consumes more land use compared to crop cultivation of fibers. In the weighting score (Table 15.4), biobased composite confirms that it had a lesser impact during the manufacturing process.

In the next phase, cradle to grave assessment was performed for both biobased and glass-based composites. Four EoL scenarios are assumed for biocomposites. EoL scenario for biocomposites, incineration (100%), landfill (100%), recycling (50%) and landfill (50%), and reuse (50%) and landfill (50%) are used as grave boundary. From the results, scenario 4 reuse (50%) and landfill (50%) had a negative impact on many indicators. Scenario 3, recycling (50%) and landfill (50%) is the second alternative scenario for biocomposites because it has a lesser impact after scenario 3. When going for recycling and reuse (scenario 3 and 4), biobased composites show better results in environmental indicators. Biobased composites show a better result in the landfilling scenario. From La Rosa et al. [6] study, landfill shows better results than incineration. In Figure 15.6, the landfill had a lesser impact than incineration.

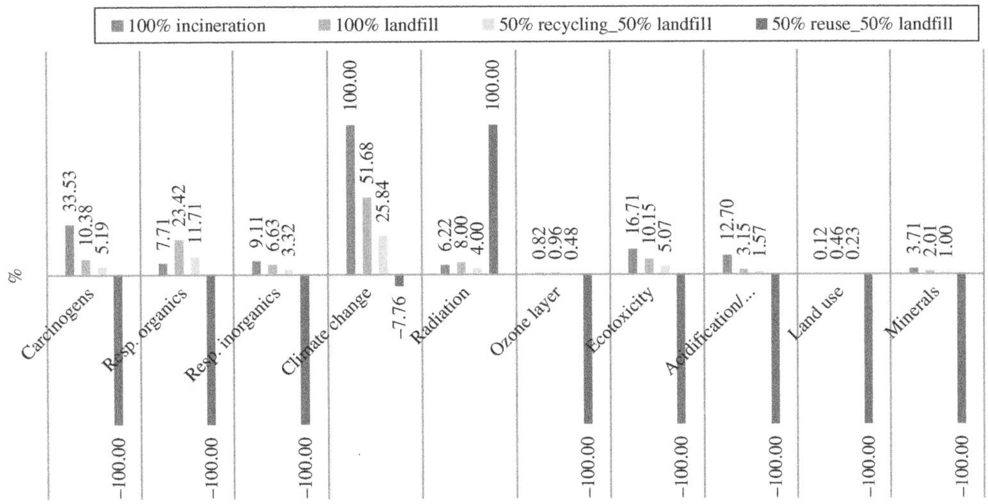

Figure 15.6 End-of-life scenario comparison of a biocomposite.

15.9 Conclusion

Using LCA methodology, the environmental impact of biobased and traditional composites are assessed. Cradle to gate and cradle to grave analysis was chosen as a system boundary. The results express that biobased composites are a good alternative material in an environmental perspective. Eco-indicator 99 (I) was used to find an environmental impact. In cradle to gate assessment, PLA production produces more impact that fiber production. While comparing both biobased and traditional composites (Figure 15.5), biocomposites provide lesser environmental burdens compared to traditional composites. In cradle to grave assessment for biobased composites, four scenarios are assumed. From the results (Figure 15.6), scenario 4 reuse (50%), and landfill (50%) had a lesser impact than other scenarios.

References

1 Mansor, M.R., Mastura, M.T., Sapuan, S.M., and Zainudin, A.Z. (2019). The environmental impact of natural fiber composites through life cycle assessment analysis. In: *Durability and Life Prediction in Biocomposites, Fibre-Reinforced Composites and Hybrid Composites*, 257–285. Woodhead Publishing.

2 Beigbeder, J., Soccalingame, L., Perrin, D. et al. (2019). How to manage biocomposites wastes end of life? A life cycle assessment approach (LCA) focused on polypropylene (PP)/wood flour and polylactic acid (PLA)/flax fibres biocomposites. *Waste Manag.* 83: 184–193.

3 Korol, J., Burchart-Korol, D., and Pichlak, M. (2016). Expansion of environmental impact assessment for eco-efficiency evaluation of biocomposites for industrial application. *J. Clean. Prod.* 113: 144–152.

4 Le Duigou, A., Deux, J.M., Davies, P., and Baley, C. (2012). PLLA/flax mat/balsa bio-sandwich – environmental impact and simplified life cycle analysis. *Appl. Compos. Mater.* 19 (3-4): 363–378.

5 Batouli, S.M., Zhu, Y., Nar, M., and D'Souza, N.A. (2014). Environmental performance of kenaf–fiber reinforced polyurethane: a life cycle assessment approach. *J. Clean. Prod.* 66: 164–173.

6 La Rosa, A.D., Recca, G., Summerscales, J. et al. (2014). Bio-based versus traditional polymer composites. A life cycle assessment perspective. *J. Clean. Prod.* 74: 135–144.

7 Günkaya, Z. and Banar, M. (2016). An environmental comparison of biocomposite film based on orange peel-derived pectin jelly-corn starch and LDPE film: LCA and biodegradability. *Int. J. Life Cycle Assess.* 21 (4): 465–475.

8 Molins, G., Álvarez, M.D., Garrido, N. et al. (2018). Environmental impact assessment of polylactide (PLA)/chicken feathers biocomposite materials. *J. Polym. Environ.* 26 (3): 873–884.

9 Khoshnava, S.M., Rostami, R., Ismail, M., and Rahmat, A.R. (2018). A cradle-to-gate based life cycle impact assessment comparing the KBFw EFB hybrid reinforced poly hydroxybutyrate biocomposite and common petroleum-based composites as building materials. *Environ. Impact Assess. Rev.* 70: 11–21.

10 Singh, V., Dincer, I., and Rosen, M.A. (2018). Life cycle assessment of ammonia production methods. In: Exergetic, Energetic and Environmental Dimensions, 935–959. Academic Press.

11 Xu, X., Jayaraman, K., Morin, C., and Pecqueux, N. (2008). Life cycle assessment of wood–fibre-reinforced polypropylene composites. *J. Mater. Process. Technol.* 198 (1–3): 168–177.

Index

Biobased Composites: Processing, Characterization, Properties, and Applications, First Edition.
Anish Khan, Sanjay Mavinkere Rangappa, Suchart Siengchin and Abdullah M. Asiri.

Printed and bound by CPI Group (UK) Ltd, Croydon, CR0 4YY